Measurement of Weak Radioactivity

Measurement of Weak Radioactivity

Páll Theodórsson

Science Institute
University of Iceland, Iceland

Published by

World Scientific Publishing Co. Pte. Ltd.

P O Box 128, Farrer Road, Singapore 912805

USA office: Suite 1B, 1060 Main Street, River Edge, NJ 07661

UK office: 57 Shelton Street, Covent Garden, London WC2H 9HE

Library of Congress Cataloging-in-Publication Data
Theodórsson, Páll, 1928–
Measurement of weak radioactivity / Páll Theodórsson.
p. cm.
Includes bibliographical references and index.
ISBN 9810223153 (alk. paper)
1. Low-level radiation -- Measurement. I. Title.
QC476.S6T54 1996
539.7'7--dc20 96-38855
CIP

British Library Cataloguing-in-Publication Data
A catalogue record for this book is available from the British Library.

Printed in Singapore by Uto-Print

Preface

It is now nearly half a century since the measurement of weak radioactivity emerged as a separate branch in radiometry — when elaborate measures were for the first time taken to achieve maximum signal and minimum background. The technique of low-level counting was invented and named by Williard F. Libby in the late 1940s when he, together with his collaborators, developed radiocarbon dating. The technique has since expanded from a single radioisotope to cover a large number of radioactive nuclides. The number of detectors has increased dramatically and the scope of applications has become highly varied, ranging from environmental control to cosmological problems.

This book on the measurement of weak radioactivity — on low-level counting — will hopefully help all those who are using the technique or are improving on it. The sources of background — environmental radioactivity and cosmic rays — and methods to suppress their effects are described. The basic principles of the detectors used are explained, systems with each of these detectors are described and their characteristic background effects discussed. Four of the most important application areas of the low-level technique are described. In order to make the book more self-contained, the basic principles of nuclear physics are also described — a brief description of the nucleus is given, radioactivity is discussed and the interaction of radiation with matter is explained in some detail.

As with all other modern measuring techniques, low-level counting has continuously been expanded and improved: new detectors have been added, older ones improved and their associated electronic equipment has kept pace with the revolutionary progress in the electronic technique. Some ten years ago, important developments were initiated to make possible studies of some rare radioactive processes of theoretical importance. Resources were made available for this work on a scale never seen before in low-level counting. This work has not only brought us greatly improved germanium spectrometers, but also a wealth of information vital to all sectors of low-level counting.

However, most of the low-level techniques were developed when the experimental methods and equipment were primitive compared to modern standards. For decades researchers had to be content with various recipes, which were not sufficiently supported experimentally, coming mainly from the early pioneers. Today, when it is much easier to study these recipes, many of them are still considered good enough. Now that we better understand the effects of the basic processes of the background and their impact on our measurements, the time is ripe for a critical review the whole field of low-level counting.

This book is not simply a review of the present state of the low-level counting technique, it also includes an analysis of a large mass of scattered material that has been published since the birth of the technique and has appeared in many scientific journals, reports and conference proceedings. Information has been collected from various sources. By piecing it together we can frequently extract a more coherent picture of the phenomena of interest. Writing this book has therefore been like putting together a large jigsaw puzzle, where bits of information from a large number of articles and reports represent the pieces. Sometimes they fit well, sometimes only nearly, but frequently not at all; and a number of pieces are no doubt missing. The author is well aware that these pieces have not always been put together correctly. In some cases there has been doubt about how to interpret the results of certain measurements. I have preferred to do my best to give a plausible explanation, even though it may well be wrong, rather than leave the question open. This will hopefully urge others to find a better explanation.

We still have a good deal to learn about our low-level systems in order to be able to design still better ones. Most of the relevant studies are relatively easy to make, especially if they are carried out through a systematic collaborative effort of a number of laboratories.

Although the main source of information on which this book is based comes from the technical literature, a large number of colleagues have given me valuable information.

I gratefully acknowledge the invaluable help of Dr. Christopher Evans for reading over the manuscript, the efficient assistance of Sveinn B. Sigurdsson and the very thorough work of Gerlinde Xander.

Without the excellent service of the library of Risö National Laboratory this book could not have been written.

Páll Theodórsson

Reykjavik, Iceland
August 1996

Contents

Chapter 1

Low-level counting

1.1 Introduction

The technique of low-level counting is used to solve a large variety of problems, ranging from daily control of contamination around nuclear plants to studies of fundamental processes in physics and astrophysics. For this work a variety of detectors and systems are used, from small semiconductor devices to large liquid scintillation counters.

Low-level counting was invented and given a name by Williard F. Libby in his pioneering work at the University of Chicago when he developed the radiocarbon (^{14}C) dating method in the late 1940s, for which he was awarded the Nobel prize in 1960. He had concluded from measurements of cosmic-ray produced neutrons in the stratosphere that ^{14}C would be produced and subsequently mix into the whole atmosphere. First he demonstrated the presence of ^{14}C in natural carbon by measuring isotopically enriched biomethane. He thereafter developed a very sensitive counting system for the determination of ^{14}C in natural carbon without enrichment. This system is shown in Figure 1.1. The most important elements of the technique are already found here: heavy shielding, anticoincidence counters, selection of radiopure materials and sample arrangement securing optimum counting efficiency and maximum sample size. The technique developed quickly. Libby's screen wall counter was replaced by gas proportional counters (requiring smaller samples), improved anticosmic counters were developed, an inner shielding layer of mercury and a neutron absorbing layer were added.

For almost a decade, low-level counting was limited to radiocarbon dating. In the mid 1950s, huge quantities of radioactive materials spread all over our globe from tests with hydrogen bombs in the atmosphere. Monitoring the fall-out created a compelling demand for expanding the low-level counting technique to a large variety of radioisotopes that required new types of detectors. The need for high sensitivity intensified in the 1960s with increasing interest in cosmogenic radioactivity in surface layers of the Earth and of meteorites, and in studies preparing for the measurement of prospective lunar samples.

In this introductory chapter, various general aspects of the low-level counting technique will be discussed. Most of the topics are treated in more detail in later chapters.

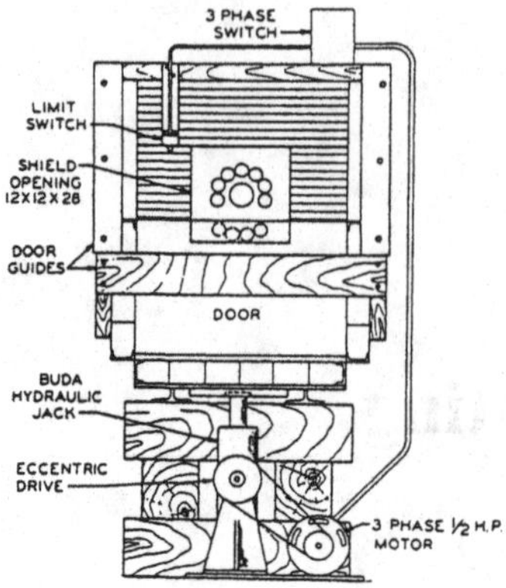

Fig. 1.1. Low-level counting started with this system. Libby's radiocarbon system with which he and his co-workers established the radiocarbon dating technique in 1947–1949.

1.2 Fields of application

The field of low-level counting started with radiocarbon dating where only gas counters were used. New radiation detectors were invented during this early period: scintillation counters with organic and inorganic crystals, liquid scintillation counters and, 15 years later, semiconductor radiation detectors. When these new detectors had reached maturity they were incorporated into systems based on the low-level counting technique, inherited from radiocarbon dating. To these radiometric methods, were later added mass spectrometric methods, where individual atoms of the radionuclides are counted rather than the few emitted decay particles, and sensitive optical methods (Chapter 16).

With new detectors and improved systems, the field of low-level counting gradually expanded. Today it is applied to a wide variety of studies, such as:

1. Radiocarbon dating.
2. Tritium in natural water.
3. Fall-out.
4. Contamination due to nuclear accidents.
5. Environmental control around nuclear establishments.
6. Distribution of primordial radioactivity in our environment.
7. General work with radioactive tracers.

8. Neutron activation analysis.
9. Tracing in hydrology and oil field exploration.
10. Mixing in the atmosphere and oceans.
11. Geochemical studies using primordial radioactive tracers.
12. Cosmogenic isotopes and extraterrestrial radioactivity.
13. Double beta decay and other rare processes.

This book is addressed to scientists working in these areas, except the last one.

It is difficult to define the field of low-level counting. Our interest starts when some special measures must be taken to increase the sensitivity of a system or a method, and it extends to sophisticated counting systems in general low-level laboratories. Expensive ultralow-level germanium spectrometers operating underground are on the border of the coverage of this book. They are, however, of interest to us as their development has brought us valuable information about the background sources of all low-level systems. Other, more specific systems, for example those for solar neutrino measurements, are not discussed here.

1.3 Detectors and systems

Today, the following detectors are of the greatest importance in low-level counting:

1. Gas proportional counters with internal samples.
2. Gas proportional and Geiger counters for external solid samples.
3. Liquid scintillation counters.
4. NaI scintillation counters.
5. ZnS scintillation counters.
6. Ge-diodes.
7. Si(Li) diodes.

These detectors are the core of low-level counting systems where various measures have been taken to increase their sensitivity, mainly to reduce their background count rate. For alpha detecting systems, the radiopurity of materials close to the detector is of primary importance. These materials are therefore carefully selected. For beta and gamma detection systems, various further measures are taken:

1. The detector is surrounded by a thick shielding layer, usually 10 cm of lead.

2. An extra inner layer of very pure material may be added: old lead, completely free of ^{210}Pb contamination, or electrolytically refined copper.
3. Sometimes there is a 5–10 cm inner layer of paraffin/boron to thermalize and absorb neutrons, produced by cosmic-ray protons in the shield.
4. Frequently the sample counter is surrounded by a system of anticosmic counters for suppressing the background contribution of cosmic-ray muons and protons.
5. Background pulses are sometimes eliminated through pulse shape or time sequence analysis.

1.4 Selection of detection system

When a radioisotope is to be measured we can frequently choose more than one method. Many aspects then come into consideration. The following points can be considered as a useful check-list:

1. Counting efficiency.
2. Background.
3. Energy resolution.
4. Number of detectors.
5. Accuracy needed.
6. Mass of sample available.
7. Size of counting samples.
8. Radionuclide enrichment in counting sample.
9. Number of samples.
10. Equipment available.
11. Cost of a new system.
12. Cost and work per measured sample.
13. Personal experience.

Let us take radon (^{222}Rn) as an extreme example. Radon is either measured directly or its progeny are separated and counted. When radon is measured, it is usually in secular equilibrium with its decay products (Section 17.8). The main methods for determining its concentration are the following:

1. The alpha activity of radon and its progeny can be measured in (1) a ZnS coated scintillation chamber (Lucas cell) or (2) in a liquid scintillation counter.

2. Radon decay products can be electrically precipitated on the window of (3) a gas proportional counter, or (4) on the window of a Si(Li) diode.

3. The radon decay products in air can be collected on a filter paper and the alpha or alpha+beta activity determined with (5) a thin window gas proportional counter. (6) with a ZnS scintillation counter, (7) with a liquid scintillation counter.

4. Radon can be adsorbed on charcoal in a canister and (8) the gamma activity determined, usually with a NaI scintillation unit. The radon can also be released from the charcoal and its alpha activity determined with a (9) a gas proportional counter, (10) in a Lucas cell or (11) by a liquid scintillation counter.

Available low-level methods for other radioisotopes are much more restricted. Two general examples will be discussed here, the measurement of (a) alpha active samples and (b) radioisotopes emitting both beta and gamma radiation.

Alpha active samples. In the measurement of alpha activity we can generally choose among four different detectors:

1. *Si(Li) diodes.* The radioisotope must be carefully separated from the matrix of the collected sample and plated out in a thin layer on a metal disk in order to secure high energy resolution (about 15 keV FWHM, see Section 8.5.3). The absolute detection efficiency is about 30%, mainly determined by the source-diode geometry. The background is very low, a few pulses per day. The diameter of the diodes is limited to about 60 mm (area 28 cm^2).

2. *Liquid scintillation counters.* The detection efficiency is practically 100%, and the energy resolution can be about 250 keV. The background can be very low, of the order of one pulse per hour. Chemical separation is necessary, the final step being dissolving the sample in the scintillating liquid.

3. *ZnS screen scintillation counting.* Scintillations of alpha particles impinging on a ZnS screen are detected with a photomultiplier tube. This method gives no information of the alpha energy and is mainly used for gross alpha counting of large, thick samples, for example filter papers. The background depends on the diameter of the ZnS screen, it can be about one count per hour for a 7 cm diameter screen.

4. *Window gas proportional flow-counters.* The samples are of the same type as those for ZnS scintillation counting and the sensitivity is similar. The counting system, frequently of the multidetector type, is simpler and more compact than the ZnS scintillating system. Good gas proportional counters have a background comparable to that of a ZnS system.

Beta/gamma active samples. Beta emitting radionuclides usually leave their daughter nuclei in an excited energy level. The nuclei get rid of the excessive energy through emitting a gamma-photon, usually instantly. These radioisotopes can be measured by three different methods:

1. *Germanium spectrometer*, detecting the gamma radiation. This is usually the preferred method, as large samples, in the order of one kilo, usually needing little or no preparation, can be measured because of the penetrating power of the photons. The high energy resolution leads to low background, gives positive identification of the radiation and allows the simultaneous determination of the concentration of individual nuclides in a mixture of radioisotopes. The detection efficiency of the germanium diodes is, however, rather low, typically a few percent.

2. *Liquid scintillation counter.* The beta detection efficiency is usually over 90% and the background of a low-level system is typically 0.5 cpm.

3. *Thin window gas proportional counter.* With thin samples the detection efficiency is typically 35–45%. The background depends on the diameter of the counter window, for a diameter of 25 mm it is about 0.2 cpm. These counting systems are usually of the multidetector type, where 4–8 samples are measured simultaneously.

1.5 Sensitivity comparison

When selecting a system for a particular job, one sometimes needs to compare quantitatively the sensitivity of two or more different systems, based on their background, counting efficiency and number of detectors. Earlier, the number of detectors was not taken into consideration, but with increasing use of multidetector systems, it is now natural also to include this parameter. Systems can be compared by the time needed to reach a given percent accuracy when a standard reference sample is measured. The relative standard deviation ε of net sample count rate is given by (see Chapter 18)

$$\varepsilon^2 = [(S+B)/T_S + B/T_B]/S^2 \qquad (1.1)$$

where S and B are the net standard sample and background count rates, T_S and T_B the counting times of sample and background. The background count rate B is generally measured regularly and it will, after proper corrections have been made, stay constant during long periods. It is therefore usually known with better accuracy than is sought for the unknown sample. The term B/T_B in (1.1) can then be omitted. The sample counting time is then

$$T_S = (1 + B/S)/(\varepsilon^2 \cdot S) \qquad (1.2)$$

The number of samples that can be measured per unit time to a given accuracy, i. e., the system's counting capacity, is a natural measure of the merit of the system. This number is inversely proportional to the counting time, so the *factor of counting capacity*, FCC, is defined by (Theodorsson 1991):

$$FCC = S/(1 + B/S) \tag{1.3}$$

The system giving the highest value of FCC is obviously most suitable. When the net count rate S of an unknown sample is large compared to the background B, FCC can be approximated by S (or rather by $S - B$). When, however, S is small compared to B, S^2/B gives a good approximation for the FCC. The most suitable system is then the one that gives the highest value of S^2/B. This is another form for the conventional *figure of merit* (FOM) of low-level systems, which is based on $1/\sqrt{T}$ rather than on $1/T$ or

$$FOM = S\sqrt{B} \tag{1.4}$$

For dimensional reasons, it is an unfortunate choice to base the widely used FOM on $1/\sqrt{T}$ rather than 1/T.

When comparing a single sample detector system with a comparably priced multicounter system, it is natural to take into account the number of detectors N and not base the comparison on S and B of a single detector element. In this case the FCC of the multidetector system will have N times the FCC value of a single detector in the system.

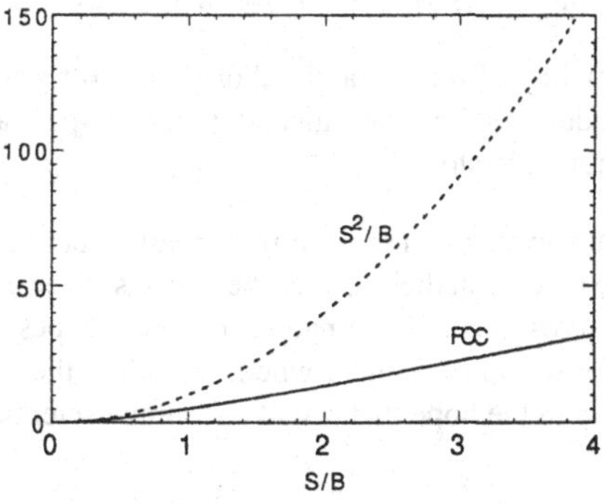

Fig. 1.2. Factor of counting capacity.

1.6 Recipes in low-level counting

In the design of low-level counting systems, as well as in their use, we depend on a number of recipes that we have inherited from our predecessors. Some of these were established by scientists short of time, setting up new low-level counting systems, wanting to start their studies with the least possible delay. Furthermore, the equipment used was expensive and cumbersome and offered limited possibilities for detailed data collection and processing compared to present day systems. Experimental work in low-level counting is usually time consuming because of the long counting times needed. Under these circumstances the scientists frequently had to settle on a procedure that seemed good enough, rather than to spend valuable time to find the best one. Various sectors of the low-level counting technique therefore still depend on lore and recipes. The description of the situation regarding scintillations cells for radon counting, as described by Cohen et al., (1983) is still valid. According to their description on the state of the technique, little information was available on these cells in the literature.

> "This does not imply that such information has not been collected, but such work has generally been done by groups with practical development goals and little taste for publication in the scientific literature. As a result the use of scintillation cells is surrounded by a lore propagated largely by private conversations and heavily influenced by isolated experience and hearsay evidence. While this evidence has been valuable, it is also somewhat variable among different practitioners and not always reliable."

We should take a critical attitude towards many of these recipes. Today, procedures would be simpler and systems better if the policy of the research laboratories of the Philips concern had guided development work in the low-level counting technique. The director of the laboratories, Hendrik Casimir, described it the following way:

> "It was an accepted policy of the research laboratories to try to really understand empirical procedures and not be satisfied with a recipe that worked well in practice, but was not understood."

Lore and recipes are, however, an inevitable part in all scientific work. In low-level counting we should show extra care in their use for the reasons given above. Five examples will be described in order to demonstrate the nature of these recipes, how they have been introduced, and what the consequences can be when we follow them uncritically. This is discussed in some detail here in the hope that it will encourage scientists to study critically some of our recipes.

1. Libby's group reported that contamination in the lead tested for the shield contributed 30 cpm to the background of the screen wall counter. Iron was therefore chosen and a 20 cm thick layer reduced the background to 10 cpm (Anderson et al., 1951). For almost twenty years, all dating laboratories using gas counters followed Libby's example and chose iron for the main shield, although lead has distinct advantages over iron, as it is easier to stack, offers greater flexibility and has higher density. With information we have today (Section 10.4.1), we can estimate crudely that Libby's lead must have had a contamination of about 8000 Bq of ^{210}Pb per kg of lead in order to contribute 30 cpm in a counter with a cathode area of 600 cm^2. This contamination level is improbable, as the ^{210}Pb concentration in recent lead is generally 100–500 Bq/kg,

although higher concentrations are met. A more probable explanation of these 30 cpm can be found in the paper describing the system (Anderson et al., 1951). Discussing the suppression of the background component due to external natural gamma radiation, the paper states:

> "Two inches of lead are sufficient for this purpose, but, unfortunately, the laboratory lead bricks themselves have been found to contribute about thirty cpm of gamma impurities. Various other materials were considered; iron was chosen as the best compromise between price, radiochemical purity, and high atomic number."

Two inches (5 cm) of lead are far from enough to suppress the contribution of external gamma radiation. Lead has a high atomic number and is very efficient in absorbing X-rays and gamma-photons with energies below one MeV, where photoelectron absorption is strong or predominating. For natural gamma radiation above 1 MeV, Compton scattering is the dominating gamma attenuating process, even in lead, and it is independent of the atomic number of the absorbing medium (see Figure 5.9). Actually, 12 cm of lead are needed to give the same gamma absorption as 20 cm of iron. It can be estimated at 300 cpm for Libby's unshielded counter from the information given in the paper. When the counter is shielded with 5 cm of lead, this count rate is reduced by a factor of about 12, so we should expect this component to have been 25 cpm, not far from the 30 cpm that was ascribed to contamination in the lead. Today we know that lead is actually the best shielding material, provided that lead with high ^{210}Pb contamination is avoided.

2. The second example of a hasty recipe is the procedure recommended by the Chicago group that developed the technique of assaying low levels of tritium in natural waters. Before the measurement, the water samples were enriched by a factor of 100 or more by electrolysis. In the description of the work it was stated that the volume reduction of the samples had been kept below 10 in order to avoid corrosion of the anode at high electrolyte concentration. It is easy to understand that scientists, needing to measure a limited number of samples in establishing the existence of cosmogenic tritium in natural waters, preferred to play safe in order to avoid possible complications due to corrosion. The samples were therefore enriched in three stages when the water was distilled between stages to remove excessive electrolyte. This recipe was followed for more than 20 years. Some laboratories, however, evaded one of the usual three stages by periodically adding water to the cells. Later it was shown that a volume concentration factor up to 300 did not affect the enrichment process and caused only mild corrosion.

3. The third example concerns the 5–10 cm thick layer of paraffin/boron inside the main shield in gas proportional counting systems, introduced by de Vries (1955) in the Groningen radiocarbon laboratory, in order to lower the neutron background component. This extra layer adds 30–40% to the mass of the main shield and therefore does not

come cheap. The Groningen laboratory only had a thin roof, giving minimum attenuation of the flux of cosmic protons that produce the neutrons in the shield. As discussed later, there is a serious doubt about the benefit of this layer in laboratories that have a few floor plates above them. Gas proportional counters in the dating laboratory at Trondheim, with four storeys above, have no neutron shield, but their background is lower than that of other counters of similar volume.

4. Graded shielding has been recommended for a long time for low-level counting systems, i.e., lining the inner surface of the main lead shield with a thin layer of cadmium and copper to absorb characteristic X-rays produced in the lead by cosmic rays. Most low-level germanium spectrometers have a shield of this type. It has recently been shown (Section 12.8) that this only helps when photons below 100 keV are being measured, as this layer increases the background in the energy range 100–500 keV by about 30%.

5. The final example concerns the anticosmic counter system inside the shield enclosing the sample detector. It eliminates the muon background contribution. It was pointed out by Scholz (1961), based on solid experimental work, that anticosmic counters on the outer sides of the shield would give the lowest background, as it would eliminate, in addition to the muon component, also a large part of the secondary radiation produced by muons and protons in the main shield. Furthermore, external anticosmic counters would reduce the mass of the main shield drastically as it makes the neutron absorbing layer superfluous and removes the anticosmic counters from the inner shield (Theodorsson and Heusser, 1991). Over 20 years elapsed until this proposal was implemented. This arrangement of guard counters is now used in a number of low-level and ultralow-level germanium spectrometers, giving high background reduction.

These examples show that we should not be satisfied with a recipe that works well in practice, but is not understood. We should continuously strive for a better understanding of all parameters and use new information for improving and simplifying our systems and procedures.

1.7 Reporting low-level systems

Today, there are still many loose ends in the low-level counting technique. As long as we cannot give a good quantitative estimate of all background components, we cannot design optimal systems.

Systematic studies, carried out with the aim of increasing our understanding of factors affecting background and counting efficiency of detectors, are naturally a major source of information. Descriptions of low-level counting systems, where their use is the main focus, are another important source of information. It is frustrating when the usefulness of good work of this type is seriously limited by not giving the value of important parameters, or when the results are described in a way that may be good enough for the purpose of the author, but not to those interested in the technique itself. It would help us in advancing the low-level counting technique if authors gave somewhat more detailed information about their systems. The following points can be considered as a check list for authors of articles on low-level counting to consult before they write their reports:

1. Overburden, mwe.
2. Thickness of all shielding layers.
3. Total mass of shield.
4. Dimensions and mass of detector.
5. Detection efficiency.
6. Background count rate.
7. Counting time.
8. X-axis of background spectra preferably linear.
9. Y-axis of background spectra in cpm/keV (not counts/channel), X-axis in keV (not channel number).
10. Match number of spectrum channels to energy resolution.
11. Quantitative reporting of ^{210}Pb in lead, Bq/kg.

A few remarks will explain the purpose of this list.

- A few floor plates above the laboratory room will significantly attenuate the nucleonic component of the cosmic rays. In order to be able to compare the background count rate of similar systems, we must know their overburden.
- In background comparisons, the mass of a germanium diode is more significant than its relative efficiency.
- The basis of the minimum detection-limit should be given, i.e., sample size, background and detection limit; the reader may want to calculate the detection efficiency using other values for the parameters or compare different systems or methods.
- The pulse height spectra of liquid scintillation counting systems are usually displayed on a logarithmic (usually uncalibrated) energy scale. This usually makes it difficult to use the information.

- Liquid scintillation systems usually have multichannel analyzers with 1000 channels. This does not match their poor energy resolution and often leads to a small number of pulses in each channel, with high scattering from one channel to the next. Fewer channels will display the results more comprehensively. It would be useful if the software offered the option of combining, for example, every four channels into a single one.

- The Y-axis of energy spectra is frequently not calibrated, but given in counts per channel. In order to make direct comparison possible the Y-scale should give cpm per keV. It is sometimes useful to know the width of the channels (in keV). It would also help if scientists would stick to one time unit, preferably minutes, not seconds, hours or days.

- Often it is useful to display the count rate of the spectra on a logarithmic scale, but only when the numbers span a wide range. In other cases it is better to use a linear scale.

- When the background of germanium spectrometers is reported, a spectrum gives the fullest information. Often we have to compress this information into a few numbers. It is then highly desirable to do this in a standardized way in order to facilitate comparison of the performance of different systems. It is recommended that the following count rates are given:

  (i) Total count rate from 50–1500 keV, in cpm
  (ii) The peak at about 150 keV, in cpm/keV
  (iii) The count rate at 500 and 1000 keV, cpm/keV.
  (iv) The fast neutron broadened peak at 691 keV, cpm.

- When scientists compare their spectra with those of other systems, an easy transfer of a spectrum into some general computer spreadsheet, such as EXCEL, is desirable. The multichannel analyzer software generally used today does not make this simple operation easy.

Finally, it should be noted that the recommendations given here will not significantly increase the length of the reports, but will substantially increase their usefulness.

1.8 Literature on low-level counting

A book was published in 1964 describing the technique of low-level counting. Today, it is naturally obsolete. Various review articles have been written, usually with main emphasis on a certain sector of the technique. Some of these are listed below, especially the more recent ones. The technique and its applications have been the subject of a number of specialized conferences, starting with a meeting of experts and scientists interested in radiocarbon dating in Copenhagen in 1954. Radiocarbon conferences have been held regularly ever since, now every three years, and their proceedings volumes have been published, in later years in Radiocarbon. The last Radiocarbon conference was held in 1994.

The first general conference on low-level counting and its applications was held by the International Atomic Energy Agency in Vienna in 1960. It has been followed up by a number of both general conferences as well as meetings for specialized sectors of the field. In most of these conferences, both techniques and applications have been discussed. Three general low-level counting conferences were held in Czechoslovakia, the first in 1975 and the last in 1989. The proceedings of these conferences have been published.

Conferences have regularly been held on the liquid scintillation counting technique, the last one in 1994. Low-level systems have been discussed at these meetings. The proceedings of the two last conferences have been published as special issues of Radiocarbon.

The International Committee for Radionuclide Metrology has held a number of conferences on various aspects of radioactivity measurements. Valuable information about low-level counting has been presented at some of these conferences. The latest, the Conference on Low Level Measurement Techniques, was held in October 1995, and its proceedings volume will be published in a special issue of Applied Radiation and Isotopes.

The University of Sevilla has arranged three times an international summer school for scientists using techniques of low-level counting. The lectures given there have been published by World Scientific.

One of the most active fields of low-level counting is the study of primordial and man-made radioactivity in our environment. A number of conferences have been held on the subject. Finally, health physics unions regularly hold conferences, usually centred on some main theme. For example, in 1993 the German Fachverband für Strahlenschutz e.V. held a conference on environmental radioactivity, radioecology and effects of radiation.

Recommended Reading

- *High Sensitivity Counting Techniques* D. E. Watt, D. Ramsden, Pergamon Press, 1964.
- *Low Level counting techniques*, H. Oeschger and M. Wahlen, 1975, Ann. Rev. Nucl. Sci., 25, 423-463. A broad general review of the technique, with emphasis on gas proportional counting.
- *Methods of Low-Level Counting and Spectrometry*, Proceedings of an IAEA symposium, IAEA 1981.
- *Radon Monitoring in Radioprotection, Environmental Radioactivity and Earth Sciences*. Proc. of an international workshop 1989, World Scientific 1990.
- *Proceedings of the ICRM Symposium on Low-Level-Radioactivity Measuring Techniques and Alpha Particle Spectrometry*, held in Monaco June 1991, ed. W. B. Mann, Pergamon Press, 1992.
- *Low-Level Measurements and their Applications to Environmental Radioactivity*, proceedings of the First International Summer School, held in Huelva, Spain, 1987. Ed. M. Garcia-Leon and G. Madurga, World Scientific, 1988.
- *Low-Level Measurements of Man-Made Radionuclides in the Environment*, Proceedings of the Second International Summer School, held in Huelva, Spain, 1990. Ed. M. Garcia-Leon and G. Madurga, World Scientific, 1991.
- *Low-Level Measurements of Radioactivity in the Environment*, Proc. of the Third International Summer School, held in Huelva, Spain, 1993. Ed. M. Garcia Leon and R. Garcia-Tenorio, World Scientific, 1994.
- *Umweltradioaktivität Radiokologie Strahlenwirkungen*, proceedings of the meetings of Fachverband für Strahlenschutz e.V. September 1993. Verlag TÜV Rheinland.
- *Low Radioactivity Background Techniques* G. Heusser, 1995. Ann. Rev. Nucl. Sci. A review article with a detailed analysis of background components and description of methods to suppress the background and description of some modern ultralow level studies. Gives 162 references.

Chapter 2

History

In order to set the material discussed in this book in perspective, a brief history of nuclear research, development of detectors and electric instruments and the technique of low-level counting is presented. Some of the milestones of interest in the present context are listed in Tables 2.1 and 2.2. In the case of radiation detectors it can be difficult to give a year of invention or the name of an inventor, as a new technique is frequently the result of many development steps. In these cases only a period is given.

2.1 Nuclear research

The beginning of modern physics is frequently traced back to the discovery of X-rays by Wilhelm Röntgen. He had begun experimenting with electric discharge between two metal electrodes in a glass tube with air under low pressure. One day in November 1895 he noticed that a screen covered with barium sulfate, lying near the tube, unexpectedly emitted weak fluorescent light when electric current passed through the tube. He traced its source to a penetrating radiation, coming from the tube, which he gave the name X–rays. Further study showed that this radiation could expose photographic plates and make air conducting.

The news of Röntgen's work spread quickly and there was intense interest in these new rays. A few months later Henri Becquerel decided to investigate whether fluorescence could produce X–rays, as these were emitted from a fluorescent spot on the glass tube where the cathode rays hit the glass. Instead of using barium sulfate Bequerel fortuitously used a sulfate of uranium, which he had at hand and knew would fluoresce strongly when illuminated by ultraviolet light. He put a small quantity of the uranium salt on black paper that was wrapped tightly around a photographic plate and exposed the salt to the sun for several hours. To his satisfaction, a black spot could be seen on the film after it had been developed. A few days later he discovered that the dark spot appeared on the film even without exposing the uranium salt to sunlight, and that pure uranium metal had the same effect. Evidently, uranium was continuously emitting a previously unknown type of penetrating rays.

Strangely enough, little work was devoted to this new phenomenon for a year and a half, until Marie Curie, shortly later joined by her husband Pierre, took up the study of the strange

Table 2.1. Historic Events in Nuclear Physics

1869	Mendeleev's periodic system of the elements
1873	Maxwell's theory of electromagnetic radiation
1888	Hertz generates and detects electromagnetic waves
1895	Röntgen discovers X-rays
1896	Discovery of radioactivity (Becquerel)
1897	Thomson measures charge-to-mass ratio of cathode rays (electrons)
1898	Isolation of radium (M. Curie and P. Curie)
1899	Discovery of alpha and beta radiation (Rutherford)
1900	Discovery of gamma radiation (Villard)
1900	Thomson's "plum pudding" model of the atom
1905	Special theory of relativity (Einstein)
1909	α particles shown to be He nuclei (Rutherford and Royds)
1910	Soddy establishes existence of isotopes
1911	Nuclear atom model (Rutherford)
1913	Discovery of stable isotopes (Thomson)
1913	Planetary atomic model (Bohr)
1914	Nuclear charge determined from X-rays (Moseley)
1919	Artificial transmutation by nuclear reactions (Rutherford)
1922	Compton effect discovered
1925	Heisenberg's first paper on quantum mechanics
1926	Schrödinger's wave mechanics
1932	Discovery of positron (Anderson)
1932	Discovery of neutron (Chadwick)
1932	First nuclear reaction using accelerator (Cockcroft and Walton)
1934	Discovery of artificial radioactivity (I. Curie, F. Joliot)
1934	Theory of β radioactivity (Fermi)
1935	Meson hypothesis (Yukawa)
1937	Discovery of muons in cosmic rays (Neddermeyer, Anderson)
1938	Discovery of nuclear fission (Hahn and Strassmann)
1940	Production of first transuranium element (McMillan and Seaborg)
1942	Controlled fission reactor (Fermi)
1945	Fission bomb tested
1947	Development of radiocarbon dating (Libby)
1947	Discovery of π meson (Powell)
1952	Thermonuclear bomb tested
1955	Discovery of antiproton (Chamberlain and Segré)
1956	Experimental detection of neutrino (Reines and Cowan)

Table 2.2. Historic Events in Nuclear Technique

1886	Crooke's discharge tube
1898	Piezoelectric electrometer (Marie and Pierre Curie)
1902	Luminescence of ZnS (Crook; Elster and Geitel)
1908	Spinthariscope for quantitative counting of alpha scintillations
1911	Wilson cloud chamber
1919	Development of mass spectrometer (Aston)
1928	Geiger-Müller counter invented
1928	Coincidence counting invented (Bothe)
1929	Electronic detection and counting of coincidences (Rossi)
1930	Electronic amplifiers for pulse ionization chambers
1930	Scale-of-two developed
1931	Electrostatic accelerator (van de Graaf)
1931	Linear accelerator (Sloan and Lawrence)
1944	Phase stability for synchrotron (McMillan, Veksler)
1945	ZnS alpha scintillations detected with a photomultiplier tube
1945	Scintillation in anthracene discovered (Kallman)
1947	Scintillation in organic liquids discovered (Kallman)
1947	First proton synchrocyclotron, 350 MeV (Berkeley)
1947	Improved particle photographic emulsions (Powell's group)
1947	Radiocarbon dating developed (Libby)
1948	Scintillation in NaI(Tl) discovered
1948	First linear proton accelerator, 32 MeV (Alvarez)
1948	Transistor invented (Shockley, Bardeen, and Brattain)
1949	Development of scintillation counters (Kallmann, Coltman, Marshall)
1952	Gas proportional counters introduced in radiocarbon dating
1952	Proton synchrotron, 2.3 GeV (Brookhaven)
1959	26 GeV proton synchrotron (CERN)
1960	Improved commercial liquid scintillation counting systems
1961	Benzene synthesis for radiocarbon LSC dating improved
1962	Semiconductor detectors invented, Ge(Li) and Si(Li)
1962	Integrated circuits invented
1971	Microprocessor invented
1971	Proton-proton collider (CERN)
1972	500 GeV proton synchrotron (Fermilab)
1975	High purity germanium detectors introduced
1983	Operation of proton–antiproton collider at 300 GeV (CERN)
1984	PC–computers
1984	Quantulus liquid scintillation counter introduced

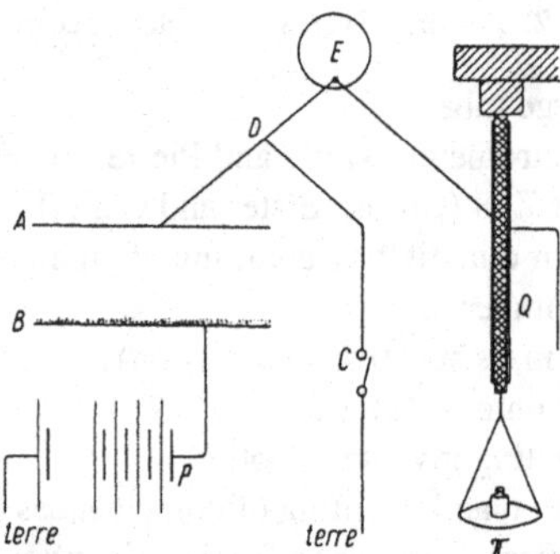

Fig. 2.1. Experimental set up of Marie and Pierre Curie for the measurement of radioactivity. The ionization chamber is in form of a parallel plate condenser. The mass on the scale H produces a small electric charge on the piezoelectric crystal Q that compensates for the charge collected through the ionization current. The electroscope E is used to indicate zero potential. The radioactive material is on plate B. (Figure from M. Curie's thesis (1903).)

radiation. She coined the name now used, radioactivity. Marie Curie discovered a uranium ore that was more radioactive than pure uranium, and in 1898 the Curies discovered two radioactive elements in this ore, which they gave the names polonium and radium.

The same year Rutherford discovered, through absorption studies with thin metal foils, that uranium and thorium emit two different types of rays, which he called alpha and beta rays. Two years later Villard discovered the third type, gamma rays. A breakthrough in these studies came in 1902 when Rutherford and Soddy discovered the exponential decay of radioactivity. They developed the theory of the instability of the atoms, exponential decay and the transformation of elements in this process. According to their theory, uranium and thorium atoms change serially to new elements through radioactive decay. Later studies of the penetration and backscattering of alpha particles impinging on very thin gold foils showed that some of the particles were strongly deflected in passing through the films, and even thrown back. This was contradictory to current ideas about the atom and led Rutherford to a new model of the atomic nucleus in 1911, according to which almost all the mass of the atom is concentrated in an extremely small nucleus. Eight years later Rutherford discovered that bombardment of nitrogen nuclei with alpha particles would in some cases lead to a nuclear transformation with the emission of a proton.

Rutherford's model could give no satisfactory explanation of the behaviour of electrons inside the atom. Bohr solved this dilemma in 1913 with a new atomic model, based partly on Planck's quantum hypothesis. This paved the way to modern quantum mechanics (Heisenberg 1925 and Schrödinger 1926). A serious obstacle to a satisfying model of the atomic nucleus was removed in 1932 when the neutron was discovered by Chadwick. Two years later, artificial (or induced) radioactivity was discovered by Juliot and Curie through the bombardment of aluminium with alpha particles. Fermi at once grasped the possibility of using the neutron as a particle to transform atomic nuclei and produce induced radioactivity. He subsequently discovered radioactive isotopes of a large number of elements.

Cosmic radiation in the atmosphere was discovered by Hesse in 1912. This radiation was evidently extremely energetic as it penetrated through all the atmosphere and even deep into the earth. The nature of cosmic rays was for a long time uncertain, but studies of their penetration were generally held to indicate that they consisted of very energetic gamma rays. Their particle nature emerged in the 1930s after the introduction of Geiger counters and counter triggered cloud chambers.

Accelerators of various types — linear, van de Graaf and cyclotrons — were developed in the thirties in order to produce an intensive beam of particles for the study of nuclear transformations and for the production of radioisotopes. Cockcroft and Walton succeeded in producing the first artificial nuclear disintegration through proton bombardment of lithium in 1932.

Bombarding uranium with neutrons brought Fermi in 1934 a result that was difficult to explain, a puzzle that was first solved four years later. The scientific community was taken by surprise in 1938 when Hahn and Strassman discovered that uranium nuclei were being split by the neutrons into two parts of similar size with a simultaneous release of a large quantity of energy. This led to the huge atomic energy program in the United States during World War II, resulting in the first nuclear reactor in 1942 and the atomic bomb in 1945. In 1952–54 the first hydrogen bombs were tested.

In 1947 Libby discovered ^{14}C (radiocarbon) in nature and subsequently invented the radiocarbon dating method. In order to measure the very weak concentration of natural radiocarbon he developed a new measuring technique that is the subject of this book, low-level counting. This technique has since expanded to the measurement of a large number of alpha-, beta- and gamma-active radioisotopes using a variety of new detectors.

2.2 Detectors and instruments

At the turn of the century, photographic plates and ionization chambers were used to detect and measure radiation from X-ray tubes and radioactive elements. Important progress was made when individual transformations of nuclei could be detected. Crooks noted in 1903 that alpha particles produce fluorescence in zinc sulfite (ZnS). Scintillations from individual alpha particles can be seen with bare eyes in a dark room, but in quantitative work they were usually observed through a magnifying glass (spinthariscopes, 1908).

The first primitive ionization pulse counters (Rutherford and Geiger 1912) were invented a few years later when current peaks could be observed in the deflection of an electroscope recording the current from an ionization chamber.

The cloud chamber (Wilson 1911), where small water droplets visualize the path of alpha particles, brought nuclear research a powerful tool. The large bubble chambers of modern high energy research are their descendants.

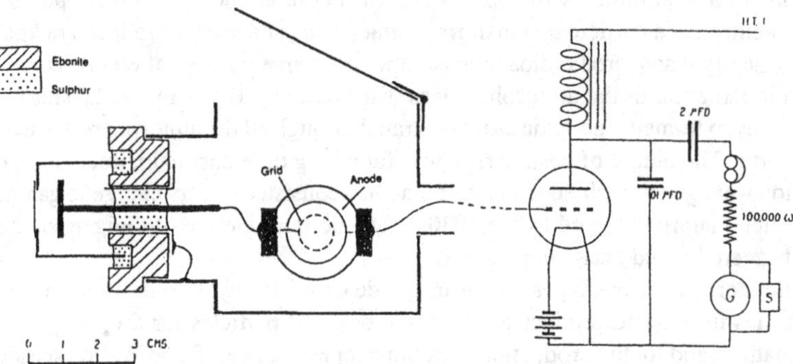

Fig. 2.2. The use of radiotubes for amplifying and counting pulses from nuclear detectors ushered in a revolution in the study of radioactivity. The figure shows one of the earliest applications of this type. A radiotube amplifies the alpha pulses from an ionization chamber. A second amplifying tube drives the galvanometer coil and headphones.

The early ionization pulse counters could only record pulses from the densely ionizing alpha particles and their sensitive volume was very small. The Geiger counter, invented by Geiger and Müller in 1928, brought nuclear research an invaluable tool. It could be made in large sizes and it detected not only the densely ionizing alpha particles, but also the weakest ionization of beta-particles and electrons because of the large internal magnification of the primary ionization. Even a single free electron triggers the counter. A continuation of this development later brought us the proportional counter.

During the late twenties and early thirties, the electron tube was adopted from the fast expanding radio technique. It was first to design fast coincidence circuits for cosmic ray research with the new Geiger counters (Rossi 1929). A year later a "scale of two" was designed in order to count Geiger pulses coming at a rate that was too fast for mechanical counters. The first electronic amplifiers, used to amplify the weak pulses of proportional chambers, appeared at the same time.

Photoelectric tubes were developed during the 1930s for use in the newly invented television technique. A surprisingly long time elapsed before they were used to detect the faint flashes produced when alpha particles strike a thin layer of ZnS. This came about first in 1945. The same year, Kallman discovered that beta particles and electrons, ejected from atoms by gamma rays, produce faint flashes in various organic crystals and four years later he discovered the scintillation of organic liquids. In 1948 Hofstadter discovered scintillations produced by gamma rays in NaI activated by thallium.

Internal liquid scintillation counters (LSC) soon became a powerful tool for the measurement of pure β-emitters such as tritium, ^{14}C and ^{32}P. These radioisotopes were used as tracers at a rapidly increasing rate in chemistry, biochemistry and medicine, as reactor-produced isotopes became readily available in the late 1950s.

The NaI scintillation counter brought scientists not only the first high efficiency gamma counter, but also a detector that gave electric pulses proportional to the energy deposited by gamma photons in the crystal. The need now arose to sort pulses into channels according to their size, in addition to counting them. Multi-channel pulse-height analyzers were developed. This development lead to the birth of gamma spectrometry and, later, also to alpha spectrometry (with Si(Li) diodes).

The transistor revolutionized electronic techniques during this period. It was invented in 1947, but more than 10 years elapsed until it could compete seriously with electron tubes in general scientific instruments. In the middle of the 1960s transistors replaced tubes completely in a few years. A second revolution followed a few years later when integrated circuits were introduced, and a third revolution came with the invention of the microprocessor (1971) and its gradual introduction in complex electronic circuits, especially computers. The semiconductor industry brought us not only transistors, but indirectly also the modern silicone and germanium diode detectors. They are a spin-off from the industry that developed the difficult technique of producing extremely high purity crystals, the technique of controlled doping, and an understanding of the electric properties of these crystals. Large area Si(Li) crystals and Ge(Li) diodes (doped with lithium) were developed in the early sixties for high resolution alpha- and gamma-spectroscopy. Some 10 years later it became possible to produce high purity germanium crystals, which made lithium doping superfluous. This paved the way for larger and better germanium diodes. During the last few years, large area Si diodes, with very low dark current, have been developed for detecting weak light signals. These diodes will probably replace the more complex photomultiplier tubes in some scintillation equipment.

With the introduction of reasonably priced minicomputers in the seventies, the powerful computer technique was applied to complex radioactivity measurements. In the eighties, when a large variety of inexpensive PC-computers appeared on the market, general purpose computers gradually became an integral part of a large number of radioactivity measuring systems. In the near future all radioactivity measuring systems, except the simplest ones, will probably consist of a computer with a small additional interface unit for the radiation detector.

2.3 History of low-level counting

The technique of low-level counting was invented and given a name by Williard F. Libby in his pioneering work in the late 1940s at the University of Chicago when he developed the radiocarbon (^{14}C) dating method. He had shown, together with his co-workers, that ^{14}C produced in the upper atmosphere by cosmic ray neutrons was found in recent biological material in a concentration of about 10 dis/min per gram of carbon. He realized that this might be used for age determination in archaeology and geology.

He demonstrated the presence of radiocarbon in recent biomethane by measuring enriched samples, using an unshielded gas proportional counter. Libby realized that a detailed study of radiocarbon in nature would never be a practical possibility if each sample had to be enriched isotopically. Before the war he had developed a sensitive Geiger counter, the screen-wall counter, for the measurement of solid internal samples of low specific activity. By smearing the carbon samples (black carbon), in a thin layer on the large inner surface of the counter wall (the cathode) he could measure a considerable amount of carbon. His counter maximized both sample size and geometrical counting efficiency.

However, the background count rate was a serious problem. Libby shielded his counter with a 20 cm thick layer of iron (Anderson et al., 1951), but his background was still about 100 cpm. The pulses evidently came mainly from the energetic muons of cosmic rays, which were only slightly attenuated by the thick iron layer. "This problem remained for some weeks until the idea of anticoincidence shielding occurred." (Libby 1967). We do not know who made this important contribution to the technique. The sample counter was now surrounded with an array of long cylindrical Geiger counters of the type used extensively at that time for cosmic-ray research. In this arrangement they are called anticoincidence, anticosmic or guard counters. Pulses from the guard counters block the pulses that muons induce in the sample counter. This reduced the background dramatically, and after making various refinements it had been reduced to 6 cpm. Libby's first counting system is shown in Figure 1.1. One cannot but admire the ingenuity and the high technical state of this first low-level counting system. Figure 2.3, a historic figure, demonstrated the foundation of the radiocarbon dating technique. It shows the results of measurements, made by Libby's group, of the specific concentration of ^{14}C in samples of known historic age. The count rate follows closely the line drawn with the measured half-life of radiocarbon. This work proved beyond doubt the basis of the radiocarbon dating method. Interest in the new dating method was immediate and widespread and in the ensuing years dating laboratories were established all over the world and various improvements in the systems were made.

In the 1950s the technique of low level counting spread to other isotopes and new detectors were added.

Gas proportional counter systems. The most important early improvement in radiocarbon dating was the introduction of gas proportional counters using carbon dioxide, methane or acetylene. Hassel de Vries pioneered in the use of CO_2 as a counting gas after he succeeded in purifying it sufficiently (de Vries and Barendsen, 1952), reducing the O_2 concentration below 1 ppm. He discovered that cosmic-ray neutrons gave a significant background contribution, which he suppressed by inserting a layer of boron/paraffin in the iron shield. Instead of using an array of single Geiger counters for anticosmic shielding he applied a compact guard proportional counter of new design introduced by Reath et al. (1951). The annular space between two cylinders of different diameters, with equidistantly spaced anodes between them, served as a guard counter. Kulp and Tryon (1951) added an extra absorbing layer of mercury between the guard and sample counter for partial absorption of gamma-rays from contamination in the main shield and secondary photons formed there by charged particles of the cosmic rays. De Vries simplified the inner shield by replacing mercury with radiopure lead.

The background components of gas proportional counters were analyzed at various lab-

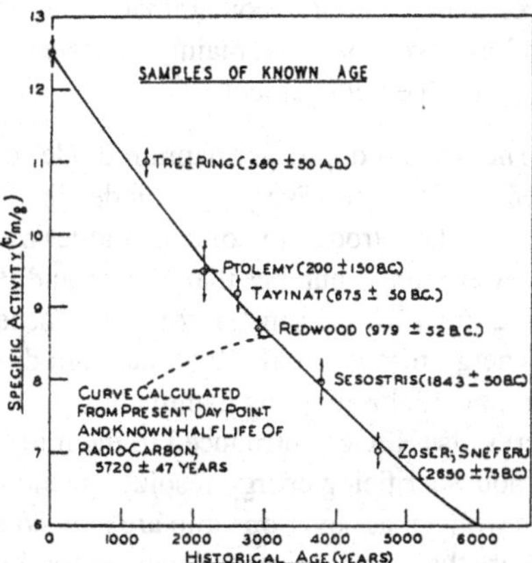

Fig. 2.3. A historic figure, from an article of Arnold and Libby (1949), demonstrating that the specific activity of ^{14}C in old organic material of known age followed the known decay of ^{14}C.

oratories. The dating laboratory in Trondheim (Nydal et al., 1976; Gulliksen and Nydal, 1979) and the University of Bern (Oeschger and Loosli, 1975) were most active in this field. In Bern a low-level laboratory was installed 25 meters (70 meters water equivalent, mwe) below the University building, and the background of various counters was measured in a basement and underground laboratory and also in a tunnel with about 3000 mwe overburden (Oeschger et al., 1982).

NaI scintillation systems. The NaI scintillation detector, invented in 1948, ushered in a revolution in gamma ray counting and brought us gamma spectroscopy. Geiger counters had up to this time been the main detector for measuring gamma radiation. Their detection efficiency was only about 1% and they gave no information about the energy of the radiation. The detection efficiency of a NaI crystal for photons passing through it can approach 100% for photons of energy below 500 keV. Equally important, the size of the pulses from the photomultiplier tube are proportional to the deposited energy. Photons that lose all their energy in the crystal give a peak in the pulse height spectrum, corresponding to the energy of the respective gamma line. This gave birth to gamma spectroscopy and started the development of multichannel pulse height analyzers.

As time went on, NaI crystals were made in even larger sizes and in a variety of shapes and these detector units were incorporated into systems based on the same principles as the radiocarbon systems. For almost 20 years NaI scintillation counters were the most important detectors in low-level gamma counting. They had, however, two weaknesses: the energy resolution was often insufficient and the glass tubes of the photomultipliers were contaminated with primordial radioactivity. In the late 1970s, NaI scintillation counters were gradually replaced by germanium diodes that offered an energy resolution almost 40 times better and which were practically free of inherent radiocontamination. Nevertheless, NaI crystals still play an important role in low-level counting, mainly in systems with large crystals in various coincidence or anticoincidence arrangements.

Germanium spectrometers. The lithium doped germanium diode for the measurement of gamma radiation was invented in 1962. However, these diodes first reached their full usefulness in low-level counting after the introduction of large diodes made of high purity germanium. Germanium has a lower atomic number than iodine and its linear photoelectric absorption coefficient is about 2 times lower than that for NaI. This is, however, more than offset by the 40 times better energy resolution of the germanium diodes. The low narrow gamma peaks allow the separation of closely lying gamma lines.

High purity germanium crystals (HPGe), introduced in the mid 1970s, have been produced in ever larger sizes without sacrificing energy resolution, and their price has continually been declining. Crystals with a mass over one kilo are now common.

Today, germanium diodes are the most important detectors for the low-level technique, not only for their high sensitivity but also for the wide range of weak radioactive samples measured with them. Background studies, carried out with ultralow-level germanium spectrometers, have brought us valuable information about the sources of background. This work, discussed in later chapters, has been transforming low-level counting from an art to science.

Liquid scintillation counters. The liquid scintillation counter was invented in 1947. It was tested for the measurement of natural ^{14}C as early as 1953, but it was not until the introduction of the two tube coincidence technique and the invention of a practical benzene synthesis in the early 1960s, that this technique could compete with gas counting in ^{14}C dating. Moderately priced commercial liquid scintillation counting systems were developed to a high technical standard and produced in thousands for tracer work in the biomedical field. Despite high background compared to gas proportional counters, they were in most cases used without any modifications in dating work, and after 1970 most new dating laboratories chose this method. Although used for the measurement of weakly radioactive ^{14}C samples, these counters can hardly be classified as low-level systems.

During the 1970s and early 1980s, specially constructed systems, with heavy shielding and guard counters, were constructed at various research institutes, but without sufficient success to warrant the large effort. A successful low-level liquid scintillation counter was first introduced in 1984, the Quantulus, produced by the Finnish firm Wallac. It has a thick lead shield, 7–12 cm, and a large annular liquid scintillation guard counter. Packard has also introduced a low-level liquid scintillation counter, where significant background reduction is obtained through pulse shape analysis.

Liquid scintillation counters give quite useful energy resolution for alpha particles, 250–350 keV FWHM in favorable cases. Furthermore, alpha pulses can be distinguished from beta pulses through pulse decay analysis, giving a background reduction up to two orders of magnitude, a few counts per day. This has greatly increased the use of liquid scintillation counters in alpha measurements in recent years.

Si(Li) alpha spectrometers. Silicone diodes where residual impurity conductivity is compensated for by lithium, for the detection of alpha and other charged particles, were developed in the early 1960s. They have been produced in ever larger sizes and are now available with diameters up to about 6 cm. Earlier they were of the surface barrier type, but they are now gradually being replaced by diodes of the ion implanted type, which has a lower leakage current and thinner surface windows that can easily be cleaned.

Multicounter systems. With the fall in prices of electronic equipment, following the introduction of integrated circuits in the mid 1960s, it became economic to increase the number of detectors in low-level work. Multisample systems with flow counters were developed in 1974 (Bötter-Jensen et al., 1977) and a system with 9 identical tightly packed CO_2 detectors was constructed at the radiocarbon dating laboratory in Heidelberg (Schoch et al., 1982), were a minicomputer replaced most of the electronic units of conventional systems. Multicounter systems with Si(Li) diodes are now available, and a multicounter system with 4 liquid scintillation detectors has recently been presented (Einarsson, 1992).

Anticosmic counter systems (guard counters). Libby used cosmic ray counters in his guard system. A few years later a dedicated guard counter was developed, as described above. Counters of this type in different sizes were used in important background studies at the radiocarbon dating laboratory in Trondheim in Norway (Nydal et al., 1977).

Flat Geiger and gas proportional counters of the flow type were introduced as guard counters in the late 1950s with window sample counters for the measurement of solid samples and larger guard counters of this type were used in multidetector systems of similar type (Bötter-Jensen et al., 1977). Larger guard counters were used in the Heidelberg radiocarbon system mentioned above. Some years later, a greatly improved arrangement with guard counters was introduced. Large flat guard counters were applied to low-level germanium spectrometers, where they were placed on the outer sides of the lead shield (external guard counters).

Scintillation counters, using either liquid or plastic scintillators, have been used as guard counters for all types of low-level counting systems. As mentioned above, an annular liquid scintillation detector is a part of the Quantulus system, and a similar counter is used as guard detector for a set of three gas proportional counters in a radiocarbon dating system, giving a very low background (Mäntynen et al., 1987). Finally, flat, box-shaped scintillating guard counters have been used as external guard counters in low-level germanium spectrometers, using either a liquid or plates of plastic as the scintillating medium (Reeves and Arthur, 1988).

Large NaI crystals have also been used as guard counters, in spite of their high price. A large annular NaI crystal with an outer diameter of 30 cm and inner diameter of 10 cm was used in a system with 12 small gas proportional counters giving a very efficient background (Otlet et al., 1986).

Electronic systems. Development in the electronic technique — radio tube, transistor, integrated circuits, large scale integrated circuits, microprocessors and PC–computers — has been revolutionary. Practically any electronic function, in signal or data processing, is now cheap. But the full power of this technique is recent and much of the low-level technique, as we have it today, was developed when scientists only enjoyed a small part of this technique. One example will be given. Although multichannel analyzers have been used since late 1950s it was only in late 1980s, after the introduction of PC–computers, that multichannel analyzers are gradually becoming common in systems with all types of detectors. Modern electronic technique has made pulse shape analysis and time sequence analysis easy, and their importance is continuously increasing.

Chapter 3

Atomic nuclei and binding energy

3.1 Atomic nuclei

Until the beginning of the present century scientists firmly believed that atoms were the smallest entity of matter and that atoms belonging to the same chemical element were identical in all respects. Investigations of electric discharge in glass tubes containing low pressure gas gave the first indication that these views had to be modified. Studies of rays emitted from the cathode of such tubes culminated in the work of W. Röntgen and J. J. Thomson. Thomson showed that these rays consist of subatomic particles carrying negative charge, all having the same ratio of charge to mass (e/m) and that the particles had a mass of about 1/2000th that of the hydrogen atom. The particle was given the name *electron*, a name that had been proposed some years earlier.

On the basis of his experimental work Thomson described the atom as a sphere of positively charged uniform mass in which electrons were evenly spread. However, studies of the scattering of alpha-particles in very thin metal films, performed some ten years later by two of Rutherford's collaborators, Geiger and Marsden, paved the way for a new model of the atomic nucleus. The expected result of their experiment was that the alpha-particles would pass through the metal film with only a minor change in direction, but they discovered that some of them were deflected by a large angle, some even had their momentum reversed, they were scattered in the backward direction. Rutherford, using their results, proposed a model of the atomic nucleus in 1911 and thus laid the foundation of our present nuclear theory. He postulated that practically the whole mass of the atom and its positive charge is concentrated in an extremely small volume with a diameter of the order of 10^{-14} m, or only about 1/10,000th of the diameter of the atom. Later work confirmed Rutherford's model and extensive studies with various scattering particles, supported by use of other methods, have shown that nuclei have a radius of

$$r = 1.2 \cdot 10^{-15} \sqrt[3]{A} \quad \text{m} \tag{3.1}$$

where A is the atomic *mass number*.

At this time it seemed most probable that the negatively charged electrons were revolving around a positive nucleus in the same way as the planets revolve around the sun. According to classical theory, electrons in such an orbit would emit electromagnetic radiation because of the constant radial acceleration. They would thus rapidly lose their kinetic energy. Niels Bohr solved this contradiction in 1913 by assuming that classical electromagnetic theory was not valid in the micro-cosmos of the atoms. Using Planck's quantum theory, developed for an unrelated phenomenon and still not well founded, Bohr postulated that electrons revolve around the nucleus in certain stable orbits determined by a given quantum condition, without losing energy. Bohr's theory was subsequently expanded and used by scientists to explain a large number of atomic phenomena. A more complete quantum theory was proposed by Heisenberg in 1925 and by Schrödinger in 1926, using a different mathematical formulation.

The scattering experiments gave Rutherford only a crude value for the positive charge of the nucleus, but in 1912 Moseley's X-ray diffraction studies showed that a nucleus has a number of elemantary charges that is equal to the element's atomic number Z in the periodic table.

Now it had to be assumed that there were $A - Z$ electrons in the nucleus in order to give it a correct electric charge. The new quantum therory gradually made this assumption increasingly more difficult to accept. This problem was solved in 1932 when Chadwick, still another of Rutherford's co-workers, discovered the neutron in the atomic nucleus. The neutron turned out to have almost the same mass as the proton, but no electric charge. According to the refined nuclear model that now emerged, the nucleus has Z protons and N neutrons, and a total of A *nucleons* (a collective name for protons and neutrons). Z is the *atomic number* of the nucleus and A its *mass number*.

Table 3.1. Physical properties of nuclear particles.

Name of particle	Symbol	Charge [e]	Rest mass [10^{-27} kg]	Rest mass [amu]	Equivalent energy [MeV]
Neutron	n	0	1.6748	1.00866	939.5
Proton	p	1	1.6728	1.00728	948.2
Alpha-particle	α	2	6.64250	4.00157	3727.2
Electron (Beta-particle)	e^- β^-	-1	0.000911	0.00549	0.511
Positron	β^+	1	0.000911	0.00549	0.511
Muon	μ	1, -1	0.189	0.114	105.
Gamma-particle	γ	0	0	0	0
X-ray		0	0	0	0
Neutrino	ν	0	0	0	0

Table 3.1 shows some properties of the particles discussed above as well as the corresponding properties of a few other particles.

3.2 Isotopes

After ten years of studying radioactivity, scientists had found a large number of radioactive species that were being considered new chemical elements. Many of these apparently had the same chemical properties as known elements. At the time it was very difficult for chemists to accept the idea that atoms, belonging to the same element, could behave differently. Fajans and Soddy solved this puzzle by stating that each element could have atoms of different weight and with different radioactive properties, and Soddy introduced in 1914 the name *isotope* for different atomic species of an element.

In 1913 J. J. Thomson discovered that an accelerated beam of neon atoms split into two beams, corresponding to atoms having a mass number 20 and 22, when it passed through a crossed electric and magnetic field. His instrument was the forerunner of the mass spectrometer. After World War I, Aston resumed Thomson's work and developed mass spectrometers that could be used to measure very precisely the mass of atoms as well as their relative abundance for individual elements. His patient and laborious work demonstrated that most elements have more than one isotope. Fluorine has, for example, only one stable isotope, ^{19}F, whereas tin has ten, with abundances ranging from 0.4 % to 28.5 %.

Atomic species with the same number Z of protons, and therefore belonging to the same chemical element, are called isotopes. Nuclei that have the same number $A = Z + N$ of nucleons are called *isobars*. A ***nuclide*** is a species of atoms with a specified number of protons and neutrons in their nucleus. If the mass number of a nuclide is to be specified, it is, according to convention in nuclear physics, indicated by a superscript on the left side of the chemical symbol, for example, ^{14}C and ^{238}U. The atomic number Z is indicated in a similar way by subscript, such as $_6$C and $_{92}$U. Generally one can describe an arbitrary nuclide by $^A_Z X$, where A is the atomic mass number, Z is the atomic number, or number of protons, and X stands for the chemical symbol of the element.

3.3 Atomic masses and binding energy

As Aston's studies progressed, he discovered that the total mass of each atom was less than the sum of the masses of its constituents, protons, neutrons and electrons. This could be explained by Einstein's theory of relativity (1905), using the most widely known equation of physics,

$$E = mc^2, \tag{3.2}$$

stating the equivalence of energy E and mass m. c denotes the speed of light.

The high energy of alpha particles emitted from naturally radioactive nuclei indicated that the energy binding the nucleons together was very large. The binding energy of a nucleus with Z protons and N neutrons, $B(Z,N)$, is the energy needed to separate all the nucleons and end up with free protons and neutrons. According to Einstein's theory, the difference of the mass of the nucleus $M_{Nu(A,Z)}$ and the sum of the mass of its protons and neutrons is equal to its binding energy:

$$B(Z,N) = [ZM_p + NM_n - M_{Nu(N,Z)}]\, c^2 \tag{3.3}$$

where M_p is the mass of a free proton, M_n is the mass of a free neutron and $M_{Nu(N,Z)}$ is the mass of the nucleus.

Frequently mass and energy are given by the same unit and in that case the factor c^2 in Equation 3.2 can be left out. The mass difference on the right side of Equation 3.3 is called the mass defect. The masses of various particles of interest is given in Table 3.1, both in units of mass and energy.

In practice, atomic masses, which are measured in mass spectrometers, are used rather than nuclear masses. In radioactive decay it is not only the nucleus which changes. Rather one atom is transformed into another. When the binding energy difference of a mother and a daughter atom is calculated, only a negligible error is introduced if atomic masses are used in Equation 3.3.

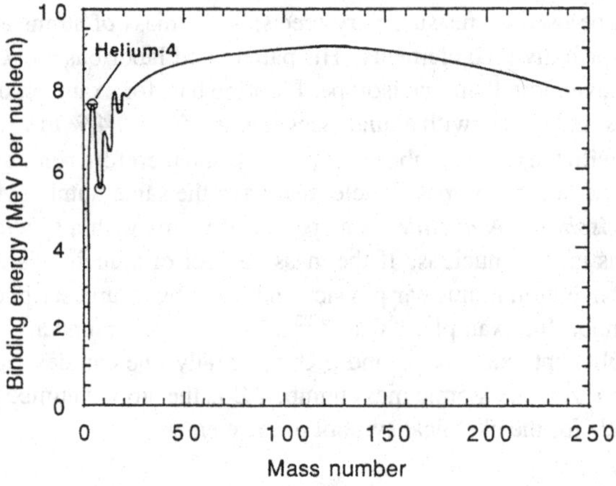

Fig. 3.1. Binding energy vs. mass number N

The binding energy is a measure of the stability of a nuclide. If either of its two isobars has lower binding energy, it may be unstable. The difference in binding energy will be released through the radiation emitted. These transformations of unstable nuclei, radioactive decay, will be discussed in the next chapter. Figure 3.1 shows the binding energy per nucleon in stable isotopes as a function of the mass number. It has a maximum at N about 70. The nuclei in this region are therefore most stable.

Chapter 4

Radioactivity

4.1 Unstable nuclides and radioactivity

Radioactivity is the transformation, also called decay or disintegration, of unstable nuclei in which they change through the emission of alpha, beta and gamma radiation by one of the processes discussed below. We begin this chapter by discussing briefly stable and unstable nuclei. Each transformation process will be described separately in the following sections. Some nuclides, radionuclides, transform spontaneously into another species belonging to a different chemical element, with a different number of protons and neutrons. We say that these nuclides are unstable or radioactive. They transform through:

1. Change of a neutron into a proton with simultaneous emission of a negative electron, called a beta particle in this case.

2. Change of a proton into a neutron with simultaneous emission of a positive electron (positron) or through capture of an orbiting electron.

3. Emission of an alpha particle.

The daughter nucleus is often left in an excited state and it gets rid of the excess energy through emission of gamma radiation. But which nuclei are stable and which unstable? There are two general conditions for stability: (a) the ratio of neutrons to protons must lie in a certain range (which depends on the mass number), (b) the total number of protons must not be higher than 82, the atomic number of lead. Of all known nuclear species, less than half are stable.

If all stable isotopes are plotted as points on a graph where the x-axis gives the number Z of protons in a nucleus and the y-axis the number N of neutrons, the points will fall into a narrow region as depicted in Figure 4.1. Nuclides lying below this region have an excess of protons and seek stability by reducing their number, transforming a proton into a neutron by radioactive decay. Nuclides above the region have an excess of neutrons and one of these can be transformed into a proton. The transformation processes can conveniently be summarized by *decay scheme diagrams* such as the two shown in Figure 4.2.

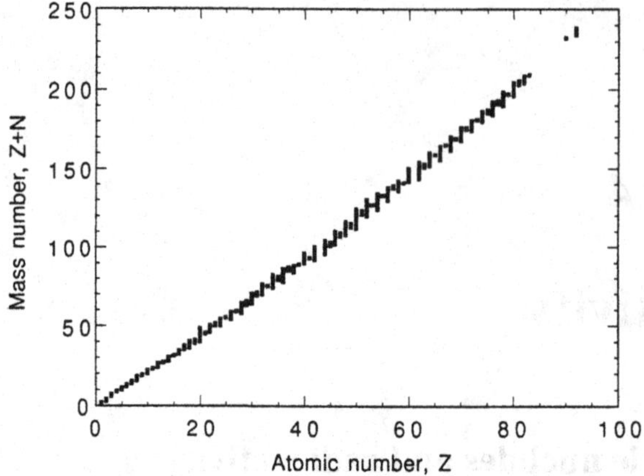

Fig. 4.1. A plot showing stable isotopes. Each point represents a stable radionuclide with a given number of protons (Z) and the total number of nucleons (A).

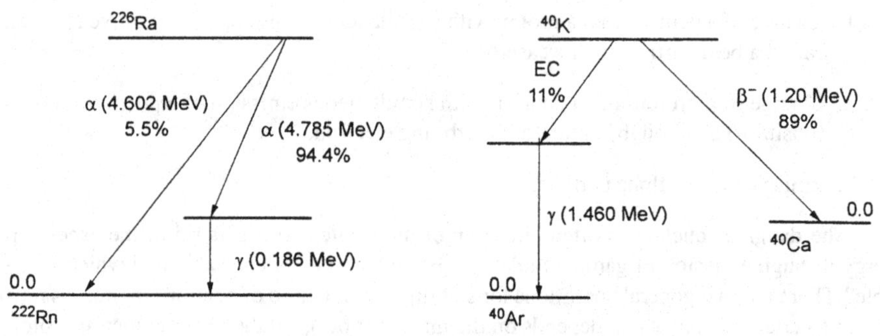

Fig. 4.2. The decay scheme of ^{226}Ra and ^{40}K

An arrow slanting downwards to the left represents a process where the atomic number of the daughter nucleus is lower (alpha emission, positron emission or electron capture) and downwards to the right a beta particle emission. Emitted gamma photons are represented by vertical wavy lines.

The ground state of the mother and daughter nuclei are shown by horizontal lines, as are the excited energy levels of the daughter nucleus. Various types of information about the transformation is usually given on the diagram: energy and relative yield of different groups of particles, half-life, etc.

The various radiation processes occurring in nuclear decay will be discussed in the following sections.

4.2 Alpha decay

Almost all alpha active radionuclides belong to the heaviest elements. The repulsive forces of the protons can make the nuclei unstable. The emission of an alpha particle, removing two protons and two neutrons, is then an energetically favorable process.

The alpha particles are emitted with discrete energies, each energy group representing a disintegration that ends in a given energy level of the daughter nucleus, which can either be the ground state or one of the excited states. In the latter case gamma rays are emitted (usually within 10^{-10} s) and the nucleus will fall to the ground level, either directly or in steps through intermediate excited levels.

^{226}Ra emits alpha particles of two different energies. 94.3% of the disintegrations end in the ground state of ^{222}Rn when an alpha particle with an energy of 4.777 MeV is emitted. Alpha particles of somewhat lower energy, 4.591 MeV, are emitted in 5.7% of the disintegrations resulting in an excited daughter nucleus which emits simultaneously a 0.186 MeV gamma photon. In 35% of excitation events no photon is emitted but the energy is transferred to one of the orbital electrons through internal conversion (see Section 4.7).

Because of the relatively large mass of the alpha particle, the nucleus recoils. The total energy of transformation (corresponding to the mass difference of the mother and daughter nuclei) is divided between the alpha particle and the nucleus, the latter receiving 50 to 100 keV, the exact amount depending mainly on the energy of the alpha particle. The recoil energy is important in some cases. It makes it possible for atoms to escape from small grains in the soil, where the recoiling atom has a range of about 50 nanometers. The atoms may be picked up by ground water or they may diffuse into the atmosphere. The recoil frequently causes disequilibrium in the natural radioactive series, as atoms that have been thrown out of the gitter structure of its crystal have a higher probability of being washed away with ground water and eventually precipitated at a different place. ^{234}U, a long lived decay progeny of ^{238}U, is frequently removed from its crystal, and their activity ratio in nature is therefore far from unity, it may range from 0.5 to 100. Radium is frequently in disequilibrium with its source, ^{238}U. Atoms in counting samples that have decayed through alpha emission may be thrown out of the sample by recoil and deposited on the detector window. If this daughter atom is radioactive and with a long half-life, this may increase the background of the detector.

4.3 Beta emission

All chemical elements have radioactive isotopes, where the ratio of neutrons to protons is outside the region of stable isotopes, as discussed in Section 4.1. If a nucleus lies above this range, it has a surplus of neutrons and will compensate for this by transforming a neutron into a proton, with the simultaneous emission of a negative beta particle (identical to the electron):

$$n \rightarrow p^+ + e^- \tag{4.1}$$

This increases the atomic number by one unit, but leaves the mass number unchanged, i.e., the mother and daughter nuclei are isobars. The process is described by the following general equation,

$$^A_Z X \rightarrow {}^A_{Z+1}Y + e^- \tag{4.2}$$

The decay of the well known radioisotope of carbon ^{14}C (C–14 or radiocarbon) is an example:

$$^{14}_{6}C \rightarrow {}^{14}_{7}N + e^- \tag{4.3}$$

It should be noted that the energy corresponding to the mass difference of the mother and daughter nuclei is not only transformed into kinetic energy of the beta particle and the neutrino (see below), but also into the mass of the beta particle, which is equivalent to an energy of 511 keV. The mass difference between two isobars must therefore surpass this minimum difference in order to permit β^- decay.

As each disintegration results in the same mass decrease for a given radionuclide, the amount of energy emitted should be the same. Studies in the 1920s showed, however, that this was not the case. Beta particles turned out to have a continuous energy spectrum, from zero to a maximum energy E_{max}, when the electron receives all the disintegration energy. The continuous spectrum confronted physicists in the twenties with a serious dilemma as it contradicted the firmly established law of conservation of energy. Pauli solved the problem in 1929 by postulating that another particle, a neutrino, was emitted simultaneously and shared the disintegration energy with the electron. He assumed that the neutrino had no charge and practically no mass. This hypothesis was further supported by Fermi who used the hypothesis to develop a theory that described correctly the experimentally measured beta energy spectrum in 1934. According to Fermi's theory equation 4.3 should be written

$$^{14}_{6}C \rightarrow {}^{14}_{7}N + e^- + \overline{\nu} \tag{4.4}$$

where $\overline{\nu}$ denotes the antineutrino. Figure 4.3 shows the energy spectrum of ^{14}C.

The neutrino is a very evasive particle because of its extremely weak interaction with matter. Its existence was not proved experimentally until 1953.

The maximum energy of beta particles spans a broad range, from 18 keV for ^{3}H (tritium) to 3.24 MeV for ^{214}Pb. The average energy of beta particles is about 1/3 of the maximum energy. The decay schemes can be quite complex, as shown by that for the ^{134}Cs (Figure 4.4), which decays into a number of excited energy levels of the daughter nuclide, ^{134}Ba. Further, the decay of the excited nuclei can follow different routes to the ground level, giving

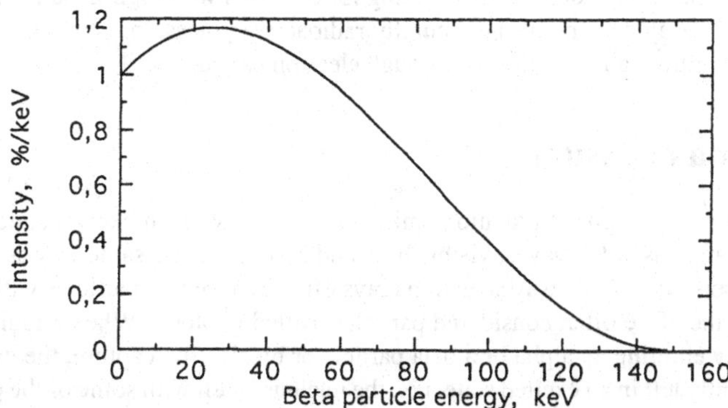

Fig. 4.3. The beta energy spectrum of ^{14}C.

a considerable number of gamma lines each with a given relative probability (gamma per disintegration).

A nucleus with a surplus of protons will decay into a more stable state through the transformation of a proton into a neutron with simultaneous emission of a positive beta particle (positron, a negative electron) and a neutrino:

$$p \rightarrow n + \beta^{+} + \nu. \tag{4.5}$$

This process is described by the following general equation,

$${}^{N}_{Z}X \rightarrow {}^{N}_{Z-1}Y + e^{+} + \nu. \tag{4.6}$$

The atomic number decreases by one unit. The energy spectrum of the emitted beta particles is similar to that of negative beta particles. It can be shown that the mass difference between the mother and daughter nuclides must exceed two electron masses for making positron decay possible.

4.4 Electron capture, EC

A nucleus with an excess of protons can also reach a stable state by capturing an electron from one of the inner shells, whereby a proton is converted to a neutron:

$$p + e^{-} \rightarrow n + \nu. \tag{4.7}$$

Most of the disintegration energy is carried away by the neutrino. Electron capture can occur in cases where the emission of a positron is energetically not possible. In some cases a nucleus can disintegrate in two different ways, by emission of a beta particle or through electron capture. The latter process is dominating for elements with high Z numbers. Figure 4.2 shows the decay scheme of the naturally radioactive potassium isotope, ^{40}K, that decays either by emitting a β^- particle or through electron capture.

4.5 Gamma emission

Gamma rays are electromagnetic radiation emitted by a nucleus in an excited state. They are of the same nature as radio waves, visible light and X–rays. In the same way as all particles behave in some cases like waves, gamma rays often behave like particles with mass. Gamma rays are therefore often considered particles, called photons. When a radioactive nucleus decays by emitting an alpha or a beta particle or by electron capture, the daughter nucleus is frequently left in an excited state, i.e., the nucleus is left with some of the disintegration energy. Each nucleus has its characteristic excited energy levels. These are similar to those in atoms and molecules, but with much higher energy. The excited nucleus will, in most cases simultaneously (within 10^{-10} s), emit gamma rays of discrete energies, corresponding to the energy difference between the higher and lower energy levels. If left in one of the higher excited energy levels, it can fall to the ground level in a single step or in two or more steps. A decay scheme, illustrating how a large number of gamma photons with discrete energies are emitted, has already been shown (Figure 4.4).

The energy of the photons spans a broad range, from about 50 keV to 3 MeV. According to historical tradition, the name gamma rays was only used for photons emitted from a nucleus during radioactive decay. Today it is also used for electromagnetic radiation produced by cosmic rays in the air and solid earth and by high energy particles from modern accelerators.

4.6 Delayed gamma emission

In some cases an excited nucleus, formed by nuclear decay, may be left in a metastable state where it does not get rid of the energy excess immediately, but after some delay. The emission from a metastable excited level is called an isomeric transition (IT), and obeys the law of radioactive decay (discussed in Section 4.11). It has a given half-life, which may extend from a fraction of a second to several hours. An excited nucleus in a metastable state is indicated by the letter m following the atomic number in the superscript on the left side of the chemical symbol. Figure 4.4 shows the decay scheme of ^{137}Cs, which disintegrates into the metastable ^{137m}Ba. ^{99}Mo, with a half-life $T_{1/2} = 3.6$ d, decays into an important metastable daughter, ^{99m}Tc with a half-life $T_{1/2} = 6$ h. ^{99m}Tc is used extensively in medical work. The mother element is bound to an ion exchanger in an *isotope generator* where the daughter can after a proper delay, be repeatedly separated by a simple elution with a suitable solution.

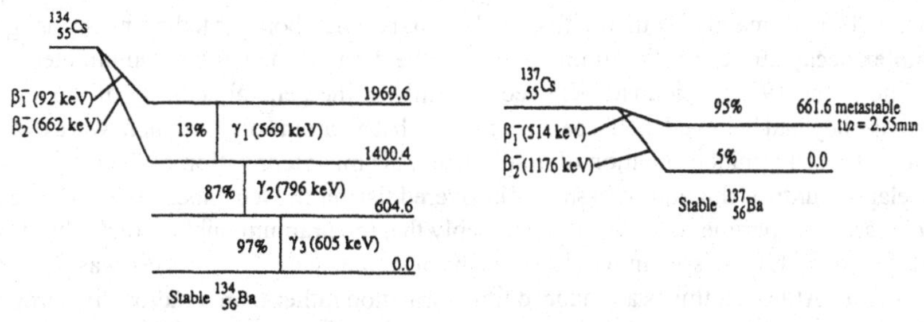

Fig. 4.4. The decay schemes of ^{134}Cs and ^{137}Cs.

4.7 Internal conversion, IC

A gamma emitting nucleus has an alternative way of losing its excitation energy. One of the orbiting electrons, usually a K electron, can absorb the excessive energy through direct interaction with the excited nucleus. The electron is then ejected from the atom with a kinetic energy E_e equal to the excitation energy E_γ minus the binding energy E_b of the electron

$$E_e = E_\gamma - E_b \tag{4.8}$$

This process is called *internal conversion* (IC). A vacancy is left in one of the innermost atomic electron shells. When this is filled by one of the outer shell electrons, X-rays with a characteristic energy are emitted.

These X-ray photons may be absorbed by one of the outer electrons, which are emitted with an energy corresponding to the difference in the two electronic energy levels minus the binding energy of the outer electron. These low-energy electrons are called *Auger electrons*. The internal conversion coefficient α is the number of conversion electrons N_e per gamma-ray photon N_γ,

$$\alpha = N_e/N_\gamma \tag{4.9}$$

In the case of the delayed gamma emission of ^{137}Cs (Figure 4.4) α is 0.11.

4.8 X-rays following IC and EC

When an electron has been removed from one of the inner shells by internal conversion or electron capture, the vacant site will immediately be filled with an electron from one of the outer shells and an X-ray photon is emitted. The photon energy will be equal to the difference in the energy levels of the two electron orbits.

4.9 Fission

Fermi irradiated a large number of elements with neutrons and studied the radio isotopes that were formed, mainly by their half-life. Uranium, when bombarded by neutrons, gave a complex decay curve that Fermi interpreted as the decay of new transuranium elements. Until late in the 1930s, scientists believed as firmly in the general stability of the atomic nucleus as they had believed, half a century earlier, that all atoms of an element were exactly identical. From Fermi's irradiation of uranium and his misintepretation of the results, four years elapsed until Hahn and Strassman discovered fission in December 1938. The result of their careful experiments showed indisputably that the uranium nucleus (actually it was the nucleus of ^{235}U) was split into two fragments of similar size. This process was given the name *fission*. Although this is an induced transformation rather than a radioactive process, it will be discussed here. Uranium nuclei can decay spontaneously by fission, but this is a very rare process. Fission normally occurs when a heavy nucleus captures a neutron or is strongly excited by some other particle.

When a ^{235}U nucleus captures a thermal neutron, violent oscillations are induced in the nucleus. These result in splitting it into two fragments (Figure 4.5). A total energy of 195 MeV is released in the splitting of each ^{235}U nucleus, most of which appears as kinetic energy of the fission fragments. Simultaneously 2–3 neutrons are emitted, which can induce fission in more uranium nuclei, eventually giving rise to a sustained chain reaction.

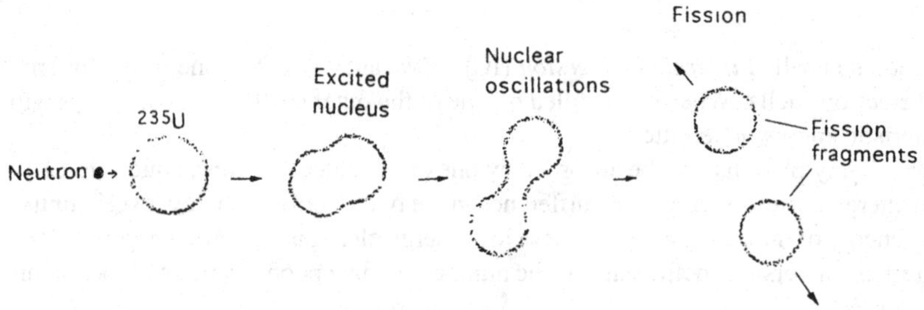

Fig. 4.5. The fission of a uranium nucleus. The capture of a neutron starts violent oscillations that result in splitting the nucleus into two nuclei, the fission products

The *fission fragments* are radioactive as they have an excess of neutrons. About 5 MeV of the fission energy appears as beta radiation and the same amount as the gamma radiation of the fission products. The radionuclides belong to a large number of chemical elements in the middle of the periodic table, and they have half-lives from a fraction of a second to thousands of years. Transuranium elements are also formed in atomic bombs and in reactors.

Less than a year after the discovery of fission, World War II broke out and a huge program was started in the USA with the aim of developing atomic bombs, and the first was tested in July 1945. After the war a large number of nuclear reactors were built in various countries, both for the production of plutonium-239 for bombs and for driving nuclear electric power plants. The development of atomic bombs was continued, resulting in a far more powerful bomb, the hydrogen bomb, based on the fusion of light nuclei. Extensive testing of the new bombs in the atmosphere, mainly from 1950 to 1962, released huge amounts of fission products that spread over the whole surface of the globe as radioactive fall-out.

Today, fission products are produced on a much larger scale in the fuel elements of nuclear electric power plants. After 2–3 years of reactor operation the fuel cells must be removed from the core of the reactors as the intense neutron radiation and heat gradually produces crystalline changes in the cladding of the fuel elements. Thereafter they either are stored indefinitely or sent to a reprocessing plant. In the latter case they are first stored for about a year before reprocessing, in which the uranium and plutonium are recovered for reuse and the highly radioactive fission products separated and stored under strict control.

The reprocessing plants will inevitably release substantial amounts of radioactivity, both as gases and dissolved in water. A considerable part of weak radioactive sample measurements are studying the distribution in the environment of fission products and other radioactive isotopes produced by the nuclear industry.

4.10 Origin of radioactive nuclei

Radioactive nuclides can be classified into three categories:

1. Primordial radioactivity, created in the nuclear syntheses in which all the matter of our solar system was originally formed.

2. Cosmogenic activity, induced both in terrestrial and extraterrestrial matter by cosmic rays.

3. Man made activity, produced in nuclear reactors, atomic bombs and in accelerators.

The primordial radioactivity, discussed in the next chapter, involves radioisotopes that have a half-life that is comparable to the age of earth (4.5×10^{-9} years) or longer. Also falling into this category are short-lived isotopes that belong to one of the three radioactive series, headed by long lived radionuclides: ^{232}Th,^{235}U and ^{238}U.

The most important cosmogenic radionuclides are tritium (^{3}H) and radiocarbon (^{14}C). Both are produced by cosmic rays in the stratosphere. Their use in geophysics is discussed in Chapter 13. In addition to these a considerable number of other cosmogenic radioisotopes, which are important in geophysical and astrophysical research, are found.

The world wide dispersal of fission products and transuranium elements coming from atomic bombs and nuclear facilities has already been mentioned. Reactors are also used for radioactivity analysis and offer, in many cases, the most sensitive analysis method. Using low-level detectors in this work can lower the sensitivity limit substantially.

Radioisotopes are used widely in scientific and technical research. Isotopes produced by neutrons, usually in reactors, emit negative beta particles. In some cases no suitable isotope of some element can be produced in this way. The production of a radioisotope emitting a positron may solve this problem. These can be produced in accelerators by bombarding material with high energy positive ions, usually ^{1}H (protons), ^{2}H (deuterons) or ^{4}He.

4.11 Kinetics of radioactive decay

Trivial as it may seem today, it was quite hard for scientists at the beginning of the century to accept the purely statistical nature of radioactive decay. The exponential decay of radioactivity can be explained by a single assumption, i.e., that the probability P of a given atom disintegrating during a time interval dt is proportional to dt.

$$P = \lambda\, dt \tag{4.10}$$

The proportionality constant λ, the *decay constant*, is characteristic of the given radionuclide. If we have N atoms of this radionuclide, the number d_N of atoms that decay during time interval dt is

$$-d_N = N\,\lambda\, dt. \tag{4.11}$$

Solving this differential equation leads to

$$N(t) = N_0\, e^{\lambda t} \tag{4.12}$$

where N_0 is the number of atoms at $t = 0$. Figure 4.6 shows the decay curve. Some atoms live a very short life, other a long one. The mean lifetime of the atoms τ is given by

$$\tau = 1/\lambda \tag{4.13}$$

After one decay time the number of atoms that have not decayed has fallen to N_0/e. Unfortunately, the rate of decay is, for historical reasons, usually described by the half-life $t_{1/2}$, given by

$$t_{1/2} = \tau \ln 2 = 0.693\,\tau \tag{4.14}$$

In our time of inexpensive electronic calculators it is often more convenient to use the following, rarely given, form of the decay:

$$N(t) = N_0\, 2^{t/t_{1/2}} \tag{4.15}$$

The unit of radioactivity is *Becquerel* (Bq), where one Bq is the disintegration rate of a radioactive source where one nucleus transforms per second. The older unit, based on the activity of one gram of radium, is the *Curie* (Ci), which is equivalent to $3.7 \cdot 10^{10}$ Bq.

4.12 Serial transformations

The daughter nucleus of a radioactive atom can be unstable, and it can be followed by a series of radionuclides. We thus meet radioactive series with the half-lives of individual nuclides ranging from a fraction of a microsecond to thousands of years. The three natural radioactive series, discussed in the next chapter, are the most important of these, but series of this type also occur in fission products. Here we will limit ourselves to the simple case where a long lived radioisotope A decays into a short lived one B and whose decay product C is stable:

$$A \xrightarrow{\lambda_A} B \xrightarrow{\lambda_B} C$$

If we start with pure isotope A ($N_B = 0$), the number of atoms B decaying per second at time t is

$$\lambda_B N_B = \lambda_A N_A (1 - e^{-\lambda_B t}) \tag{4.16}$$

where N_A and N_B denotes the number of atoms of type A and B.

Equation (4.16) shows that the activity of isotope B will grow, approaching asymptotically the disintegration rate of A (Figure 4.6). The two isotopes will end in a *secular equilibrium*, also called radioactive equilibrium, where the same number of both types of atoms decay per second. After 6.6 half-lives radioisotope B has reached 99% of its final disintegration rate.

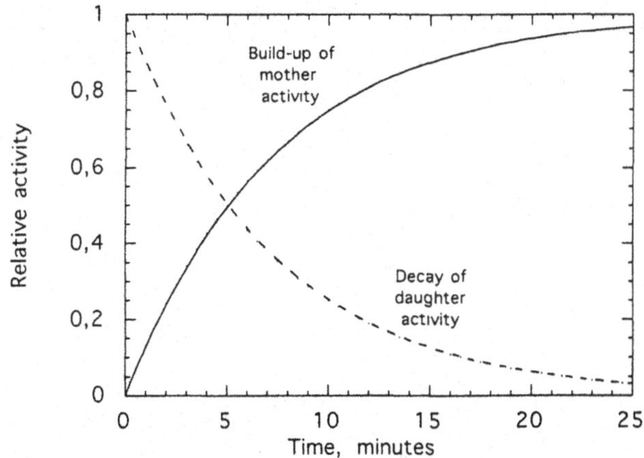

Fig. 4.6. The building up of the radioactivity of radioisotope B, half-life 5 min (curve B). Curve D, decay of radioisotope B after separation from isotope A.

4.3 Stand Transformations

For the propagation of an enlarged image... the medium... forward by a transformation S. The linear displacement... height is determined...

$$\text{Note} = R\left[\frac{F}{\eta}\right] - SRF$$

where A and S are ...

Figure 4.2 ...

Chapter 5

Interaction of radiation with matter

It is useful for scientists engaged in measuring weak radioactivity to be familiar with the interaction of radiation with matter in order to understand: (a) the response of their detectors to the radiation they want to measure and (b) how the shield attenuates interfering radiation which comes from both radioactive materials and cosmic rays.

The interaction depends on the type of radiation and on the detecting medium. The following types of radiation will be discussed here:

1. Heavy charged particles.
2. Electrons, positive and negative.
3. Photons, gamma-rays and x-rays.
4. Neutrons.

Energy is transferred from this radiation to the atoms of the medium through a number of processes. The probability of each process depends on the type of radiation, its energy, and the type of absorbing matter.

5.1 Charged particles

5.1.1 Ionization of heavy particles

Heavy charged particles, for example muons, protons and alpha particles, interact with matter mainly through electrostatic forces during collisions with orbiting electrons. These are either brought to a new orbit with less binding energy (*excitation*), or torn away from the atom (*ionization*). Primary ionization, that is, electrons removed from atoms by the particle itself, accounts for about 1/3 of the total ionization, the remaining 2/3 is due to the more energetic primary electrons, known as *δ-rays*.

The average energy loss per unit length, $-dE/dx$, is called the *stopping power* or the specific energy loss, and is usually expressed in MeV per cm. Bethe derived the following expression for the stopping power of heavy particles in 1930:

$$-\frac{dE}{dx} = \left(\frac{e^2}{4\pi\epsilon_0}\right)^2 \frac{4\pi z^2 N_0 Z \rho}{m_e c^2 \beta^2 A}\left[\ln\left(\frac{2m_e c^2\beta^2}{I}\right) - \ln(1-\beta^2) - \beta^2\right] \tag{5.1}$$

where

z	=	number of elementary charges of the ionizing particle
Z	=	atomic number of target material
A	=	mass number of target material
N_0	=	Avogadro's number
m_e	=	mass of the electron
e	=	charge of the electron
c	=	speed of light in vacuum
ϵ_0	=	permittivity constant
β	=	v/c, where v is the velocity of the heavy particle
ρ	=	mass density of the stopping material
I	=	mean excitation energy (including ionization)

The last two terms inside the bracket almost cancel each other and only give a small contribution to the total value inside the bracket, so Equation 5.1 can be simplified, giving

$$-\frac{dE}{dx} = \left(\frac{e^2}{4\pi\epsilon_0}\right)^2 \frac{4\pi z^2 N_0 Z \rho}{m_e c^2 \beta^2 A}\left[\ln\left(\frac{2m_e c^2\beta^2}{I}\right)\right] \tag{5.2}$$

Equation 5.1 shows that the stopping power does not depend on the mass of the particle, only on its velocity (through β). Substituting the respective values into 5.2, we get

$$-\frac{dE}{dx} = 0.306z^2\left(\frac{Z}{A}\right)\frac{\rho}{\beta^2}\left[\ln\left(\frac{1.02\beta^2}{I}\right)\right] \qquad MeV/cm \tag{5.3}$$

where I is in MeV and ρ in g/cm^3.

The stopping power depends on the medium traversed via $(Z/A)\rho$ and the mean excitation energy I. It should be noted that $(N_0\,(Z/A)\rho)$ is the electron density in the absorbing material in units of electrons per cm^3. As the ratio of atomic mass number and atomic number (Z/A) varies only slightly, from 0.5 for light elements to about 0.4 for heavy, this term only contributes a maximum variation of about 20% to the stopping power.

Inner electrons are more tightly bound in high Z atoms, giving a higher mean excitation energy, which can be approximated by

$$I = 10 + 10Z \quad (eV) \tag{5.4}$$

The stopping power depends only weakly on Z, and only through I, which appears in a logarithmic factor.

The stopping power for alpha particles, protons, muons and electrons in materials with low atomic number is shown in Figure 5.1. When the particles have a velocity close to the speed of light, or rather, when their kinetic mass is at least twice the rest mass, the stopping power curve enters a broad minimum, about 1.9 MeV per g/cm^2, irrespective of the rest mass of the particle. The electron reaches this value at about 1 MeV, the muon at about 200 MeV, and protons and alpha particles (or one should rather say helium nuclei) at still higher values. Above the minimum, the stopping power rises slowly when radiation loss becomes significant (Section 5.2).

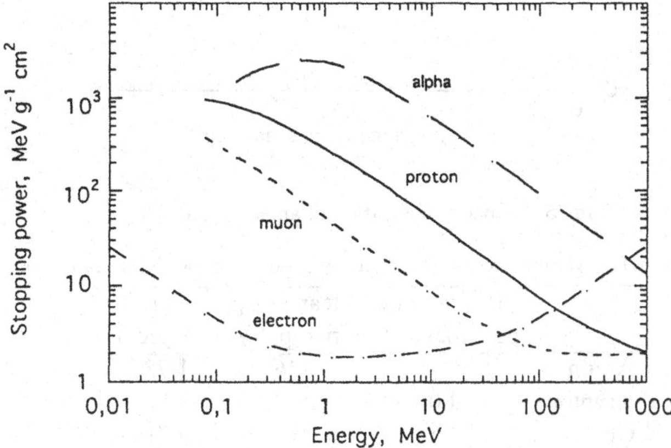

Fig. 5.1. Stopping power of electrons, muons, protons and alpha particles in low Z materials.

Specific ionization s, describing the ionization density, is defined as the average number of ion pairs formed per cm in the track of a charged particle. It is related to stopping power by

$$-dE/dx = s\,w \tag{5.5}$$

where w is the net energy required to produce an ion pair, including both energy loss due to ionization and excitation. Figure 5.2 shows the specific ionization of an alpha particle, with an initial energy of 5 MeV, in air. As it loses energy, the specific ionization increases and reaches a maximum close to the end of the track where it then falls rapidly to zero. In this range, the alpha particle is continuously exchanging electrons with gas atoms and consequently has a net charge between 2 and zero. Equation (5.1) is then no longer valid.

The mean energy w required to form an ion pair in gas is nearly independent of the energy of the particle, its charge and mass. The value of w for gases commonly used in gas counters is shown in Table (5.1). As the ionization potential is only about half of w, it is clear that about half of the energy must be lost through the excitation of the atoms.

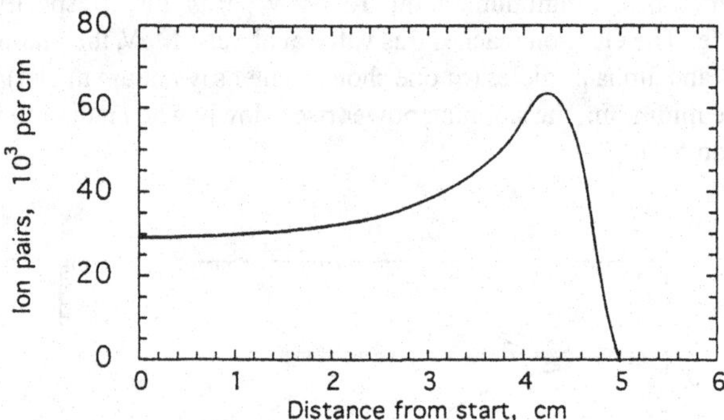

Fig. 5.2. Specific ionization of an alpha particle.

Table 5.1. Average energy w lost in the production of an ion pair (ip).

	Ionization potential, eV	Mean energy per ip, eV	Density mg/cm^3
Argon	15.7	26	1.78
Krypton	14.0	24	3.74
CH_4	14.5	27	0.716
CO_2	14.4	33	1.98
Air		34	1.20

5.1.2 Ionization of electrons

Electrons, including beta particles, lose energy in a similar way as alpha particles, that is, through collisions with electrons bound to the atoms of the absorbing material. Because of the electron's small mass it can lose a large part of its energy in a single collision. The stopping power of electrons is described by an equation similar to that for heavy particles:

$$-dE/dx = \left(\frac{e^2}{4\pi\epsilon_0}\right)^2 4\pi N_0(Z/A)\frac{\rho}{(m_e c^2\beta^2)} \cdot \left[\ln\{m_e c^2\tau\sqrt{(\tau/2+1)}/I\} + F(\beta)\right] \tag{5.6}$$

where $\tau = T/m_e c^2$, i.e., the kinetic energy of the electron in terms of its zero energy (0.511 MeV). The term $F(\beta)$ only gives a small contribution to the sum in the bracket, and can for most practical purposes be omitted. Substituting the respective values into (5.6) gives

$$-dE/dx = 0.170\rho(Z/A)/\beta^2\left[\ln\{0.511\,\tau\sqrt{(\tau/+1)}/I\}\right] \tag{5.7}$$

where I is in MeV.

Figure 5.1 shows the stopping power of electrons in materials with a low atomic number. It is high at low energy (low velocity), it then decreases and reaches a long flat plateau somewhat below 1 MeV. It rises slowly again at higher energy as a new process, radiation loss, becomes effective.

5.1.3 Range of heavy particles

Heavy charged particles only lose a small part of their total energy in each collision and travel along straight paths and have a rather well defined path length, or range. If a beam of parallel, monoenergetic heavy particles impinges on a plate of varying thickness, the number of penetrating particles is almost constant until a given thickness R is nearly reached. Then the number falls rather rapidly to zero. This is shown schematically in Figure 5.3. There is a small variation in the range of individual particles *(straggling)*, about 5% for alpha particles. Like stopping power, range, expressed in g/cm^2, depends only weakly on the

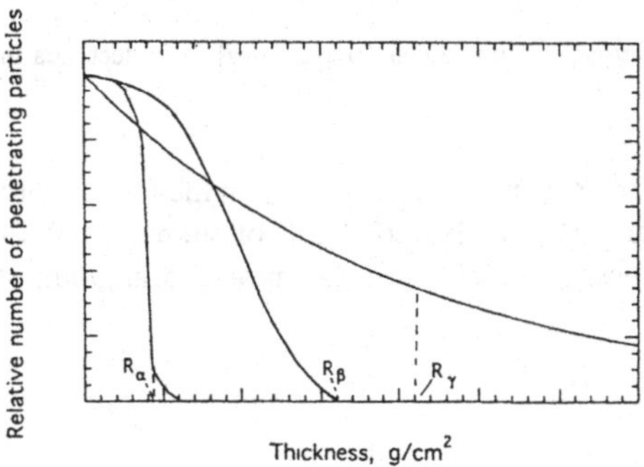

Fig. 5.3. Schematic illustration describing range determination for monoenergetic alpha particles, protons and photons.

atomic number. The range of protons, helium nuclei and electrons in a wide energy range is shown in Figure 5.4.

5.1.4 Absorption and range of beta particles

Although the average energy loss of beta particles and electrons is small per collision, it can be very large in individual collisions. The straggling of electrons is therefore much larger than for alpha particles and the track is tortuous at low energies. Beta particles have a continuous energy distribution (Section 4.3), from zero to a maximum value, E_{max}, characteristic for each radionuclide. Their absorption is a rather complicated process and must be treated empirically. Let us look at a beam of collimated beta particles incident on an absorbing foil of thickness dx, in a geometry similar to that shown in Figure 5.5. Some of

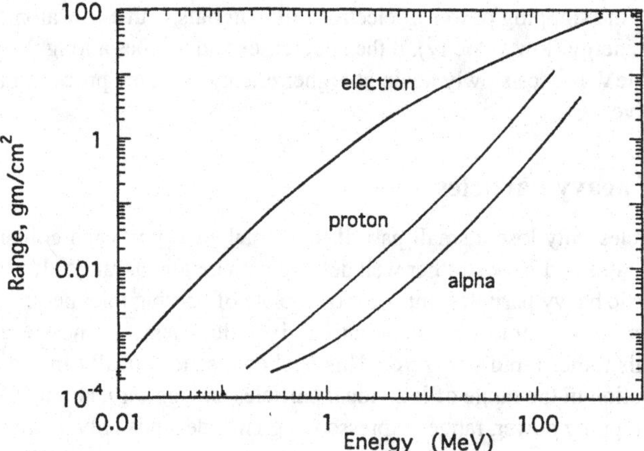

Fig. 5.4. The range of protons, helium nuclei (α-particles) and electrons in low atomic number materials.

the particles may lose all their energy and be stopped in the foil (absorbed) or they may be deflected (scattered) out of the beam. The count rate of our detector will therefore decrease with increasing foil thickness x, which is usually measured in g/cm^2. Figure 5.6 displays

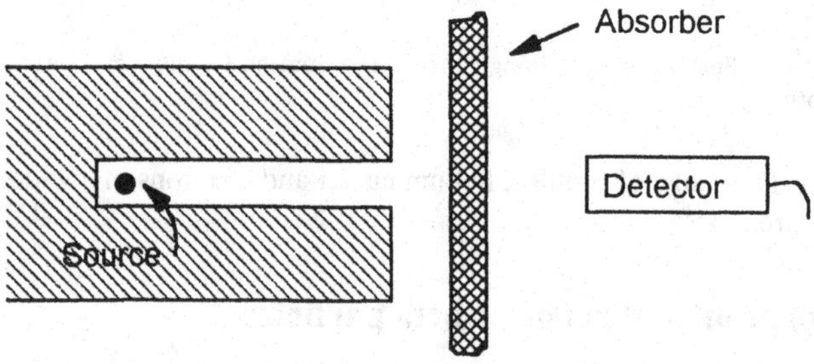

Fig. 5.5. Schematic illustration of the measurement of the absorption of nuclear radiation under good geometry.

a typical curve showing the transmission of beta particles as a function of the thickness of the absorbing layer. The form of these curves depends somewhat on the geometry of the source, absorber, collimator and detector. It is a fortunate coincidence that the combination of the continuous energy spectrum and scattering results in an attenuation curve that is approximately exponential. Thus,

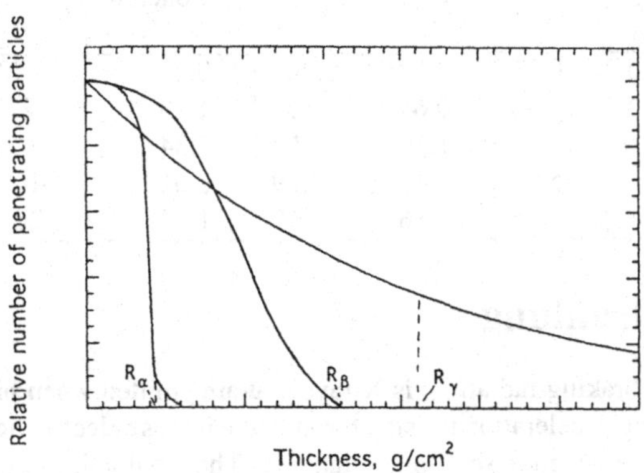

Fig. 5.6. Transmission of beta particles through aluminium.

$$I(x) = I(0)e^{-\mu x} \tag{5.8}$$

where $I(x)$ and $I(0)$ are the beta count rates with and without absorber and μ is an empirical mass absorption coefficient M, measured in cm²/g. The mass absorption coefficient μ can be expressed empirically by

$$\mu = 17E_{max}^{-1.43} \tag{5.9}$$

where E_{max} is the maximum energy of the beta particles in MeV and μ is in cm²/g.

The range of beta particles is defined by the thickness that the most energetic particles can penetrate. As for heavy particles, the range depends only weakly on atomic number of the absorbing material. Figure 5.6 shows the range in aluminium as a function of E_{max}. For $0.01 \leq E \leq 2.5$ MeV the range curve can be approximated by

$$R = 412E_{max}^{1.265-0.0954\ln E_{max}} \quad (\text{mg/cm}^2) \tag{5.10}$$

and for energy above 2.5 MeV

$$R = 530E_{max} - 106 \quad (\text{mg/cm}^2) \tag{5.11}$$

The range of energetic electrons in some materials of interest is shown in Table 5.2.

The definition of range for alpha particles and electrons are quite different. The path of the alpha particle is almost straight and the range is therefore the same as the path length. The path of beta particles is on the other hand tortuous. Their range is defined on the basis of transmission measurements of a parallel beam described above.

Table 5.2. Range of electrons in some materials.

	Range, cm				
Energy MeV	NaI	Ge	Pb	Concrete	Air
1	0.17	0.12	0.069	0.22	408
5	0.90	0.64	0.32	1.23	2270
10	1.63	1.16	0.54	2.34	4310
50	5.08	3.46	1.29	8.61	16200
100	7.45	4.86	1.67	13.4	26300

5.2 Bremsstrahlung

Bremsstrahlung, or braking radiation, is X-rays that are emitted when high speed charged particles suffer strong deceleration passing through the intense electric field close to nuclei. Let us look at the phenomenon semi-quantitatively. The Coulomb force F between an incident particle with a charge ze and the positive charge Ze on the nucleus is

$$F \sim ze\,Ze/r^2 \tag{5.12}$$

where r is the distance between the charges. The deceleration is $a = F/M$, where M is the mass of the particle. The intensity I of the radiation is proportional to a^2:

$$I \sim z^2\,Z^2/M^2 \tag{5.13}$$

This indicates that the intensity should be much higher for electrons than muons and higher for high atomic number materials. The ratio of the contribution of bremsstrahlung and collision is approximately given by

$$(dE/dx)_r/(dE/dx)_c = Z/800 \tag{5.14}$$

where the subscripts r and c refer to radiation and collision losses.

5.3 Cerenkov radiation

Charged particles, passing through a dielectric medium with a velocity greater than that of light in the medium, emit electromagnetic rays, called *Cerenkov radiation*, due to the polarization of atoms the particle passes. Its wavelength ranges from ultraviolet to visible light. The energy lost in this way is small compared to that lost in the processes discussed above. For a relativistic particle, 200 photons in the visible region are created per cm in water, corresponding to about 400 eV. The same particle will lose 1.8 MeV/cm through ionization in water.

Cerenkov radiation is sometimes used to measure radionuclides emitting high energy beta particles in liquid scintillation counters and it is very important in high energy research.

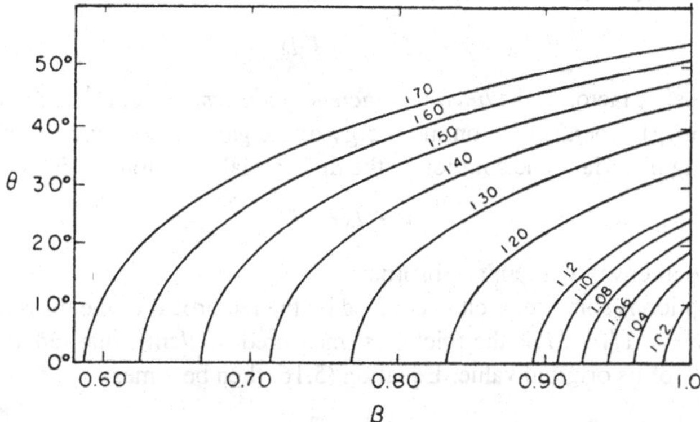

Fig. 5.7. Variation of Cerenkov angle.

The photons are emitted in a direction along a conical surface making an angle θ with the direction of movement of the particle. Figure 5.7 shows the variation of the Cerenkov angle with $\beta\,(= v/c)$ for various indices of refraction. Cerenkov radiation produced by cosmic rays in the glass of photomultiplier tubes and in vials contributes significantly to the background count rate of liquid scintillation counters.

5.4 Gamma rays

The attenuation of gamma rays is quite different from that of charged particles. Photons have no electric charge and they can travel a considerable distance with no energy loss. Probability governs how deep they penetrate into an absorbing layer without interacting. Gamma rays lose energy through three different processes:

1. Photoelectric absorption.
2. Compton scattering.
3. Pair production.

5.4.1 Attenuation of gamma rays

Let us consider a narrow beam of monoenergetic gamma rays incident normally on a slab of material (Figure 5.5). As the photons traverse the slab, some will be absorbed and others deflected. If $I(x)$ is the intensity of the beam, for example the number of photons per second, after having penetrated a thickness x, measured either in cm or g/cm^2, the number dI lost in travelling further dx is proportional to I and dx

$$dI = -\mu \, I \, dx \tag{5.15}$$

The proportionality factor μ, the *linear coefficient of absorption* (cm^{-1}), or mass absorption coefficient (cm^2/g) depends both on the energy of the gamma rays and the atomic number of the absorbing material. The solution to the differential equation (5.15) is

$$I = I_0 \, e^{-\mu x} \tag{5.16}$$

where I_0 is the intensity without an absorber.

The absorption is sometimes characterized by the reciprocal of the mass absorption coefficient, or $M = 1/\mu$. M is the thickness, measured in g/cm^2, that reduces the gamma intensity to $1/e$ of its original value. Equation (5.16) then becomes:

$$I = I_0 e^{-m/M} \tag{5.17}$$

where m is the thickness of absorber in g/cm^2. This expression has the advantage that the physical meaning of M is more easily grasped than μ. A half value layer is sometimes also used, that is, the thickness of absorber that will reduce the intensity to half its initial value. This thickness, $M_{1/2}$, is given by:

$$M_{1/2} = \ln 2 \, M = \ln 2/\mu \tag{5.18}$$

The photons can be removed from the beam through three different processes, discussed below, each having its separate absorption coefficient.

5.4.2 Photoelectric effect

A photon can lose all its energy E_γ in an interaction called the *photoelectric effect.* The photon then interacts with the atom as a whole, which results in the ejection of an electron from an inner shell (Fig. 5.8). As energy is conserved we have

$$E_\gamma = T + B_e \tag{5.19}$$

where T is the kinetic energy of the electron and B_e its binding energy. The probability for this process is highest for K-electrons (typically about 80% of all events). The vacancy left is filled by one of the electrons in the outer shells and characteristic X-rays of the absorbing nucleus are emitted.

The photoelectric absorption coefficient μ_{ph} varies approximately as

$$\mu_{ph} \sim Z^4/E^3 \tag{5.20}$$

All three absorption coefficients for lead are shown as a function of energy in Figure 5.9. The discontinuity at about 0.1 MeV, the K absorption edge, corresponds to a photon energy below which the photon does not have sufficient energy to eject a K-electron, the most tightly bound electron.

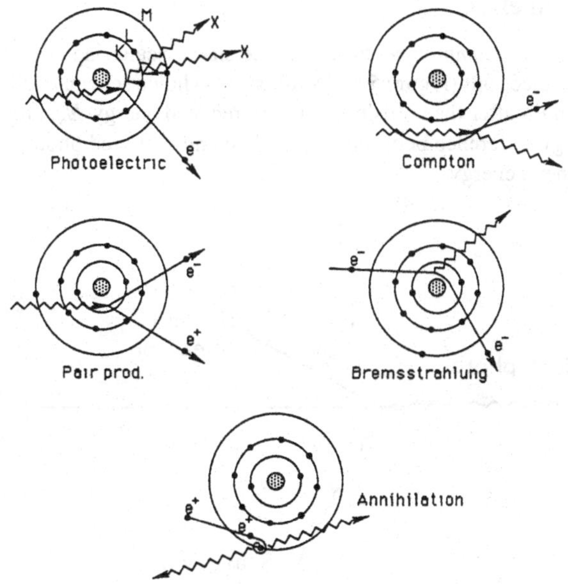

Fig. 5.8. Schematic illustration of different types of gamma interaction with atoms.

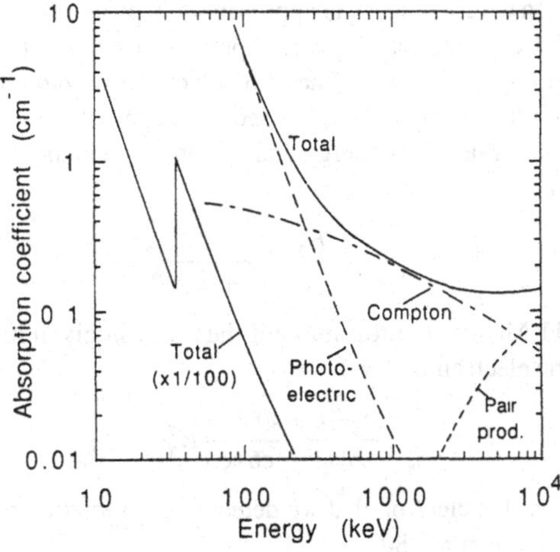

Fig. 5.9. Gamma absorption coefficients in lead

5.4.3 Compton effect

The *Compton effect*, or *Compton scattering*, is the elastic scattering of photons by loosely bound, nearly *free*, electrons (Figure 5.8). A collision is shown schematically in Figure 5.10. The photon continues after the collision, but with reduced energy $E_\gamma^{'}$, and deflected in direction. The energy difference between the incident and scattered photon is transferred to the electron as kinetic energy.

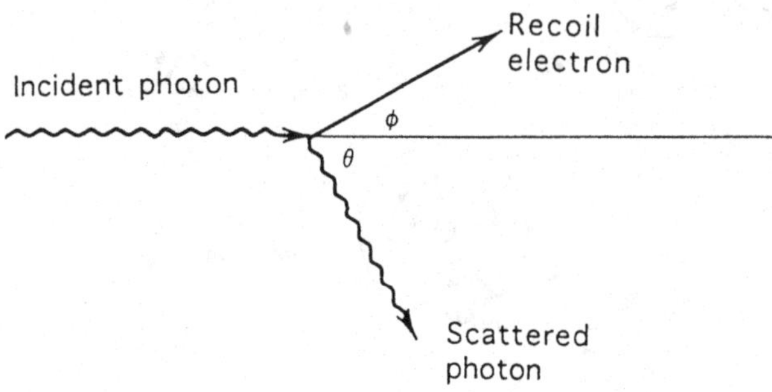

Fig. 5.10. A Compton collision.

The Compton absorption coefficient μ_C is approximately proportional to E_γ^{-1} and proportional to the number of free electrons. If μ_C is expressed in cm^2/g, it is nearly independent of the atomic number of the absorber. The photoelectric Compton absorption coefficient for lead is shown in Figure 5.9 and the cross section per electron in Figure 5.11.

Classical laws of the conservation of energy and momentum can be used to find the energy $E_\gamma^{'}$ of the scattered photon

$$E_\gamma^{'} = \frac{E_\gamma E_e}{E_\gamma(1 - \cos\phi) + E_e} \tag{5.21}$$

where $E_e = m_e c^2 = 0.511$ MeV, is the rest mass of the electron given in terms of energy. The kinetic energy T of the electron is

$$T = \frac{(1 - \cos\phi)}{E_e - E_\gamma(1 - \cos\phi)} E_\gamma^2 \tag{5.22}$$

It is the ionization caused by this electron that we detect in our radiation counters. Its maximum kinetic energy $T_{\max}$ is for $\phi = 180°$

$$T_{\max} = \frac{2E_\gamma^2}{E_e + 2E_\gamma} \tag{5.23}$$

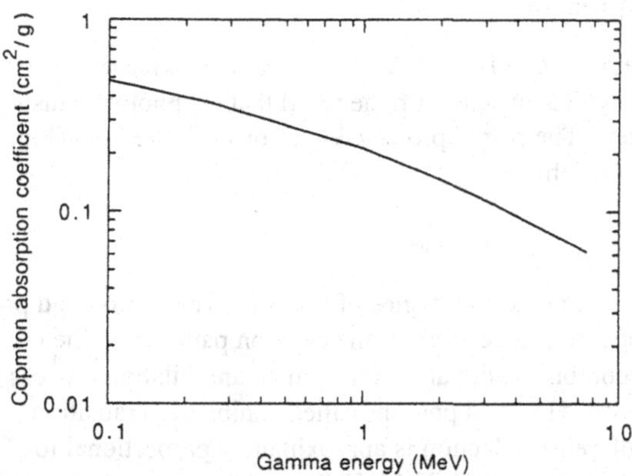

Fig. 5.11. The Compton cross section.

The Compton electrons generally lose all their energy in liquid and solid detectors. The pulse height spectra of monoenergetic photons will show a Compton spectrum that is nearly flat up to a value that corresponds to T_{max}, above which the count rate decreases rapidly with increasing energy (Compton edge).

The kinetic energy of the ejected electrons varies continuously from zero to T_{max}. The average fraction of energy which the photon loses in a Compton collision is 14% at 0.1 MeV, 44% at 1 MeV and 68% at 10 MeV.

For incident photons with energy below 0.1 MeV, the scattered photons are emitted with similar probability in all directions, also backward. With an incident energy of 1 MeV, a large fraction of the photons are scattered in the forward direction as are the majority at 10 MeV, most of them within a cone of 30°. The distribution of angular scattering is shown for different energies in Figure 5.12.

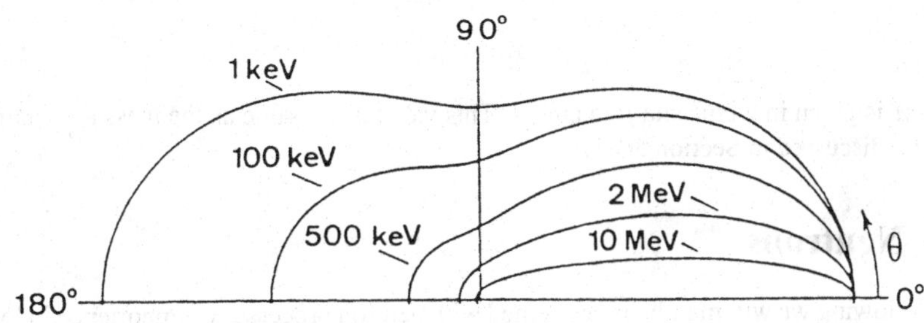

Fig. 5.12. Angular distribution of Compton scattered photons.

5.4.4 Pair production

A photon having energy above 1.02 MeV, i. e., more than twice the rest mass energy of an electron, can interact with a nucleus with the result that the photon vanishes and a positron/electron pair is created. The pair is projected in a forward direction (Figure 5.8). Conservation of energy requires that

$$T_{el} + T_{pos} = E_\gamma - 1.022\text{MeV} \tag{5.24}$$

where T_{el} and T_{pos} are the kinetic energies of the negative electron and positron (in MeV) and 1.022 MeV is the rest mass energy of the electron pair. When the positron has lost all its kinetic energy, it combines with an electron in an annihilation process and the mass of the pair appears in two 0.511 MeV photons called annihilation radiation.

The probability of pair production is approximately proportional to $Z^2 + Z$ and it increases rapidly with energy above 1.02 MeV. Pair production is therefore important in absorption in high Z materials.

5.4.5 Combined effects

According to equation (5.16), a narrow beam of monoenergetic gamma rays is attenuated exponentially

$$I = I_0 e^{-\mu x} \tag{5.25}$$

where μ is the total absorption coefficient. It is composed of coefficients due to the three effects discussed above:

$$\mu = \mu_{ph} + \mu_C + \mu_{pair}. \tag{5.26}$$

The total absorption coefficients for air, iron and lead are shown in Figure 5.13 as functions of energy from 0.1 to 10 MeV. Note that the absorption coefficient is almost the same fròm 0.5 to 5 MeV for air and iron as the Compton effect, which is independent of atomic number, is predominant in this range.

The gamma photons have no range like alpha and beta particles. Their mean range M can, however, be found by integrating (5.25), giving

$$M = 1/\mu \tag{5.27}$$

where M is given in g/cm^2 and μ in cm^2/g. This value is the same as the mass attenuation thickness discussed in Section 5.4.1.

5.5 Neutrons

In the following we will mainly be concerned with neutron processes of importance in low-level counting and those occurring in neutron detectors. We are therefore only interested in one source, i.e., cosmic rays neutrons, released in high energy nuclear collisions (Chapter 7).

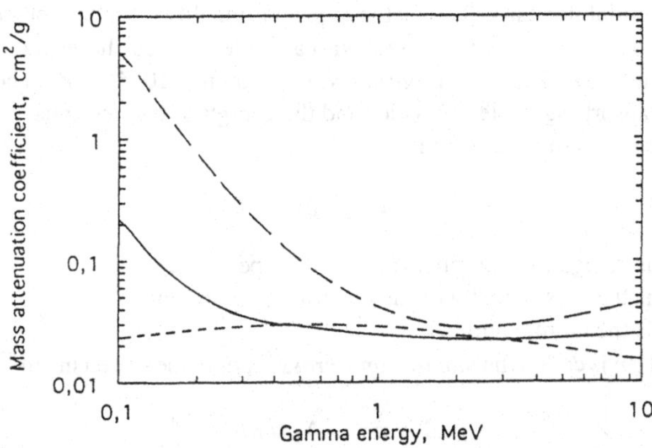

Fig. 5.13. Total absorption for gamma radiation in air, iron and lead.

Neutrons are frequently classified according to their energy. *Thermal neutrons* are in thermal equilibrium with the surrounding matter. Their most probable energy is 0.025 eV and practically all of them have energy less than one eV. *Intermediate neutrons* have energy from 1 eV to 100 keV and *fast neutrons* from 0.1 to 20 MeV. Those of the highest energy, above a few hundred MeV, are called *relativistic*.

Neutrons react with nuclei through a number of different processes that are described by a special notation, such as

$$^{10}Be(n,\alpha)^7Li \quad \text{or} \quad ^{65}Cu(n,2n\gamma)^{64}Cu \tag{5.28}$$

where the target nucleus is given first, then the projectile, then the emitted particle or particles, and finally the end form of the hit nucleus. When one is only interested in the type of reaction the symbols for the nuclei are omitted, we speak for example of (n,α) and (n,2n) processes.

5.5.1 Neutron cross section

When a collimated beam of monoenergetic neutrons passes through an absorbing layer with a thickness dx, neutrons are removed from the beam (scattered or absorbed) in a similar way as in the case of gamma radiation (Figure 5.5). The number of neutrons removed is proportional to the density n of atoms in the material (atoms/cm^3) and to the thickness dx (cm). The change dI in the intensity I is then

$$dI = -I\,\sigma\,n\,dx \tag{5.29}$$

The intensity decreases exponentially with the thickness x of the slab

$$I = I_0\,e^{-\sigma n x} \tag{5.30}$$

where I_0 is the initial intensity. The factor of proportionality σ, called the *cross section*, is the probability that a neutron will interact with a single nucleus. It has the dimension area per atom. It is usually measured in *barns*, where 1 barn = 10^{-24} cm^2. The cross section depends on the absorbing nucleon species and the energy of the neutrons.

The number n of atoms per cm^3 is

$$n = N_0\rho/A \tag{5.31}$$

where N_0 is Avogadro's number (6.02×10^{-23} per mole),
ρ is the mass density of the material in g/cm^3 and
A is the atomic number

The product σn is called the *macroscopic cross section*, measured in cm^{-1}, is designated by Σ:

$$\Sigma = n\sigma = N_0\,\sigma\,\rho/A \tag{5.32}$$

Its reciprocal value, $1/\Sigma$, is equal to the mean interaction length (cm) of the neutrons, i.e., their mean free path. Neutrons can be removed from the beam by an elastic or inelastic collision, or by absorption. The cross section is the sum of the cross sections for these three processes, σ_{el}, σ_{inel}, and σ_{ab}.

5.5.2 Elastic scattering, thermalization

All cosmic ray neutrons are born fast in energetic nuclear interactions. They lose energy through elastic and inelastic collisions. The former is dominating at energies below 1 MeV. Let us look at an elastic collision of a neutron with energy E_0, where the direction of the neutron after collision makes an angle ϕ relative to the incident neutron in the center of mass system. This collision can be treated by classical mechanics, using the laws of conservation of energy and momentum. The situation is similar to that of Compton scattering (Figure 5.10). The energy of the scattered neutron is

$$E' = E_0\frac{(A^2+1+2A\cos\phi)}{(A+1)^2} \tag{5.33}$$

where A is the mass number of the absorber. Theoretical calculations show that the energy loss is nearly independent of the energy of E_0.

The scattered atom loses one or more electrons in the collision and will therefore ionize atoms along its path. The neutron energy loss can therefore be studied through the pulse height spectrum of gas proportional counters or organic scintillation counters irradiated with neutrons. The equal distribution in energy is illustrated in the pulse height spectrum of a stilbene scintillation counter (Fig. 5.14).

For a head-on collision ($\phi = 180°$) we have

$$E'/E_0 = 1 - \left(\frac{A-1}{A+1}\right)^2. \tag{5.34}$$

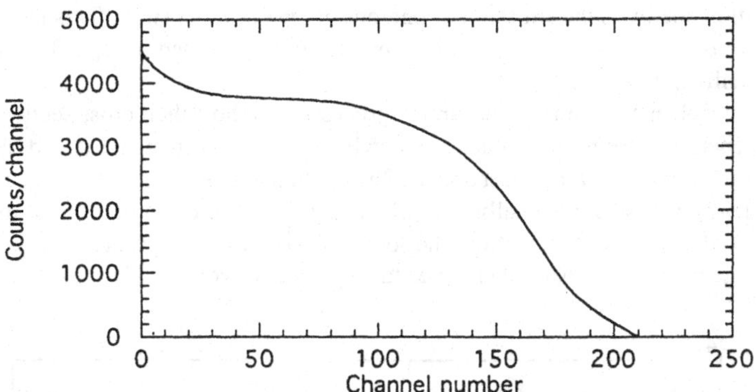

Fig. 5.14. Pulse height spectrum of 2.4 MeV neutrons in a stilbene scintillation counter.

A neutron loses all its energy in a head-on collision with a hydrogen nucleus, but only a part of it in head-on collisions with other nuclei (25% with nitrogen and 2% with lead).

In grazing collisions the energy loss is smaller. For all but the lightest atoms the average value of the ratio E'/E_0, denoted by f in the following, can be approximated by

$$\left(\frac{E'}{E_0}\right) = f = e^{-2/(A+2/3)} \tag{5.35}$$

The natural logarithm of $1/f$ is known as a legarthy in reactor physics. For lead, f has the value 0.9904, so 99.0% of the neutron energy is conserved, and only 1.0% is transferred to the lead nucleus. It is evident that it takes a large number of collisions to degrade the energy of fast neutrons by orders of magnitude in their thermalization. If n is the average number of collisions needed to thermalize a neutron with an initial energy of E_0 (eV), i.e., to reduce its energy to 25 meV (25×10^{-9} MeV), then

$$E_0\, f^n = E_0\, e^{-2n/(A+2/3)} = 0.025 \text{ eV} \tag{5.36}$$

and the number of collisions is

$$n = \frac{A + \frac{2}{3}}{2} \ln \frac{E}{0.025} \tag{5.37}$$

For an initial energy of 1 MeV (cosmic evaporation neutrons) the thermalization will take 500 collisions in iron and 1800 in lead.

5.5.3 Capture of slow neutrons

When a nucleus captures (absorbs) a thermal neutron it will have an excess energy of about 8 MeV, released by the binding energy of the neutron. The most probable process to get rid of this excitation energy is the emission of photons, in one or more steps. This is called an (n, γ) process, or *radiative capture*, and its cross section is denoted by σ_{ab}. The residual nucleus is usually radioactive.

Let us now look at some important capture processes and how their cross section varies with neutron energy. It seems plausible that a slowly moving neutron, which spends a longer time in passing by a nucleus than a fast one, will have a higher probability of being captured. This is frequently the case, especially in heavy nuclei, and the cross section is inversely proportional to the speed of the neutron, the so-called $1/v$ law. In light nuclei, σ_{ab} usually does not change much from thermal energies to megaelectronvolts.

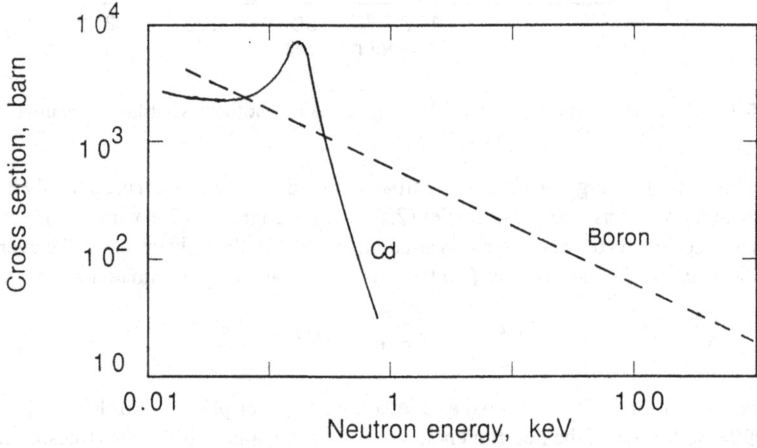

Fig. 5.15. Neutron cross sections of ^{10}B and Cd.

Boron has a high absorption cross section for thermal neutrons. Its cross section is shown in Figure 5.15. This is an exceptional thermal neutron process as an alpha particle is emitted (an (n, α) process):

$$^{10}B +^{1} n \rightarrow {}^{7}Li +^{4} He \tag{5.38}$$

In 93% of the reactions the ^{7}Li nucleus is left in an excited state and a 480 keV gamma photon is emitted. Boron is thus a useful neutron absorber in many application. For neutron thermalization and absorption, boron is frequently incorporated in a hydrogen rich material in the form of 5–10 cm thick sheets, usually of paraffin or polyethylene. A neutron absorbing layer of this type is used in most low-level gas proportional counting systems. Boron is an advantage in this case as the alpha particle carries away most of the binding energy of the neutron, and less than one photon (0.48 MeV) is emitted per captured neutron. This process is also important in neutron measurements as discussed in Section 8.16.

Absorption cross sections are frequently greatly enhanced in a narrow range close to the energy levels of the nucleus. The cross section for the ^{115}In isotope is an important example where there is a very strong resonance level at 1.4 eV.

5.5.4 Inelastic collisions

A nucleus can capture a high energy neutron and form a *compound nucleus*. As there is now more energy available than in the case of a thermal neutron capture, there are other possible processes to get rid of the excessive energy. Each of these requires a given minimum energy, the *threshold energy*, that depends on the type of nuclide. The main processes of interest are:

1. The nucleus may re-emit the neutron. The nucleus is then left in an excited state and a photon is emitted. We speak here about an $(n, n\gamma)$ process, often also written $(n, n'\gamma)$. This process is common for energies above 100 keV and up to a few MeV.

2. At higher energy, close to 10 MeV, two neutrons and gamma rays may be emitted, in an $(n, 2n\gamma)$ process.

3. At still higher energies, a positive particle may be emitted, a proton or an alpha particle, in (n, p) or (n, α) processes. In a few cases this can occur at intermediate energies, even at thermal energies (^{10}B).

Gamma peaks due to inelastic collisions with atoms are seen in the pulse height spectra of low-level germanium spectrometers. When fast neutrons collide inelastically with germanium nuclei, these will recoil giving a small pulse, usually below 50 keV. Simultaneously the nucleus deexcites by emitting gamma radiation. If the diode photoelectrically absorbs the photon, the resulting peaks in the spectrum have a characteristic shape, as they represent the sum of the energies of the photon and the varying recoil energy. The peak will have almost a triangular shape, with a sharp rise corresponding to the energy of the gamma photon, then declining slowly. This will be discussed in more detail in Chapter 12.

where there is a very strong resonance level at 1.4 eV.

5.4 Inelastic collisions

[illegible]

Chapter 6

Our natural radioactive environment

6.1 Radioisotopes in the environment

Energetic nuclear radiation is everywhere around us, coming from primordial and artificial radioisotopes, both man-made and cosmogenic. This radiation is of interest to us, both because it may give a disturbing contribution to the signal of our low-level detectors, and as it may be the subject of our study.

The source of the disturbing background radiation comes either from external or internal radioactivity. External radiation is due to potassium, thorium and uranium (K/Th/U) in materials in the laboratory, predominantly in building materials. Internal radiation comes from primordial and artificial radioactivity in the shield that surrounds our low-level detector and the materials inside it, including the detector itself.

Matter that has not been processed chemically by man generally contains natural radioactivity in substantial concentrations. Most naturally occurring radioisotopes are earth's inheritance of the original nuclear synthesis of the matter of our solar system (*primordial radioactivity*). They are survivors of the synthesis, radioisotopes with a half-life of the same order of magnitude as the age of our Earth (4.5×10^9 years) and three of them are accompanied by a series with shorted half-lives, having half-lives from less than a microsecond to thousands of years. Table 6.5 lists these primordial radionuclides.

6.2 Radiation from K, Th, U

^{232}Th, ^{235}U and ^{238}U are the first links of three chains of radioactive nuclides that all end in a stable lead isotope, ^{208}Pb, ^{207}Pb and ^{206}Pb, respectively. The isotopic abundance of ^{235}U is only 0.7%; its contribution to the total activity of uranium is therefore small and it will not be discussed further here.

Potassium contains 0.0118% of ^{40}K and only 11% of the disintegrations are followed by the emission of gamma rays. Potassium is, however, about 10^4 times more abundant by weight in bedrock and soil than thorium and uranium. Its contribution to terrestrial gamma radiation is therefore comparable to that of the uranium and thorium series.

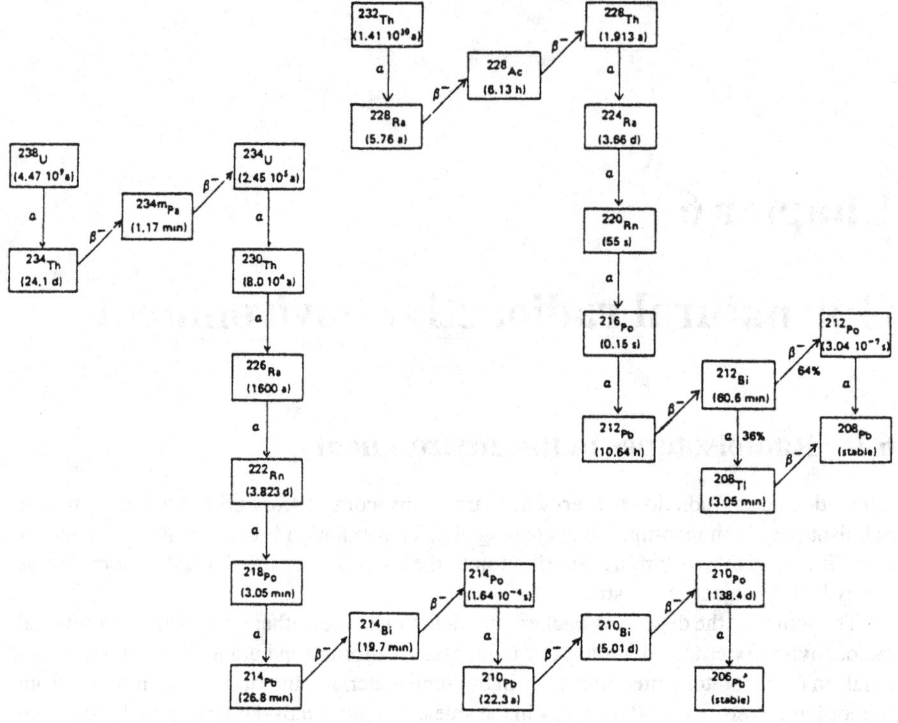

Fig. 6.1. Radioactive decay chains of the ^{232}Th and ^{238}U series.

Figure 6.1 depicts schematically the radioactive decay chains of the ^{232}Th and ^{238}U series. Table 6.1 lists the half-life, specific activity and total number of photons emitted per disintegration of a ^{40}K, ^{232}Th and ^{238}U atom.

The gamma activity of the ^{232}Th and ^{238}U series comes from a number of radionuclides. Table 6.4 lists their main gamma lines, their energies (keV) and intensities (γ/dis). Figure 6.2 gives the same information graphically.

Table 6.1. Half-life, specific activity and total number of photons emitted per disintegration of ^{40}K and the ^{232}Th and ^{238}U series.

	Half-life, years	Activity Bq/kg	γ/dis (>50 keV)
Potassium, ^{40}K	1.25×10^9	2.98×10^4	0.11
Thorium, ^{232}Th	14.0×10^9	4.02×10^6	2.08*
Uranium, ^{238}U	4.5×10^9	12.3×10^6	1.80*

* per disintegration of ^{232}Th and ^{238}U respectively, assuming equilibrium in the series.

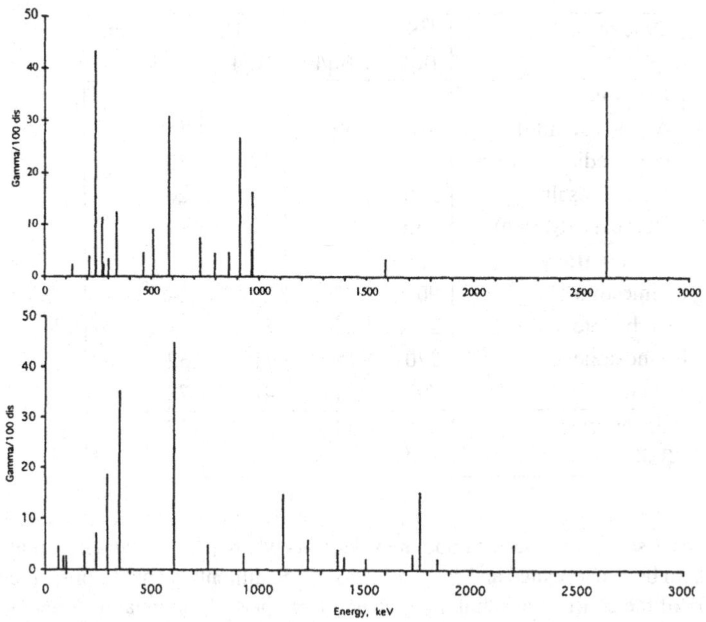

Fig. 6.2. Abundance of main gamma lines of the ^{232}Th and ^{238}U series

6.3 K, Th and U in rocks and soil

There are large variations in the concentration of potassium, thorium and uranium in nature. Their original source is the crust of our Earth and the underlying magma. As molten magma cools and solidifies, silicate minerals are formed and chemical differentiation occurs. At an early stage, "mafic" rocks, rich in magnesium and ferrous iron, are formed with rather low concentrations of thorium and uranium. The remaining magma is thus enriched in these elements. Further partial solidifications give rocks with successively higher concentrations. The radioactivity of bedrock, therefore, depends primarily on its geomorphological type. Table 6.2 gives typical concentrations of K, Th and U for various types of rocks and their mean values in crust and soil. In order to have a parameter that reflects the total gamma flux from these elements coming from a surface, it is convenient to use the absorbed air dose rate 1 meter from the surface. This is usually measured in nGy/h (10^{-9} Gray per hour). This dose rate is given in Table 6.2.

Water plays a major role in the breakdown of igneous rocks and the development of sedimentary rocks and soil. Some of the chemical constituents may be separated and washed away. Further, mechanical and biological processes can give differentiation. The product of

Table 6.2. Typical concentrations of K, Th and U in various rocks and absorbed dose rate in air 1 m above the surface (UNSCEAR, 1977).

Type of rock	^{40}K Bq/kg	^{238}U Bq/kg	^{232}Th Bq/kg	Dose rate nGy/h
Igneous				
Acidic (granite)	990	59	81	140
Intermediate (diorite)	590	23	32	53
Mafic (basalt)	330	11	11	26
Ultrabasic (durite)	150	0.4	24	17
Sedimentary				
Limestone	90	27	7	24
Carbonate	Low	26	8	21
Sandstone	370	18	11	32
Shale	700	44	44	78
Upper crust	830	34	45	77
Soil	370	25	25	43

these processes is sometimes buried under new deposits where pressure and heat may cause the sediments to be consolidated and form solid rocks. Sedimentary rocks constitute only a small fraction of the earth's crust, but they cover a large part of the inhabited land areas of the continents.

Shale is composed of fine grains of clay, silt and mud from chemical and mechanical breakdown of rocks. Sandstone is composed of medium-sized grains from the mechanical breakdown of rocks. Carbonate rocks, limestone (calcium carbonate) and dolomite, are produced by chemical precipitation from water.

The radioactivity of soil is determined by the rock from which it is formed, but the leaching of water can have diminished it. Soil may have been produced from local bedrock, but it is just as likely that it comes from material transported by rivers, winds or ancient glaciers. Table 6.3 lists mean concentrations of potassium, thorium and uranium in soil, and their photon fluxes, and air dose rate measured above the soil.

6.4 Regional and local variations

So far, we have only discussed the global distribution of primordial radioactivity as shaped by geochemical and geophysical processes. Superimposed on this are regional variations, and finally variations in individual buildings.

There are large regional variations in the distribution of natural radioactivity, and in each region considerable local variations can be found. These are well described by the distribution in the dose rate taken 1 meter above the surface. Figure 6.3 shows the results of

Table 6.3. Primordial radioactivity in soil, flux and dose rate at 1 meter above its surface at mean concentration.

	Potassium	Thorium	Uranium
Mean concentration, Bq/kg	370	25	25
Concentration range, Bq/kg	100-700	7-50	10-50
Dose rate, nGy/h per Bq/kg	0.043	0.662	0.427
Photons (>50 keV), $cm^{-2}s^{-1}$ per Bq/kg	0.00513	0.103	0.076
Photon flux above mean soil, $cm^{-2}s^{-1}$	1.9	2.6	1.9
Dose rate above mean soil, nGy/h	16	17	11

combined measurements from France, Germany, Italy, Japan, and the United States, population weighted. About 90% of the total population is exposed to a terrestrial outdoor dose rate of more than 25 nGy/h and 10% to more than 65 nGy/h. We can therefore expect a variation of about a factor of 3 in the concentration of natural radioactivity. When we select an optimum thickness for our shield, this variation in natural radiation should be taken into consideration.

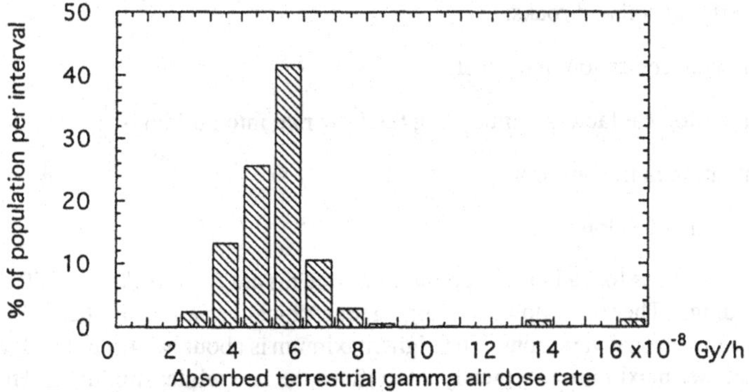

Fig. 6.3. The results of combined measurements from France, Germany, Italy, Japan and the United States, population weighted, of the dose rate 1 meter above ground (UNSCEAR, 1982).

6.5 Radon

One radioactive element, radon, with an isotope in the thorium series and the two uranium series, deserves special attention as it is a gas, belonging to the family of noble gases. These isotopes are:

^{232}Th series:	^{220}Rn	half-life	54.5 s
^{235}U series:	^{219}Rn	half-life	3.92 s
^{238}U series:	^{222}Rn	half-life	3.82 days

In building materials and in the ground a small fraction of the radon atoms can escape through recoil from the small particles or grains in which the radium mother atoms are bound, into interstitial air space or ground water. Being a gas, radon can be carried away by air or groundwater, far from its place of formation. This is particularly true for ^{222}Rn because of its comparatively long half-life. The short half-lives of ^{220}Rn and ^{219}Rn will prevent these isotopes from diffusing far. Their concentrations in air are therefore usually much lower than that of ^{222}Rn. In the following, we will concentrate on ^{222}Rn (which will be referred to as radon) and its daughter products. Most of the gamma radiation from the ^{238}U series comes from the decay products of ^{222}Rn (Table 6.4).

Because of the health hazard of radon, its concentration in outside air and in buildings has been studied extensively. Its concentration in indoor air depends on various factors:

1. Building materials.
2. Underlying soil and rocks.
3. Radon concentration in tap water.
4. Meteorological factors, influencing the flow rate into buildings.
5. Ventilation in the building.
6. The decay of radon.

Typical mean values for radon concentration in free air are 5–10 Bq/m^3 and 50–100 Bq/m^3 inside buildings. There are, however, large geographical variations in these values as well as diurnal and seasonal variations. The night maximum is about two times the day minimum and the summer maximum is typically three times the late winter minimum. Higher radon concentrations are encountered at continental sites and lower concentrations on islands and at coastal sites.

Radon enters buildings from soil and rocks, from building materials and tap water. Figure 6.4 shows the distribution of mean radon concentration in the basements of houses in 29 states in the USA. The concentration was measured in 43000 houses. This figure reflects the width of the geographical radon variations. Taking basement rooms as typical for low-level counting laboratories we can expect, if no special ventilation precautions are taken, a mean radon concentration in the range from about 40 to 160 Bq/m^3, with considerable temporal variation, depending mainly on atmospheric conditions and ventilation. These are mean values for a large area. The radon concentration in individual houses can be considerably higher.

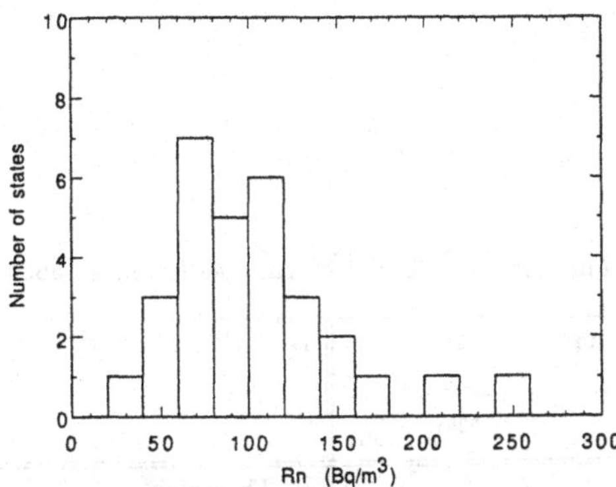

Fig. 6.4. Distribution of mean radon concentration in inside air.

6.6 Cosmogenic and man-made radioactivity

In addition to primordial radioisotopes there are two other categories of radioactivity in our environment, i.e., man-made radioactivity and activity produced by cosmic rays in the atmosphere and the surface of earth, such as tritium, ^{14}C and ^{32}Si.

Cosmogenic radioactivity is measured in various geophysical studies. During the last ten years, cosmogenic radioisotopes produced in ultralow-level germanium spectrometers have been studied extensively.

Man-made radioactivity comes mainly from fission products that have, in the last four decades, been spread all over our globe. The largest part is fall-out from tests with nuclear weapons in the atmosphere, mainly in the period 1952–1962. A major part of this activity was carried into the stratosphere by hot air masses produced in the nuclear explosions, where it was temporarily trapped because of limited mixing between the stratosphere and the lower lying troposphere. Each spring there occurs some mixing between these layers at middle latitudes. About a third of the stratospheric radioactivity then leaks into the troposphere, where, after a few months, it is mixed evenly in the whole lower atmosphere of each half of the globe. Mixing between the northern and southern hemisphere takes a few years.

During the period of high rate of fall-out, from the early 1950s to the 1970s, precautions had to be taken in some low-level work in order to avoid contamination. Now that this activity has fallen to a very low level, this risk is of much less concern.

There are other sources of radioactive fission products. Vast quantities were spread over parts of central Europe following the Chernobyl catastrophe. Reprocessing plants in England and France discharge low activity water in considerable quantities into the sea, but this has much declined in the last 15 years. Finally, radioactive materials, mainly in gaseous form, are released from reactors in smaller quantities.

Table 6.4. Gamma lines in the ^{238}U and ^{232}Th series having an abundance above 2%.

Nuclide	Energy	Intensity %, per decay of ^{238}U atom	Nuclide	Energy	Intensity %, per decay of ^{232}Th atom
^{238}U series			**^{232}Th series**		
Thorium-234	63.0	4.5	Actinium-228	129.1	2.2
	92.4	2.6		209.3	3.8
	92.8	2.6		270.2	11.2
Radium-226	186.2	3.5		463.0	4.5
Lead 214	242.0	7.1		794.7	4.3
	295.2	18.1		911.1	26.6
	351.9	35.1		964.6	5.1
Bismuth-214	609.3	44.6		969.1	16.2
	768.4	4.8		1588.0	3.3
	934.6	3.1	Radium-224	241.0	4.1
	1120.3	14.7	Lead-212	238.6	43.5
	1238.1	5.8		300.1	3.3
	1377.6	3.9	Bismuth-212	727	7.3
	1408.0	2.4	Thallium-208	278	2.3
	1509.2	2.1		583	30.7
	1729.6	2.9		860.4	4.6
	1764.5	15.1		2614.7	35.6
	1847.4	2.0			
	2204.2	5.0			

Table 6.5. Primordial radionuclides.

Nuclide	Abundance %	Half-life years	Decay mode	Energy MeV	Daughter nuclide
^{40}K	0.0117	1.25×10^{9}	β^-	1.312 (89%)	^{40}Ca
			EC	γ1.456 (10.7%)	^{40}Ar
^{50}V	0.250	$>6\times10^{15}$	EC	γ1.554	^{50}Ti
			β^-	0.79	^{50}Cr
^{87}Rb	27.83	4.9×10^{10}	β^-	0.27	^{87}Sr
^{113}Cd	12.22	9×10^{15}	β^-		^{113}In
^{115}In	95.7	4.4×10^{14}	β^-	0.49	^{115}Sn
^{123}Te	.0908	1.3×10^{13}	EC	0.79	^{123}Sb
^{138}La	0.009	1.1×10^{11}	EC		^{138}Ba
			β^-	γ0.789 (34%)	^{138}Ce
				γ1.44 (66%)	
^{144}Nd	23.8	2.1×10^{15}	α	1.83	^{140}Ce
^{147}Sm	15.0	1.1×10^{11}	α	2.23	^{143}Nd
^{148}Sm	11.3	7×10^{15}	α	1.96	^{144}Nd
^{174}Hf	.162	2.0×10^{15}	α	2.50	^{170}Yb
^{176}Lu	02.59	3.7×10^{10}	β^-	0.57	^{176}Hf
				γ0.31 (93%)	
^{187}Re	62.60	4.5×10^{10}	β^-	0.0026	^{187}Os
^{186}Os	1.58	2×10^{15}	α	2.75	^{182}W
^{190}Pt	.001	7×10^{11}	α	3.18	^{186}Os
^{132}Th	100	1.40×10^{10}	α	4.010	series
^{235}U	.720	7.0×10^{8}	α	4.401	series
^{238}U	99.27	4.47×10^{9}	α	4.196	series

Suggested reading

- Environmental Radiation Measurements, NCRP Report No. 50, National Council on Radiation Protection and Measurements, 1985.
- Ionizing Radiation: Sources and Biological Effects. United Nations, New York, 1982.
- Exposure of the Population in the United States and Canada from Natural Background Radiation, Issued by National Council on Radiation Protection and Measurements, 1987.
- Proceedings of the International Workshop on Radon Monitoring in Radioprotection, Environmental Radioactivity and Earth Sciences, eds. L. Tommasino, G. Furlan, H. A. Kahn and M. Monnin, World Scientific, Singapore,1990.
- Environmental Radon. Ed. C. R. Cothern and J. E. Smith, Plenum Press.

Chapter 7

Cosmic rays

7.1 Cosmic rays and low-level counting

The most important, and also most difficult, part of the technique of low-level counting is to suppress to the lowest practically possible level the contribution of cosmic rays to the background of our sample detectors. This component can be reduced to an insignificant value by locating the counting system deep underground, where cosmic rays are reduced by orders of magnitude. This is, however, only in rare cases economically viable. Here we are primarily interested in systems working in surface laboratories, having some shielding in floors and walls of a building, and in shallow laboratories having an absorbing layer of a few meters of rocks or soil. The total absorbing mass layer M_{ob} above the system is called *overburden* and is measured in g/cm^2 or in *meters of water equivalent* (mwe). A typical concrete floor plate is about 50 g/cm^2, or 0.5 mwe.

In order to understand the cosmic ray background component and the means to reduce it, one must have good knowledge of its complex nature.

7.2 History of cosmic ray studies

During the first decade of this century, scientists tried to find the source of a small residual current always found in well shielded ionization chambers. The prevailing opinion was that some penetrating radiation was emitted from the ground. A decisive experiment was performed by Hess in 1912 when he took an ionization chamber in a daring balloon flight to a height of 5 km. He found that the ionization current first declined a little with altitude, but it then rose at an increasing rate. Hess proposed that a very penetrating radiation, höhenstrahlung, was entering our atmosphere from above.

Millikan, who was at first a sceptic, carried out (in 1923–26) an extensive series of measurements where he used a very refined technique for automatically recording the value of the ionization current on a photographic film, eliminating the danger of manned high altitude flights and greatly reducing the cost. Millikan's studies gave conclusive evidence for the extraterrestrial origin of the radiation, generally believed to be energetic gamma rays, and he gave it the name *cosmic rays*.

As long as only the integrating effect of the cosmic rays could be measured, i.e., the ionization current, scientists could hardly discover their true nature. A technical breakthrough, both in the study of cosmic rays and radioactivity, came in 1928 when Geiger and his student, Müller, transformed the small, slow and erratic point counter into a fast and reliable detector that could be made in large sizes. This detector is called the Geiger–Müller counter or simply the *Geiger counter*. It brought about a revolution in cosmic ray studies.

Bothe, working in Geiger's institute, almost immediately started in collaboration with Kolhörster a study of cosmic rays using the new detector. He noticed that when a Geiger counter was above another one, simultaneous pulses from both counters were frequent, presumably coming only from cosmic rays. Bothe now introduced *coincidence counting*. The signals from two electroscopes, each connected to a Geiger counter, were recorded on a long strip of photographic paper and events were only counted (visually) when pulses appeared simultaneously in both channels, i.e., coincidences.

At this time, the rapidly advancing electronic technique, developed for the young radio industry, had not yet been applied in nuclear research. A few months after the work of Bothe and Kolhörster, Rossi greatly improved their technique by detecting coincidences and counting pulses through the use of radiotubes. This greatly reduced the coincidence resolving time and the work involved. Rossi thus introduced electronic techniques into nuclear research.

Cloud chambers had been of almost no use in cosmic ray studies. Random triggering of a chamber very seldom yielded a cosmic ray track because of the short sensitive time (0.01 sec) of the chamber. In 1932, Blackett and Occhialini (who had worked with Rossi) overcame this limitation when they developed the new technique of triggering the chamber by a coincidence between a pair of Geiger counters, one on each side of the chamber.

A third powerful technique for the study of cosmic-ray processes was added soon after World War II when a group led by Powell developed greatly improved photographic emulsions that could visualize the tracks of weakly ionizing particles. This made possible a very detailed study of the nuclear interactions of cosmic rays, as the emulsions could be exposed for long times and were easy to bring to high altitudes.

In 1947, the scintillation counter was invented and gradually developed into a versatile and very efficient instrument in cosmic-ray research, and in the following decades a series of sophisticated new detectors were added to the arsenal of cosmic ray physicists. In the 1960s, high energy studies could be made more efficiently with the new generation of accelerators, yielding protons with energy above GeV. A major part of the studies discussed here were, however, carried out before this more recent technique became common.

The work of Bothe gave the first strong indication of the particle nature of cosmic rays and their penetrating power. In the mid 1930s Rossi discovered through absorption measurements in lead that cosmic rays consist of two components, the *soft* component, almost completely absorbed by a 10 cm thick layer of lead, and the *hard* component, which is only weakly attenuated by the lead. Rossi further discovered, by studying coincidences between three Geiger counters not in line, that cosmic rays could create new particles through interactions in the material they traversed.

Studies with cloud chambers in a strong magnetic field, used for the momentum analysis of charged particles, gave detailed information about the cosmic ray processes. In 1933 Anderson discovered a new particle, the *positron*, in a cloud chamber photograph. The existence of a positively charged antiparticle of the electron had been predicted theoretically by Dirac in 1928.

Neddermayer and Anderson, using the same technique, identified in 1938 a particle, now called *muon*, with a mass between that of the electron and proton. It was first, incorrectly, identified with the strongly interacting *meson*, which had been proposed theoretically by Yakawa in 1935 in order to explain nuclear forces. In 1945 it was shown that this interpretation was not correct as the particle's interaction with nucleons was much weaker than theory predicted. Later studies identified the muons with the hard component of the cosmic rays.

Powell's group discovered (1947) Yakawa's meson in a photographic emulsion that had been exposed to cosmic rays at high altitude. A meson had come to rest in an emulsion and decayed into a muon. This particle is now called a π-meson or a pion. It can have either a positive or a negative charge. Some years later a neutral pion was discovered. It has a very short half-life and decays into two energetic photons that are the source of the soft component of the cosmic rays.

Table 7.1. Particles in cosmic rays of importance for low-level counting

Particle	Mass MeV	Decay time, s	Decay products
Proton	938.26		
Neutron	939.55	1013	$p + e^- + \nu$
$\pi^{\pm}$	139.6	$2.55 \cdot 10^{-8}$	$\mu^{\pm} + \nu$
π^0	135.0	$1.78 \cdot 10^{-16}$	2γ
$\mu^{\pm}$	105.7	$2.20 \cdot 10^{-6}$	$e^{\pm} + 2\nu$
$e^{\pm}$	.511	Stable	
γ	0	Stable	

A large number of other short lived particles have been discovered in cosmic rays during the last decades, but they are of less interest in the present context.

7.3 Kinematics and theory of relativity

Before going further, important kinematic parameters of particles moving with a velocity close to the speed of light will be described briefly. According to Einstein's theory of relativity the mass m of a particle changes with its velocity v according to the equation

$$m = \frac{m_0}{\sqrt{1 - v^2/c^2}} = \frac{m_0}{\sqrt{1 - \beta^2}} \tag{7.1}$$

where m_0 is the rest mass and

$$\beta = v/c \quad \text{and} \quad \gamma = \frac{1}{\sqrt{1 - v^2/c^2}} \tag{7.2}$$

The theory also expresses the equivalence of mass and energy, E, relating them by the equation:

$$E = mc^2 \tag{7.3}$$

The total energy of a particle is equal to the sum of its rest mass energy and its kinetic energy:

$$E = T + m_0c^2 \tag{7.4}$$

The mass of particles is therefore often given in equivalent energy units.

The energy of a relativistic charged particle is frequently determined by measuring the radius of curvature of its track in a strong magnetic field, where the radius is proportional to its momentum. The relation between energy and linear momentum p is given by:

$$E^2 = (m_0c^2)^2 + (pc)^2 \tag{7.5}$$

When the rest mass energy is small compared to total energy, we have $p = E/c$. The momentum for relativistic particles is therefore frequently given in units of MeV/c.

7.4 Cosmic rays in the atmosphere

The primary particles of cosmic rays entering the atmosphere are nuclei of light atoms with very high energy:

Protons	86%.
Helium nuclei	12.7%.
Heavier nuclei	1.3%.

In view of the relatively large number of protons, the heavier nuclei will be disregarded in the following.

We introduce the description of cosmic-ray processes in the atmosphere by showing the illustrative result of an early scientific rocket experiment where the pulse rate of a Geiger counter in a rocket penetrating the atmosphere was recorded (Figure 7.1). At high altitude, outside the atmosphere, the pulse rate is constant as no matter interferes. At an elevation of about 50 km, where the cosmic particles have traversed an atmospheric layer of 2 g/cm^2, the pulse rate begins to rise and reaches a broad maximum at a height of 15 km, when they have traversed about 150 g/cm^2. Below this altitude, down to the surface of earth, the pulse rate of the Geiger counter decreases rapidly. The pulse rate rise in the upper atmosphere is due to nuclear collisions of the primary particles with nuclei of gas atoms in which a number of secondary particles are produced. At the altitude of maximum count rate, production

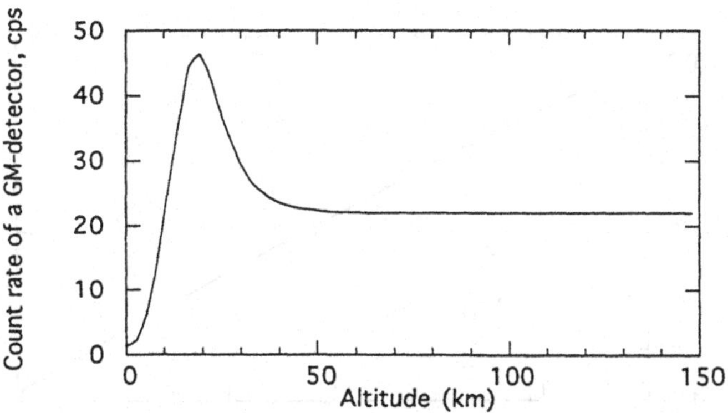

Fig. 7.1. Pulse rate of a Geiger counter in a rocket penetrating the atmosphere.

is balanced by loss of particles that have lost all their energy, mainly through nuclear collisions, but partly through ionization. Below this region, the loss of particles is higher than the production of new ones.

In cosmic-ray physics, we usually speak of depth in the atmosphere rather than height. The depth is measured by the mass per unit area of overlying gas, measured in g/cm^2 or meters water equivalent (mwe, 1 mwe = 100 g/cm^2). Sea level is at a depth of 1030 g/cm^2 or 10.3 mwe. Figure 7.2 shows the depth in g/cm^2 as a function of height.

A variety of particles are produced in the atmosphere. When we speak of secondary radiation in the following, we include all particles formed in the atmosphere and solid earth, including gamma quanta, even though the production has taken place in two or more steps.

In space, far from the Earth, cosmic rays are omnidirectional. When they come close their path is influenced by the Earth's magnetic field, which is equivalent to that of a large dipole, diverting less energetic particles back to space. At the equator a proton moving in a vertical direction must have an energy (cut off energy) of at least 15 GeV to reach earth. At middle latitudes the minimum is 1–4 GeV. At the magnetic poles there is no filtering of the vertical cosmic-ray particles.

Temporal magnetic variations in interplanetary space as well as in the vicinity of earth, caused by varying solar winds, modulate the cosmic-ray intensity. These fluctuations are discussed in Section 7.5.

Although we are primarily interested in cosmic rays in our laboratories, it is useful to know about the processes that occur in the atmosphere, as they determine the composition of cosmic rays at sea level and the same processes continue in our laboratories. The result of these processes, described in Sections 7.6 to 7.11, is illustrated schematically in Figure 7.3.

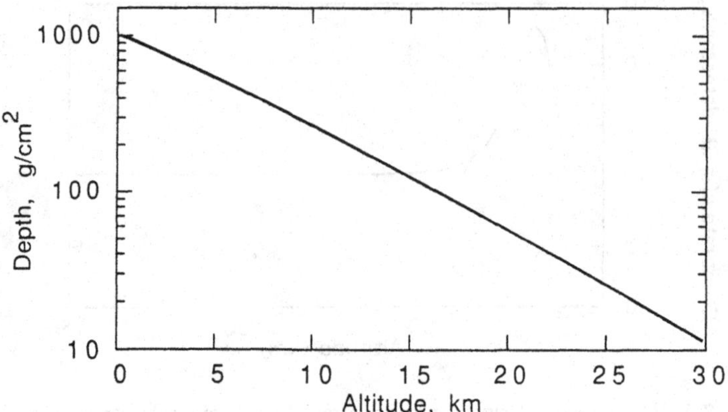

Fig. 7.2. The depth in the atmosphere as a function of height.

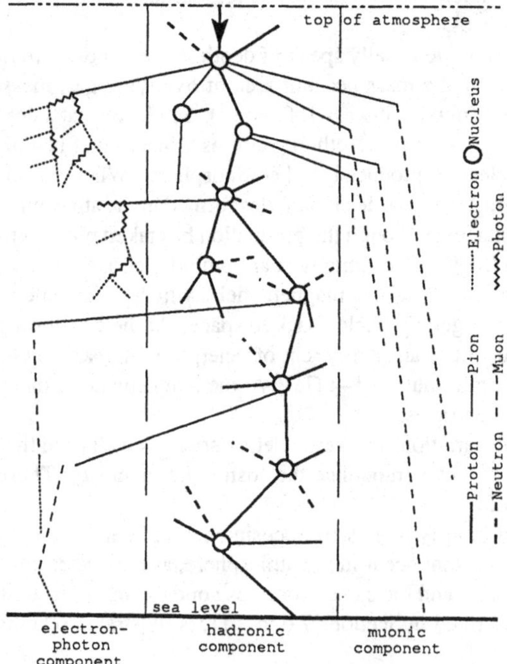

Fig. 7.3. A schematic description of the production and fate of cosmic ray particles in the atmosphere and upper layer of the Earth.

7.5 Temporal intensity variations

The intensity of cosmic rays at sea level shows small variations in time, both periodic and aperiodic, influenced by:

1. Atmospheric pressure.
2. Height of the main muon production level.
3. Extraterrestrial factors.

When atmospheric pressure increases, the cosmic rays must penetrate a denser atmosphere, resulting in a decrease ΔI in their mean intensity I_0 at sea level. We will now look at the variations of $\Delta I/I_0$ due to a deviation Δp in atmospheric pressure from its mean value for the two cosmic ray components of main interest in the present context, neutrons and muons.

The fractional change in the neutron flux due to pressure variations is

$$\Delta I/I_0 = -\alpha_N \Delta p \tag{7.6}$$

The value of α_N is 0.096 per cm Hg. This corresponds to an attenuation thickness of 142 g/cm^2 in the atmosphere. In a similar way, the fractional change in the muon flux is given by

$$\Delta I/I_0 = -\alpha_\mu \Delta p \tag{7.7}$$

where α_μ is 0.021 per cm Hg.

The mean height of muon production, usually measured by the height of the 100 mbar level (H), also influences the muon flux at sea level through the survival probability of the muons. This change is described by

$$\Delta I/I_0 = -\alpha_H \Delta H \tag{7.8}$$

where α_H has the value 0.05 per km and ΔH is the deviation in kilometers from mean height of 100 mbar pressure. As the value of ΔH is determined by the mean temperature in a vertical column of the atmosphere, the 100 mbar level varies seasonally, in most regions by 3–5 km; it is high in the summer and low in winter. It should be noted that the three intensity variations discussed above depend on local climatic factors.

There is also a periodic variation in the cosmic ray intensity with a period of 11 years. This is brought about through a magnetic modulation effect caused by solar plasma, consisting of trapped electrons and protons, moving radially from the sun to great distances. This phenomenon is related to solar eruptions and is inversely correlated to the 11 years period of solar sunspots.

In addition to this, a decrease in the cosmic ray intensity that may amount up to 10% occurs after a solar flare (Forbush effect). The total period of decreased intensity may last a few days, but full recovery sometimes takes several weeks.

7.6 Protons

The mean interaction length of the energetic protons in the atmosphere is about 80 g/cm^2, almost independent of energy. In each collision they lose, on average, about half of their energy. In traversing the atmosphere they therefore decrease both in number and energy. Figure 7.4 shows the vertical differential proton energy spectrum in space and at sea level. The number of 0.2 GeV protons at sea level, the remnants of former high energy protons, has decreased by a factor of 2×10^3. The reduction factor is 5×10^3 for 10 GeV protons.

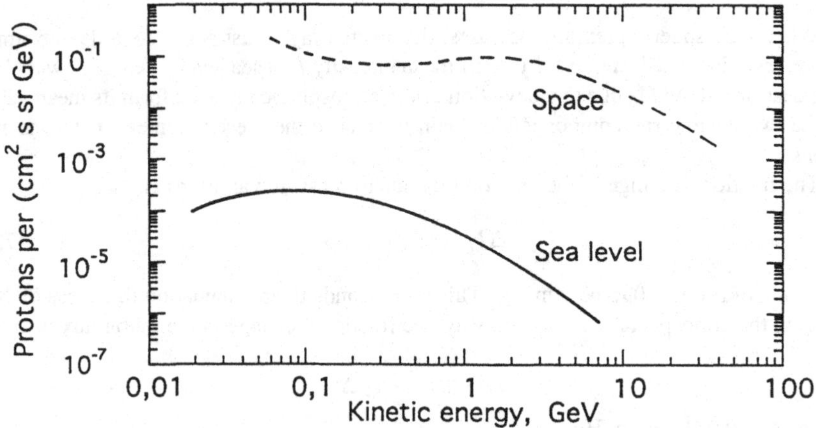

Fig. 7.4. Vertical differential proton energy spectrum in space and at sea level.

The high energy primary particles (87% protons) cause violent, explosion-like events when they collide with atomic nuclei, primarily of nitrogen and oxygen, in the atmosphere. About half of the energy of the incident proton is lost in each collision as a result.

The first phase of this process can be regarded as a collision with a single free nucleon inside the nucleus. The reactions occurring are similar for all nuclei and lead to multiple collisions, both elastic and inelastic (pion production), inside the nucleus. Several particles may be emitted—protons, neutrons and pions—with an energy of 200 to 500 MeV, mainly in the direction of the incident proton. This process is called the *nuclear cascade*. It is illustrated schematically in Figure 7.5. After this violent phase, the nucleus is left in a highly excited state and additional nucleons will escape isotropically as evaporation protons and neutrons, having a broad energy spectrum with a maximum at about 1 MeV and extending somewhat above 10 MeV.

When the energy of the incident proton is below 1 GeV, the emitted particles are mainly protons and neutrons, but above 1 GeV the number of pions is usually larger. In addition to protons, energetic neutrons and pions can initiate the same kind of collisions and their interaction length is similar. These particles form the *nucleonic* part of cosmic rays and have the common name *hadrons*.

Energetic protons have a good chance to continue after a collision and their mean attenuation length in the atmosphere is 120–130 g/cm^2, considerably longer than their interaction length. The total depth in the atmosphere is about 13 interaction lengths, so practically none of the primary protons will reach the surface of the Earth without colliding with atomic nuclei a number of times.

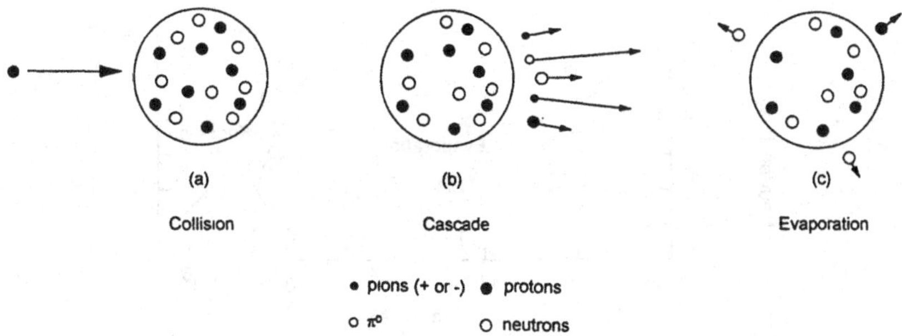

Fig. 7.5. Nuclear cascade, a collision of a high energy proton with a nucleus. Elastic and inelastic collisions inside the nucleus result, emitting knock-on (or cascade) particles and finally evaporation neutrons and protons.

7.7 Neutrons

Each primary proton produces a large number of neutrons in the atmosphere with a broad energy spectrum. These are the very high energy cascade, or knock-on, neutrons and the evaporation neutrons. The energy spectrum of the latter is expressed by

$$P(E)dE = E\,e^{-E/E_0}\,dE \tag{7.9}$$

where E_0, the "temperature" of the excited nucleus, has a value of about 1 MeV. The path of the neutrons in the atmosphere is longer than that of protons as they lose no energy through ionization.

The neutrons are slowed down through elastic and inelastic collisions with atmospheric nuclei, until they are finally captured. The root mean square value of their range is about 150 g/cm^2. As the production and slowing down processes are the same at all depths, the shape of the neutron energy spectrum does not change below 200 g/cm^2, except for its intensity. The differential energy spectrum at sea level is shown in Figure 7.6. This extends from thermal energy to above one GeV. The small bump at one MeV is due to evaporation neutrons. The integral energy spectrum, showing the flux of neutrons above a variable energy E (Figure 7.7), is often more convenient to use. It shows that about 10% of the neutrons have an energy above 10 MeV. Low-level germanium spectrometers give indirect information about the energy spectrum of the neutrons inside the lead shields (Section 12.1).

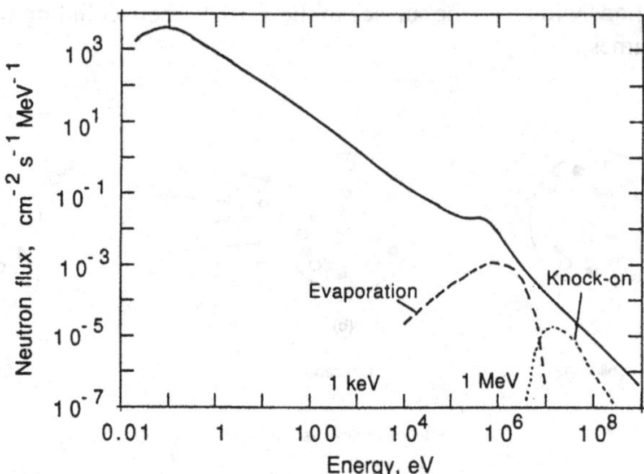

Fig. 7.6. Differential energy spectrum of neutrons at sea level, extending from thermal energy to above one GeV.

In a vertical column in the atmosphere 6.2 neutrons cm^{-2} s^{-1} are produced, about 80% by evaporation and 20% by a knock-on process. Each primary proton produces on average about 20 neutrons. These are either eventually absorbed or escape out of the atmosphere. 64% are captured after thermalization by nitrogen nuclei, producing ^{14}C through a $^{14}N(n,p)^{14}C$ reaction. 20% are captured through other processes and 16% leak out of the atmosphere.

7.8 π-mesons (pions)

In high energy cascade collisions, three kinds of π-mesons, neutral (π_0), positive (π^+) and negative (π^-), are produced with approximately equal probability. They are all unstable and have a very short mean-life, τ:

$$\pi^{\pm} \longrightarrow \mu^{\pm} + v \quad \tau = 2.6 \cdot 10^{-8} \quad s$$
$$\pi^{0} \longrightarrow 2\gamma + v \quad \tau = 1.8 \cdot 10^{-16} \quad s$$

The number of pions produced in a single collision can exceed that of emitted nucleons by a considerable factor. The charged pions interact strongly with nuclei and can give rise to further nuclear transformations. The mean-life τ_{LS} of charged pions in an earth stationary laboratory system is

$$\tau_{LS} = 2.6 \cdot 10^{-8} \cdot \frac{E}{m_\pi} \quad s \qquad (7.10)$$

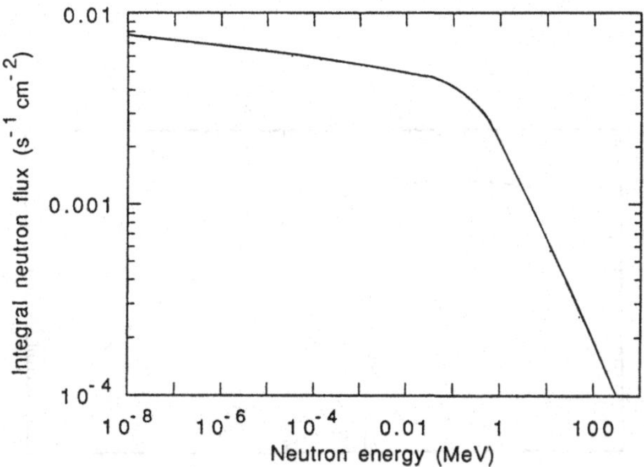

Fig. 7.7. Integral energy spectrum of cosmic ray neutrons at sea level

where m_π is the rest mass of the pion in energy equivalent units (140 MeV), and E the total energy. At relativistic energies the mean free path before decay is therefore 56 m/GeV. In air at sea level, at the highest gas density, a pion with an energy of 1 GeV traverses a mass thickness of 7 g/cm^2 in its mean path length. This is only 8% of its interaction length, so the majority of pions will decay in the atmosphere without colliding with a gas nucleus.

Because of the extremely short half-life of the π^0 meson it will only have travelled in the order of a micrometer before it decays into two high-energy gamma quanta. These photons give rise to the electromagnetic cascades (discussed in Section 7.10) that are the main source of the soft component of the cosmic rays in the atmosphere.

7.9 Muons

Muons are decay products of charged pions. Their direction of propagation is nearly the same as that of their parent pions and in average they receive close to 80% of their parent's energy. They are very penetrating as their nuclear interaction cross section is only about $2 \cdot 10^{-29}$ cm^2, or 10 μbarns. They lose energy practically only through electromagnetic interaction. The predominant process is ionization and excitation, but also knock-on collisions with electrons and, to a lesser degree, bremsstrahlung and pair creation.

The muon energy spectrum at the point of production is therefore nearly the same as that of the pion. However, muon passage through the atmosphere modifies the spectrum due to ionization losses, decay and capture of negative muons. The differential and integral energy spectra of the muons at sea level are shown in Figure 7.8 and Figure 7.9, respectively.

In air, a considerable number of the muons decay in flight where the decay electron receives about 1/3 of the kinetic energy of the muon. Neutrinos carry away the rest of the

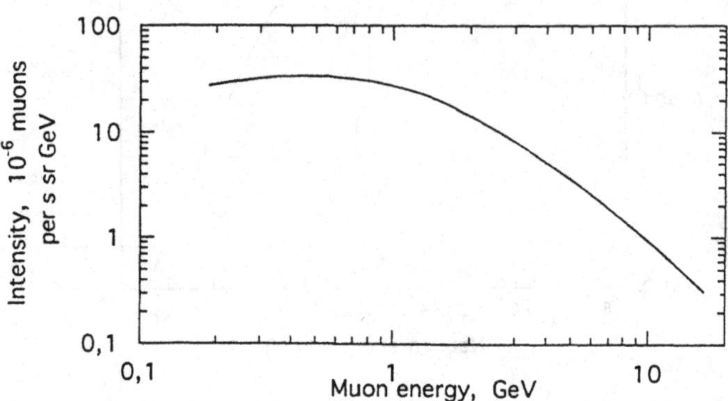

Fig. 7.8. Differential energy spectrum of muons at sea level.

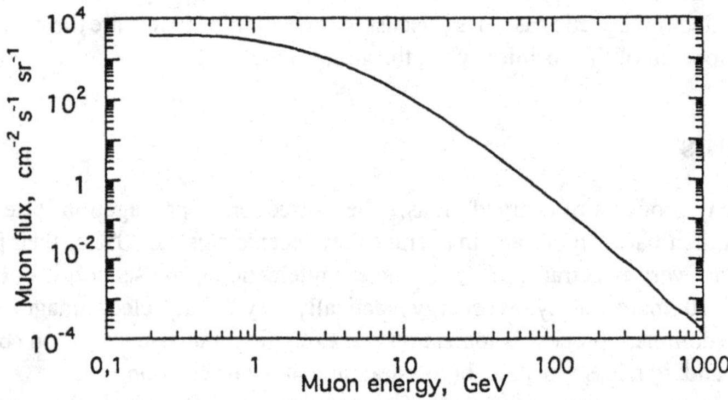

Fig. 7.9. Integral energy spectrum of muons at sea level.

energy. In dense material, most of the muons lose all their energy through ionization before decay. In this case, the decay electrons and the neutrinos share the muon rest mass energy, 106 MeV. The maximum energy of the decay electrons is 50 MeV and their mean energy about 40 MeV.

At a height of 10 km (at a depth of 270 g/cm^2), about 90% of the total number of muons produced in the atmosphere will already have been formed. Taking relativistic time dilution into account, their mean lifetime in a laboratory system is

$$\tau_{LS} = 2.2 \cdot 10^{-6 \cdot E/m_\mu} \text{ s} \tag{7.11}$$

where E is the total muon energy in MeV, and m_μ its mass (106 MeV). The probability P_s of a muon surviving at sea level after travelling vertically from a height of 10 km, is

$$P_s = e^{-10000/c \cdot \tau_{LS}} \tag{7.12}$$

where c is the speed of light in m/s. Although a large part of the muons decay in the atmosphere, a substantial part will reach sea level, 20% of those with an energy of 1 GeV and 80% at 10 GeV. Because of the high penetration power of the muons, their flux decreases more slowly with atmospheric depth than that of its source, the protons (see Figure 7.10).

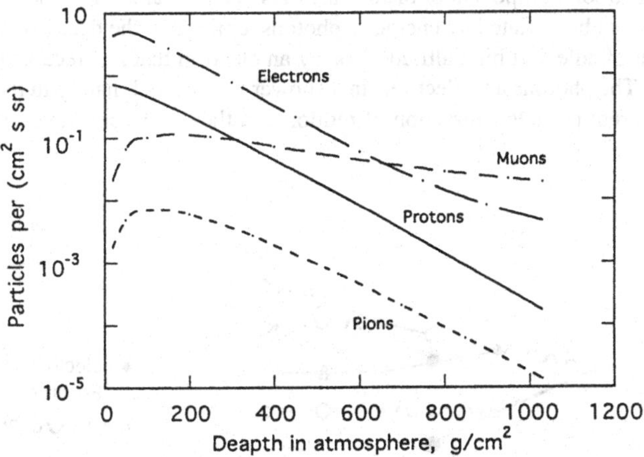

Fig. 7.10. Fluxes of cosmic ray particles in the atmosphere.

Because of their higher mass, stopped negative muons are trapped in atomic orbits lying much closer to the nucleus than electron atomic orbitals. Muonic X-rays are emitted upon trapping. Now, a second process competes with decay, i.e., capture by a proton in the nucleus

$$\mu^- + p^+ \longrightarrow n + \nu$$

Decay is more probable in materials of low atomic number, but capture is dominating for high Z nuclei. For $Z = 20$, 80% of the negative muons are captured, and 95% at $Z = 50$. The rest mass energy of the muon (106 MeV) is released and the nucleus is left in a highly excited state. It deexcites by emitting one or more neutrons. The number of neutrons emitted depends on the mass number of the capturing nucleus, it is 1.6 neutrons per muon capture in lead.

7.10 Electron-photon cascades

A cosmic-ray electron with energy E_0 loses energy mainly through bremsstrahlung, emitting photons with a rather flat spectrum, extending from zero to E_0. The photon receives in average about 40% of E_0. The mean length X_0 traversed by an electron before radiating a photon is called the *radiation length*, measured in g/cm^2. It is practically independent of E_0, but is approximately inversely proportional to the atomic number of the penetrated material, as discussed in Section 4.3.2.

Let us look at the fate of the energetic photon produced. The most probable photon interaction process is electron pair production. The average length traversed by the photon is nearly the same as the radiation length of the electron. The repeated combined effect of these two processes of high energy electrons and photons gives rise to *electromagnetic showers* or *cascades*. The soft component of the cosmic rays is identical to these electrons and photons. Showers can be initiated by energetic photons, emitted in the decay of a neutral pion (dominating at middle and high altitudes) or by an electron that has received high energy from a muon. The photons and electrons in a shower are emitted mainly in the same direction as their parent particle's direction of motion and they therefore travel inside a rather narrow cone.

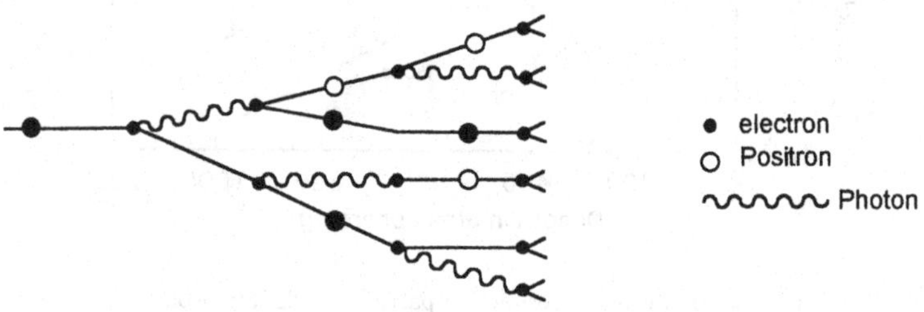

Fig. 7.11. Schematic description of the development of an electromagnetic shower.

The development of a shower can be described quite well by a simple model illustrated schematically by Figure 7.11. We start with an electron with energy E_0 and assume that it radiates half of its energy through a bremsstrahlung photon after traversing one radiation length. We assume further that that the photon, after a second radiation length, creates an electron pair, each electron receiving half of the photon's energy ($E_0/4$) and the initial electron radiates still another photon with half its energy, or $E_0/4$. After two radiation lengths we will have, in addition to the original electron, a photon, an electron and a positron, all with the same energy, $E_0/4$. This multiplying process is repeated until a *critical energy* E_c is reached where electrons lose more energy by ionization than radiation, and the photons lose more energy through Compton scattering than pair creation. The multiplication in the shower then stops.

After t radiation lengths, the number of particles is

$$N = 2^t \tag{7.13}$$

and the mean energy per particle

$$E(t) = E_0/2^t. \tag{7.14}$$

The shower reaches a maximum at

$$t_{max} = \ln(E_0/E_c)/\ln 2 \tag{7.15}$$

where the number of particles is

$$N_{\text{max}} = E_0/E_c \tag{7.16}$$

Table 7.2 gives the value of the radiation length and critical energy of some materials of interest.

Table 7.2. Radiation length and critical energy of some materials

Material	Z	X_0, g/cm^2	X_0, cm	E_c, MeV	ρ, g/cm^3
Fe	26	13.9	1.8	20.7	7.9
Cu	29	13.0	1.45	18.8	8.9
Pb	82	6.4	0.56	7.4	11.3
Concrete		26.5	11.1	49.8	2.4
Air		37.2	$3.1\cdot10^4$	81	$1.21\cdot10^{-3}$
Water		36.0	36.0	78.7	1.00

At the end of an electromagnetic shower in air, at t_{max}, the mean energy of the electrons and photons is 81 MeV. The electrons now lose energy mainly by collisions, but may still produce a few photons. They have a maximum range of about 26 g/cm^2 (Equation (5.11)). The photons gradually lose energy, mainly through Compton scattering, giving rise to a number of photons of decreasing energy. Their attenuation length in air is 55 g/cm^2 (corresponding to about 500 m at sea level) in the energy range 10–80 MeV, but decreases to 20 g/cm^2 at 2 MeV. A significant number of photons have a range far beyond that of the electrons.

At sea level we have photons and electrons with energy of tens of MeV produced in showers initiated at low altitude. The low energy electrons are being slowed down and a number of Compton energy degraded photons are produced for each photon of energy close to the critical energy. Figure 7.12 shows the energy spectra of electrons and photons at sea level.

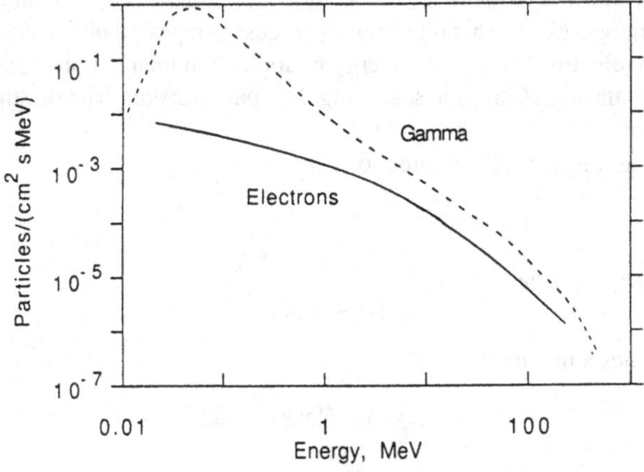

Fig. 7.12. The energy spectra of electrons and photons at sea level.

7.11 Particle fluxes in the atmosphere

Now that the processes occurring in the atmosphere have been described, we consider the relative composition of the cosmic rays as a function of depth. Figure 7.10 shows the flux of protons, pions, neutrons, muons, and electrons. The photon flux is not shown because at energies above the critical energy it will be close to that of electrons.

Protons and pions (or rather the primary particles) are absorbed exponentially with depth in the atmosphere with an attenuation thickness of 120–130 g/cm^2. The protons are continuously producing secondary particles at a rate that is nearly proportional to their flux Γ_p (particles/s cm^2). It is useful to look at the ratio of the flux Γ of the other four particles, specified by a lower index, relative to the proton flux. The pion–proton flux ratio increases slowly with depth (Figure 7.13) and is about 0.08 pions per proton at sea level.

Neutrons. The shape of the neutron flux follows clearly that of the proton flux. If we look at the ratio of neutrons/protons as a function of depth (Figure 7.14), we see that the ratio increases gradually from about 22 in the upper atmosphere to 38 at sea level. The neutron flux is higher than that of protons because the neutron path length is much longer as they do not lose energy by ionization.

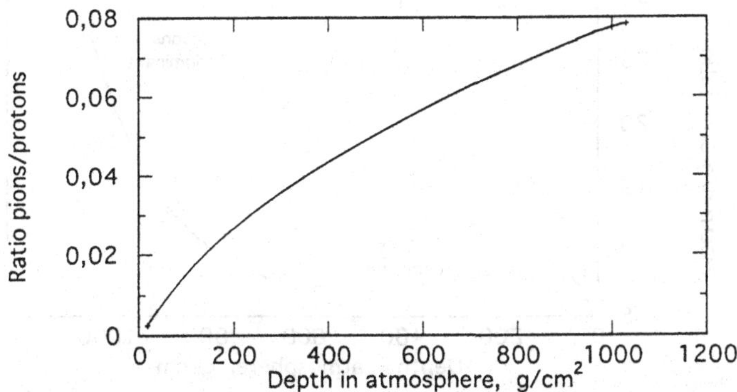

Fig. 7.13. The ratio of pions to protons in cosmic rays in the atmosphere.

Electrons. Electrons also follow approximately the same flux pattern as that of protons (Figure 7.10), but their flux decreases more slowly close to sea level. The ratio of electrons to protons (Figure 7.15) is nearly constant to a depth of about 700 meters where it begins to rise.

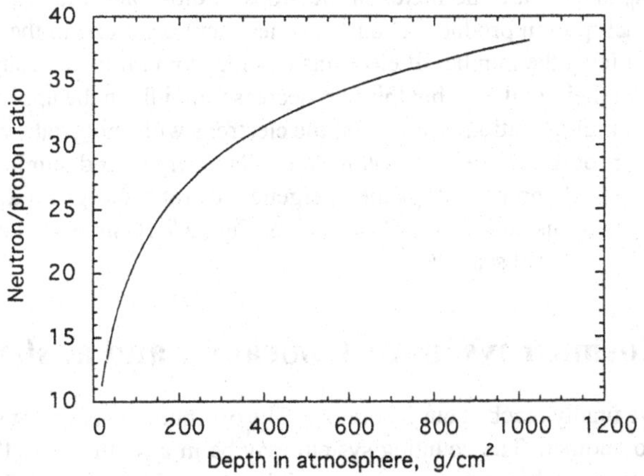

Fig. 7.14. The ratio of neutrons to protons in cosmic rays in the atmosphere.

In the upper atmosphere, electrons are produced predominantly by energetic gamma photons emitted in the decay of neutral pions, produced by protons. Below 700 meters the ratio begins to rise, mainly through muon decay electrons, when the contribution of muons becomes significant, as their flux decreases much more slowly than that of the protons. The

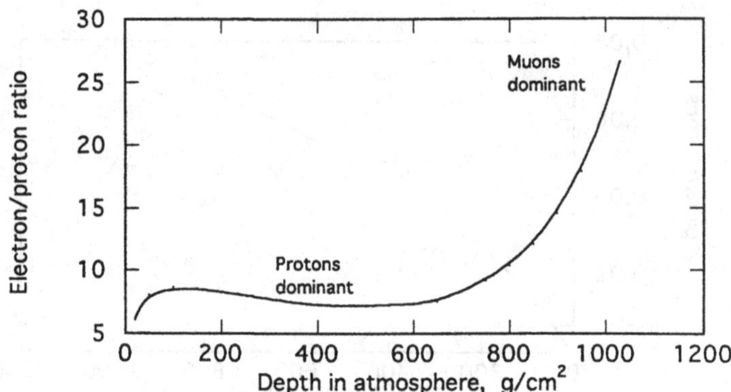

Fig. 7.15. The ratio of electrons to protons in cosmic rays in the atmosphere.

electron component can be described within 15% by

$$n_{el} = 7.1\, n_p + 0.14\, n_\mu \tag{7.17}$$

where n_{el}, n_p and n_μ are the fluxes of electrons, protons and muons, respectively. This shows that each proton produces about 50 times more electrons in the atmosphere than a muon. At sea level the number of electrons coming from protons through neutral pions is about a third of their total flux, but this part decreases rapidly in the uppermost meters of the Earth. During multiplication in a shower, the electrons with energy above critical energy in air, 83 MeV, are accompanied by photons of similar energies and numbers. When the electromagnetic cascade process stops the energetic electrons and gamma rays will continue their interaction as discussed in the last section. Figure 7.16 shows the energy spectrum of electrons and photons at sea level.

7.12 Cosmic rays in the laboratory and at shallow depth

Now we have finally reached our laboratory. The overburden varies considerably from one laboratory to another. The counting system may be in a room where there is only a thin roof above, or in the basement of a large building, possibly having 6 floor plates or more above. A few underground laboratories have been specially constructed, usually below a few meters of soil or rocks. Here, we are mainly concerned with the intensity of cosmic rays in laboratories with an overburden of 20 mwe or less.

At sea level, outside a laboratory, the cosmic rays consist of the following components, measured in particles cm^{-2} per 1000 s (NCRP Report, 1975):

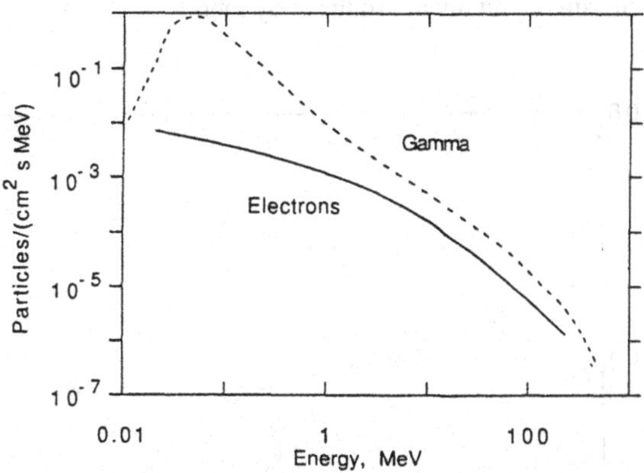

Fig. 7.16. The energy spectrum of electrons and photons at sea level

Muons	19.1
Electrons	5.5
Protons	0.17
Pions	0.0013
Neutrons	7.5

Protons and neutrons. The proton attenuation thickness in building materials is apparently somewhat higher than that in air and 1.6 mwe will be used here, but values given in literature may deviate 10% from this. The attenuation A_p is given by

$$A_p = I/I_0 = e^{-m_{ob}/1.6} \tag{7.18}$$

where I and I_0 are the proton fluxes under an overburden of mass m_{ob} and outside the laboratory.

Here we are primarily interested in the contribution of cosmic-ray components to the background of our low-level detectors. It is generally easy to get rid of the pulses caused by the direct interaction of the protons. The background contribution of the neutrons, which are mainly produced by protons in ordinary laboratories, is more difficult to suppress.

Neutrons are not only produced by protons in surface and near surface laboratories, but also by muons, mainly through capture of negative muons. Measurements made by Coccioni and Tongiorgi (1951) illustrate well the production of neutrons at shallow depths (Figure 7.17). They determined the relative number of neutrons produced in a lead layer on both sides of thermal neutron counters, embedded in thermalizing paraffin. This unit was immersed in lake water at varying depth. Although the measurements are rather coarse, they show that the neutron intensity can be decomposed into:

1. An exponentially decreasing component due to protons with a mass attenuation thickness of about 1.6 mwe.

2. A slowly decreasing component produced by muons.

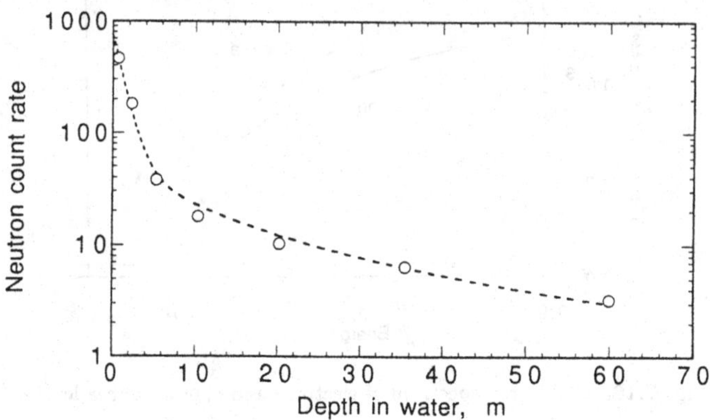

Fig. 7.17. Neutron production at different depths in water. From Coccioni et al., 1951.

The attenuation of the muon component follows Equation (7.19) quite well. These measurements indicate that with no overburden about 92% of the neutrons are produced by protons and 8% by muons. These two components have equal intensity at an overburden of about 4 mwe. A few mwe below this depth, the neutron component due to protons becomes insignificant.

At air/concrete interface, the thermal neutron flux will first increase. Figure 7.18 shows the results of measurements and Monte Carlo calculations of the relative neutron flux under a layer of concrete of varying thickness (Dep et al., 1994). It shows that the thermal flux rises to a maximum below an overburden of about 0.4 mwe. The thermal neutron flux is therefore considerably higher under the first floor plate than in the air above it.

Muons are very penetrating, their most energetic fraction penetrates hundreds of meters below the surface of earth. Figure 7.19, shows the attenuation of muons underground and is based on a large number of experimental studies.

The curve can be approximated by

$$A_\mu(m_{ob}) = I/I_0 = 10^{-1.32\log d - 0.26(\log d)^2} \tag{7.19}$$

where I_0 is the intensity at zero overburden and I below overburden M_{ob} (mwe), and d is given by

$$d = 1 + m_{ob}/10 \tag{7.20}$$

The parameter d can be interpreted as the total overburden (atmosphere included) measured in number of atmospheres. Equation (7.19) describes muon flux with an accuracy of about 5% to a depth of 100 mwe and within 10% up to 1000 mwe. It is interesting to compare

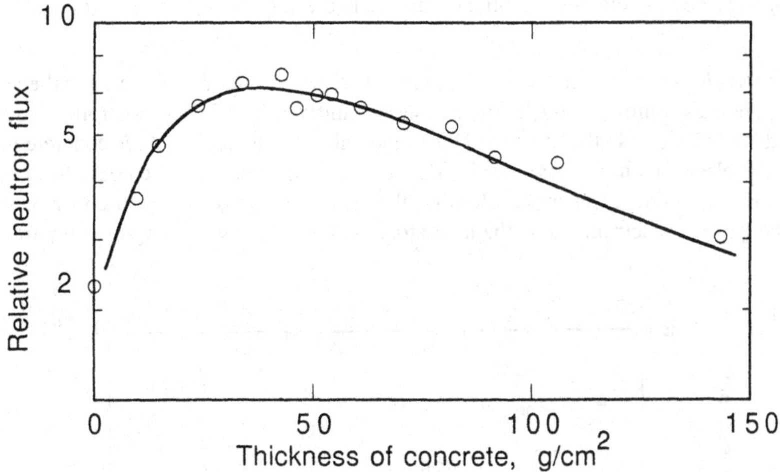

Fig. 7.18. The relative cosmic neutron flux at sea level under concrete layers of varying thickness, measured (circles) and calculated (lines) by the Monte Carlo method (Dep et al., 1994).

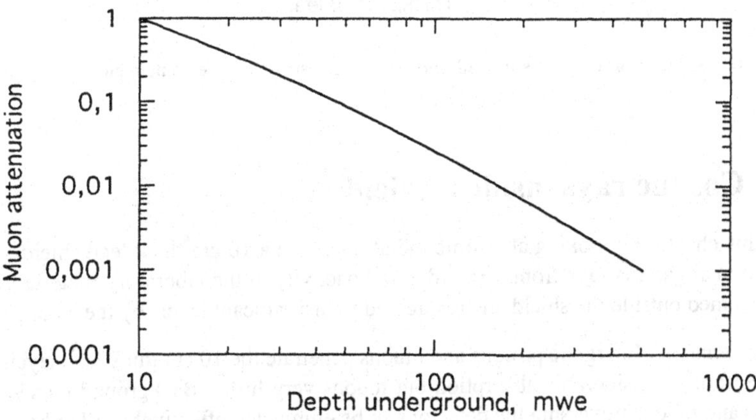

Fig. 7.19. The flux of muons underground.

this equation with measurements of muon attenuation in thick lead and iron absorbers made with a quadruple coincidence technique (Figure 7.20). The agreement with Equation (7.19) is quite good, showing that there is little difference in the absorption in rocks, iron and lead.

Photons and electrons. The radiation length in concrete is 22 g/cm^2 and the critical energy 38 MeV. The maximum electron density in a shower initiated by a 2 GeV electron will occur at a depth of 145 g/cm^2 below its start. This is equivalent to the mass in three concrete floor plates. The electron–photon cascades inside the laboratory can therefore come from both the outside and building materials. Because the critical energy is lower in concrete than in air, the gamma spectrum inside the laboratory is shifted to lower energies compared to outside air.

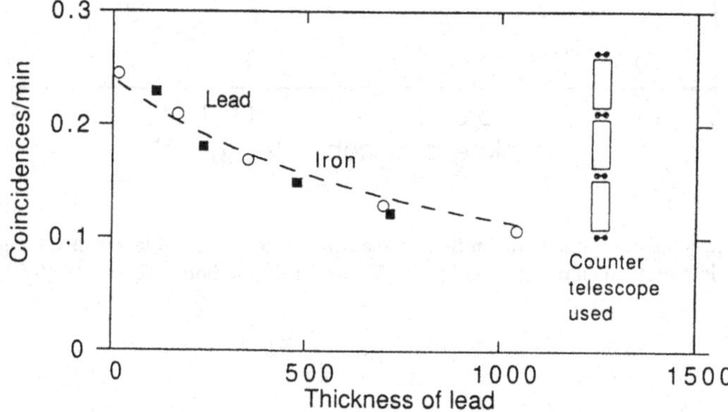

Fig. 7.20. Absorption of muons in lead and iron as measured by the coincidence technique.

7.13 Cosmic rays inside a shield

We end this chapter by looking at cosmic radiation inside a 10 cm thick lead shield. The flux of external gamma rays from primordial radioactivity in the laboratory materials and showers formed outside the shield are suppressed to insignificant levels by the lead.

Protons and muons. Energetic protons and muons penetrate the 10 cm thick lead layer, the protons suffering considerable absorption but muons very little. Background pulses due to protons and muons traversing the detector can be eliminated effectively, either by pulse height discrimination or through the use of anticosmic counters. Their most important background effect comes from the secondary radiation — neutrons, electrons and photons — they produce in the lead.

Neutrons. Lead has a low thermal absorption cross section. When there is neither a moderating nor an absorbing material inside the shield, it will be relatively transparent to neutrons. The flux and spectrum of neutrons will then be almost the same inside as outside the shield, except for additional neutrons produced in the lead. Lead, exposed to high energy protons, is a very efficient neutron producer. The proton interaction length in lead is about 170 g/cm^2 (15 cm). About 50% of the protons incident on the shield will therefore experience a collision, with a high probability of producing one or more neutrons. Arthur et al. (1988) have made a very useful study of thermal and fast neutrons inside a lead shield of varying thickness. They measured the neutrons with a BF_3 proportional counter, sensitive only to thermal neutrons. In order to determine the flux of fast neutrons the counter was surrounded by a 5 cm thick layer of paraffin. Figure 7.21 shows the net count rate of the neutron detector inside lead shields of varying thickness. The background of the detector (0.03 cpm) was determined by covering it with a 0.05 mm foil of cadmium that absorbs all external neutrons. The measurements demonstrate a copious production of fast neutrons in the lead and that these have no measurable effect on the thermal flux inside the shield, as measured with a bare counter.

The elastic neutron cross section in lead in the energy range from a fraction of an eV to 5 MeV is 5–10 barn, corresponding to a mean free path of 3–6 cm. Neutrons lose on average only 1% of their kinetic energy in each elastic collision with the lead nuclei. We can therefore conclude that the fast neutrons produced in the lead do not contribute measurably to the small thermal flux inside the shield as long as there is no thermalizing material there. This is verified by the measurements shown in Figure 7.21. This is contrary to a widespread opinion that there is a significant thermal flux inside the shield of low-level germanium detectors, due to fast neutrons being slowed down there. This is discussed further in Section 12.1.

Electrons and photons. Electromagnetic showers, the soft component of the cosmic rays, that are formed outside the shield cannot penetrate the 10 cm thick lead layer. A new generation of showers is, however, formed in the lead and these will give rise to a new flux of electrons and photons inside the shield. These showers are secondary radiation of: (1) protons through neutral pions and (2) muons, mainly through knock-on electrons. In order to estimate crudely the relative intensity of these two components we assume that the shower processes are independent of material and its density, i.e., the processes will continue to operate even in our lead shield. We can then write an equation for the electron flux Γ_{el} as a function of overburden by inserting the proton and muon flux at sea level given above into Equation 7.12

$$\Gamma_{el} = 1.20\, A_p + 2.66\, A_\mu \tag{7.21}$$

where A_p and A_μ are the attenuation factors given in Equations 7.13 and 7.14. By inserting the expression for A_p and A_μ we can calculate Γ_{el}.

According to these crude calculations the muons produce about 2/3 of the electron flux at small overburden, with 1/3 coming from protons. It can be assumed that the ratio of the gamma fluxes will be similar. The muon component decreases only slowly but the proton

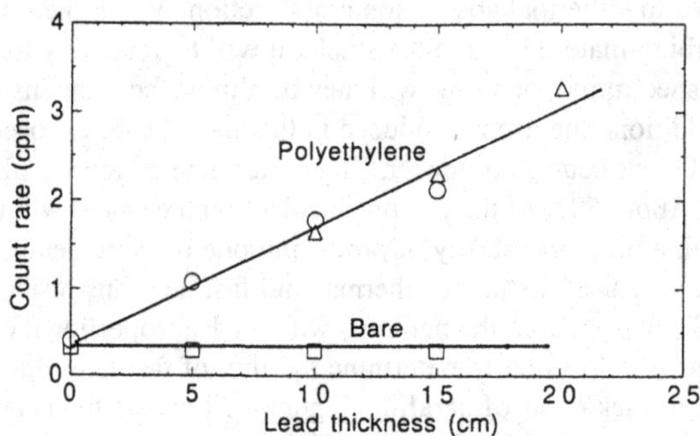

Fig. 7.21. Neutrons inside a lead shield of varying thickness.

component exponentially. At 5 mwe the proton component has fallen to an insignificant value but the muon flux to 55% of its surface value. The main trend for these two components can be assumed to be similar in lead.

Recommended reading

- S. Hayakawa, *Cosmic Ray Physics*, John Wiley & Sons, 1969.

Chapter 8

Radiation detectors

8.1 Introduction

Marie and Pierre Curie devised an accurate instrument for measuring the intensity of radioactivity: the parallel plate ionization chamber with a sensitive electrometer. A few years later, two new methods were invented for the detection of individual alpha disintegrations: (a) visual counting of light flashes, scintillations, emitted when the particles impinge on ZnS crystals, and (b) detection of electric pulses with small point-counters. Although important results were obtained with these two methods in nuclear research, they were limited to low counting rates. About two decades elapsed until a versatile nuclear detector was invented and electric methods developed to count pulses at a fast rate. This came with the invention of the Geiger counter in 1928 and the application of the new electronic techniques. The Geiger counter revolutionized the nuclear radiation detection technique and it remained for 20 years the most practical detector for radioactivity measurements.

After World War II, when atomic reactors were beginning to produce vast amounts of radioactivity, supplying scientific and industrial research with a large variety of radioisotopes, better detectors were urgently needed. Great advances made in the electronic techniques during the War had prepared the ground. Scintillation detectors, using both solid and liquid scintillators, were invented in the late 1940s, and semiconductor detectors, based on both silicone and germanium, some 15 years later.

The basic principles of the three main categories of detectors used in low-level counting will be described here. These detectors are:

1. Gas discharge detectors.

2. Scintillation detectors.

3. Semiconductor detectors.

Their application in low-level systems will be described in Chapters 12–15.

8.2 General properties

First, some general properites, more or less common to all the detectors, will be discussed. The signal is derived from the ionization and excitation of atoms, produced by charged particles travelling through the detecting medium. The original signal can come from (a) the primary ionization or (b) from a shower of photons, i.e., a scintillation that is subsequently transformed into an electric pulse by a photomultiplier tube. The electrical signal is usually nearly proportional to the energy deposited in the detecting medium. We are mainly interested in the following characteristics of our detectors:

1. Energy resolution.
2. Time resolution.
3. Detection efficiency.
4. Proportionality of signal to energy deposited.
5. Pulse shape.

Energy resolution. Let us take the measurement of ^{137}Cs with a NaI scintillation detector as an example. A significant part of the photons impinging on the crystal will deposit all their energy through the photoelectric effect. We focus on these events. The shower of photons created will in average release N_e electrons from the cathode of the photomultiplier tube. Because of the statistical nature of the processes, the distribution in the number of the electrons released at the photocathode by the monoenergetic gamma radiation, 664 keV in our case, will have nearly a Gaussian distribution, giving a peak in the pulse height spectrum, as can be seen in Figure 8.1. The resolution is defined as the full width of the peak at half maxinum (FWHM) divided by the mean size of the pulses. The lower this ratio is, the better will the detector be able to separate closely lying peaks. As a rule of thumb we can say that two peaks can be resolved when their energy differs by more than one FWHM.

Only about 10% of the energy deposited in the NaI crystal is converted to light, the rest is lost as heat. About 500 eV are needed to produce one photoelectron. In germanium crystals it only takes about 3 eV to produce an electron-hole pair, giving a much sharper photopeak, as shown in Figure 8.1, and an energy resolution that is about 40 times better than in NaI scintillation counters.

Time resolution. The time sequence of two related pulses is sometimes of interest. The time resolution is defined in a similar way as the energy resolution through the FWMH time delay of the pulses.

Detection efficiency. The detection efficiency is the probability of registering a particle emitted by a sample in a well defined geometry. It can be divided into two components: *absolute efficiency*, ε_{ab}, and *intrinsic efficiency*, ε_{in}, where

ε_{ab} = number of recorded pulses/number of particles emitted by the source,

ε_{in} = number of recorded pulses/number of particles hitting the detector.

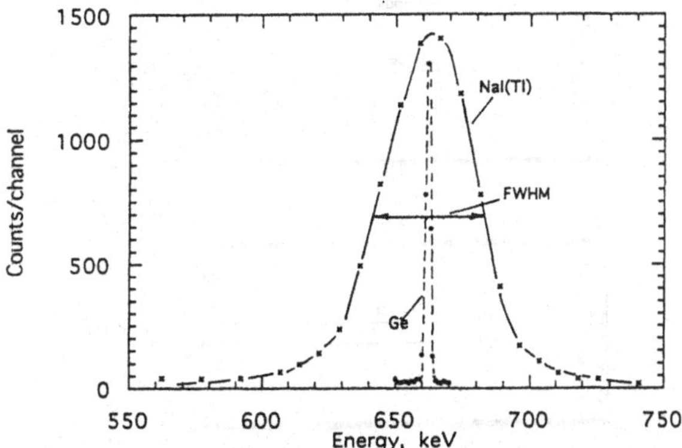

Fig. 8.1. A part of the spectrum of a ^{137}Cs measured with a NaI scintillation detector and a germanium diode. The energy resolution is measured by the full width of the peak at half maximum, either in keV or in percent.

The two efficiencies are related by

$$\varepsilon_{ab} = \varepsilon_{in}\,(4\pi/\Omega) \tag{8.1}$$

where Ω is the average solid angle of the detector as seen by the source. Frequently, we are only interested in pulses lying in an alpha or gamma peak. We then speak of *peak detection efficiency*, as opposed to *total efficiency*.

Energy proportionality. The pulse size is usually nearly proportional to the energy deposited by a charged particle in the detecting material. For scintillation detectors the ionization density of the particle influences the transformation of energy into light. The light pulse of an alpha particle is, for example, only about 1/10 of that of an electron depositing the same amount of energy in liquid scintillators.

Pulse shape. Either the rate of rise of the electric pulse or its rate of decay can, in some cases, be used to discriminate true signal pulses from interfering pulses. This is discussed in section 11.3.

8.3 Gas discharge detectors

8.3.1 Electric multiplication in gas

A gas discharge radioactivity detector consists of a container with two electrodes between which a potential difference is applied.

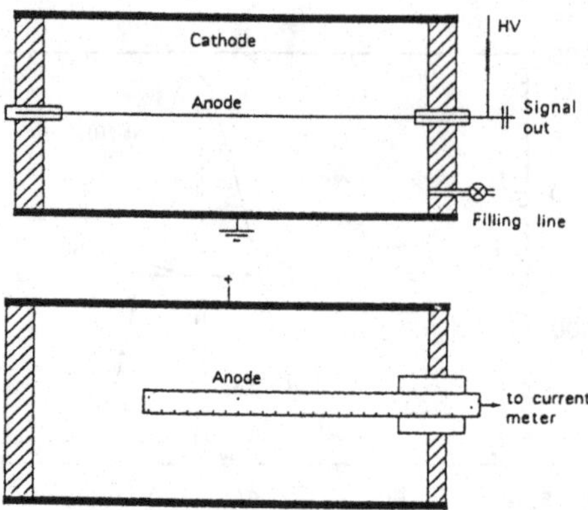

Fig. 8.2. A schematic drawing of a gas discharge detector.

Figure 8.2 shows schematically a gas proportional counter. Its geometry and the gas used depend on the application.

The primary ionization can be produced by a variety of charged particles; here we are mainly interested in alpha particles and electrons (including beta particles). For most gases the net energy required to produce an ion pair, including excitation energy, is about 30 eV, independent of the type of particle and its energy (Section 5.1.1, Table 5.1). The electrons and the positively charged gas molecules move in the electric field between the two electrodes, the electrons towards the anode and the positive ions to the cathode. Gas radiation detectors can be divided into three categories:

1. Ionization chambers.
2. Gas proportional counters.
3. Geiger counters.

Now we discuss the difference in the mechanism of their operation. The detector (Figure 8.2) is in the form of a cylinder with a thin central wire. The cylinder (cathode) is usually grounded and the wire (anode) connected to a positive high voltage supply. The electrostatic field at a point P between the electrodes is then

$$E(r) = \frac{V}{r\ln(b/a)} \tag{8.2}$$

where a = radius of the wire,
b = inner radius of cylinder,
r = distance of P from the center of the cylinder.

The diameter of the wire is usually 20–100 μm (1 μm = 10^{-3} mm). The electric field strength close to the surface of the wire is therefore very high. The free electrons move almost radially towards the anode wire. They make frequent elastic collisions with gas molecules, losing in each collision a major part of the energy they have acquired in passing through the electric field from the preceding collision.

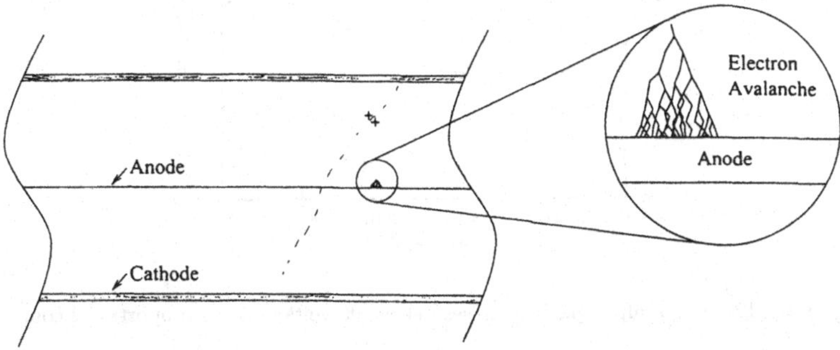

Fig. 8.3. The formation of an avalanche close to the anode wire (a) and the space charge left by the avalanche when the electrons have been collected.

Close to the wire, at a distance of a few anode radii, the electrons can acquire enough energy between the collisions to ionize gas atoms (inelastic collisions). The two electrons then continue towards the anode and both can cause further ionization. A single electron can thus initiate an avalanche, where the anode finally collects M electrons for each primary electron. M is called the *gas multiplication factor*. Typical values for M lie in the range from 10^2 to 10^5.

8.3.2 Categories of ionization gas detectors

The description of the discharge in gas proportional counters is usually simplified. Here it is described in some more detail. The approach is mainly graphical, showing how the pulse size depends on the applied high voltage. The description is based on an experiment made with a flat (1 cm thick), box shaped detector, with a thin window (1 mg/cm^2) and a 0.1 mm anode. The counting gas, a mixture of 90% argon and 10% methane, was continuously renewed (flow counter).

Let us first look at the response of the counter to the heavily ionizing alpha particles. The counter was exposed to a ray of collimated beam falling perpendicularily on the counter window and the maximum pulse size as a function of anode voltage was read on an oscilloscope. It can be assumed that each alpha particle forms about 40 thousand ion pairs in the gas.

The pulse size for a high voltage from zero to up to 800 volts is shown in Figure 8.4. At the lowest anode voltage the pulse size rises rapidly, it then levels off until it reaches a saturation size of 50 μV at an anode voltage of about 60 volts. The pulse height then remains practically constant up to about 500 volts, where it begins to rise. The first rising part of the

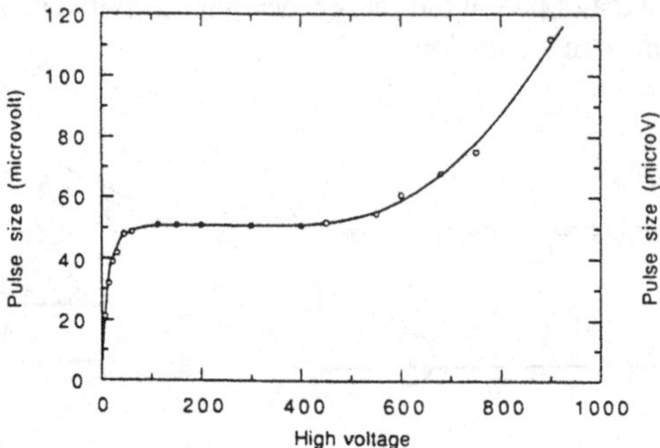

Fig. 8.4. The size of alpha particle pulses below 800 volts in a gas proportional counter.

curve belongs to the *region of recombination* where the electric field is still weak and the competing process of recombination is strong or significant compared to charge collection by the electrodes. Between 60 and 500 volts no recombination occurs and the pulse height is practically constant. This part of the curve is called the *ionization chamber region*. When the current is measured, rather than the pulse size, the current will be constant in this voltage range.

Above 500 volts, the size of the alpha pulses begins to rise due to electron multiplication close to the anode, as described above (Figure 8.4). The pulse size increases slowly at first, but exponentially from 1000 volts to 1600 volts. The pulse size is therefore shown on a logarithmic scale in Figure 8.5.

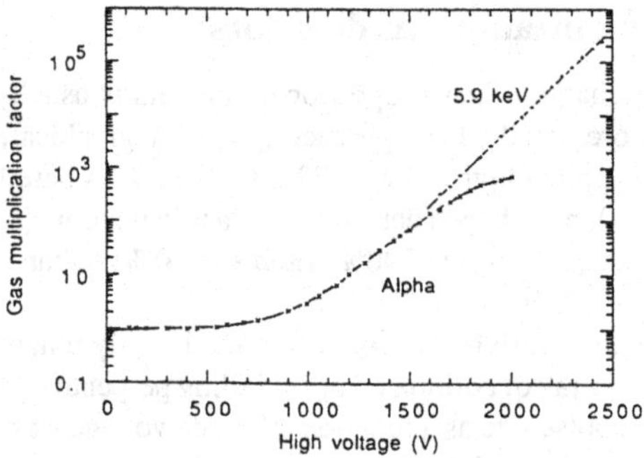

Fig. 8.5. The size of alpha particle and 5.9 keV X-ray pulses in a gas proportional counter versus high voltage.

Above 1600 volts, the pulse size begins to level off due to strong space charge effects, discussed below. The counter was then exposed to 5.9 keV X-rays from ^{55}Fe (Figure 8.5).

The pulse size first rises above the noise level at about 1600 volts and increases exponentially up to about 2500 volts. At this voltage an increasing number of spurious afterpulses makes the counter operation unreliable.

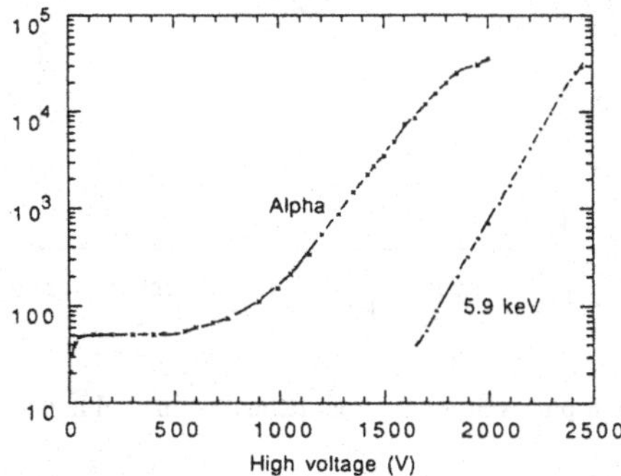

Fig. 8.6. Gas multiplication constant determined for alpha particles and 5.9 keV X-rays.

The *gas multiplication factor* M is, according to the definition given above, equal to the total pulse height $P(V)$ at high voltage V divided by the pulse height P_{pr} the primary ionization gives:

$$M = P(V)/P_{pr} \tag{8.3}$$

It is easy to find M for the alpha pulses, as P_pr can be measured, here it is 50 μV. Figure 8.6 shows the gas multiplication constant. The gas multiplication factor for the 5.9 keV X-rays cannot be determined directly. It is assumed to be the same as for alpha particles at low voltage, as shown in Figure 8.7.

It is interesting to look at the relative pulse size due to the secondary ionization only, i.e., at $(P(V) - P_{pr})/P_{pr}$, that is equal to $M - 1$. The result is shown in Figure 8.7. Here the exponential part of the curve extends down to 400 volts.

According to theory (Knoll 1989, page 170) the slope of the exponential gas multiplication line should not depend on the source of ionization. The difference in the slope of the exponential part of the curve for alpha particles and 5.9 keV photons indicates strongly that space charge affects the large alpha pulses even to the lowest values of M.

The energy linearity of the counter at fixed voltage was studied by exposing it to X-rays of varying energy. The highest photon energy that could be measured was about 30 keV. Above this limit, the photoelectrons will only lose a part of their energy as their range then exceeds the dimensions of the counter and the detection efficiency becomes very low. The ratio of the size of a 14.4 keV peak to that of the 5.9 keV peak remained constant up to 2000 volts. The ratio then began to decrease and had fallen by 15% at 2400 volts. In the high voltage region, where space charge (discussed below) has not yet a significant effect on the gas amplification, in the *region of proportionality*, the size of the pulses is proportional to the

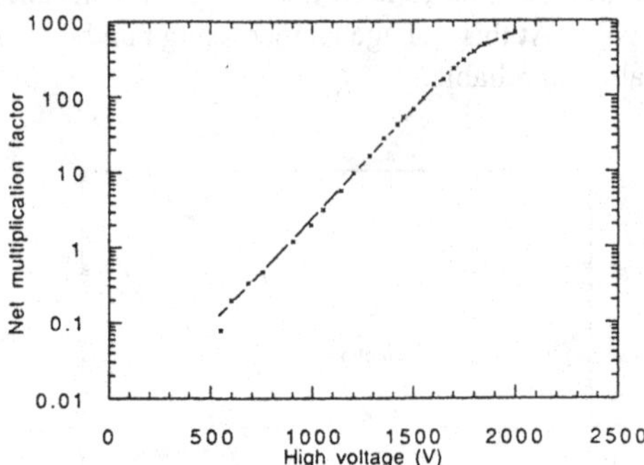

Fig. 8.7. The size of alpha pulses in a gas proportional counter due to gas multiplication only.

primary deposited energy. This voltage region ends where the space charge begins to significantly affect the gas multiplication, and the *region of limited proportionality* starts. In this region the pulse size still depends on the primary ionization, but is no longer proportional to it.

This limitation in proportionality is explained by the space charge formed in each electron avalanche close to the anode as a result of the multiplication process. The positive ions have a mass about 10^5 larger than the electrons and they will therefore hardly have moved when the multiplication process is brought to an end. They form a local cloud of positive charge very close to the anode wire, locally attenuating the electric field (Figure 8.3). The last electrons to arrive, those produced farthest from the anode, will thus pass through a significantly reduced electric field and their multiplication will be smaller than for the electrons that arrived first and passed through an undisturbed field. For particles giving dense primary ionization, or at high gas multiplication, the pulse size will not be strictly proportional to the initial ionization, i.e., the counter will be working in the region of limited proportionality.

The upper limit of the proportional region is not a fixed voltage for a given detector, as usually assumed, but it depends on the type of the radiation. Particles creating the same number of primary ion pairs may give pulses of different size, as one pulse may come from a high energy electron or beta particle that has a long track and loses only a part of its energy in the gas. Another may come from a low energy electron or beta particle that has a short track and loses all its energy. The multiplication of the former will be spread over a large part of the anode, giving low space charge density, but for the latter multiplication will be limited to a narrow range of the anode, forming a large localized space charge density. The region of limited proportionality is thus not only a question of high gas amplification, but depends also on the density of the primary ionization. At a given high voltage, a counter may therefore be in the range of proportionaity for beta particles, but in the range of limited proportionality for alpha radiation.

It should be noted that the alpha particle tracks in the measurements described above were nearly perpendicular to the direction of the anode and most of the ionization, and the following gas multiplication of each alpha particle, lies in a narrow (1–3 mm) axial region, enhancing the space charge effect. It seems probable that space charge also affects the alpha pulse height at low gas multiplication, even below 2, as the slope of the curve showing the size of the alpha pulses is considerably less than that of the 5.9 keV X-ray line in Figure 8.6.

If the argon gas used for the proportional counter discussed above, containing 10% methane, is replaced by argon with 1% propane, the proportional region is very narrow and at about 1100 volts all pulses are of equal size, irrespective of the amount of primary ionization. This is the beginning of the *Geiger region*. Even a single ion pair will trigger a pulse of the same size as an alpha particle, that produces thousands of ion pairs. Furthermore, the multiplication now spreads along the entire length of the anode through ultraviolet light emitted by the ionized atoms and the multiplication continues until the space charge halts any further secondary ionization. The detector is then insensitive for a given time, the *dead time*, which is of the order of 100 μs, until the space charge has drifted sufficiently far away from the anode.

8.3.3 Ionization chambers

Ionization chambers are mainly used for measuring environmental gamma and cosmic radiation. They are usually steel cylinders filled with argon to a pressure of 10–20 atm in order to increase their sensitivity. Ionization chambers can also be used in a pulse mode for alpha particle spectroscopy, giving a relatively good resolution. The electrodes are then in the form of parallel plates. Pulse ionization chambers, frequently called Frisch chambers, were used extensively for alpha spectrometric work late in the 1950s and in the 1960s, but they were eventually superseded by Si(Li) diodes. The advantage of these chambers is that they are of simple construction and can measure a source of large area. In recent years, they have been used in an ingenious way to measure ^{210}Po, a decay product of radon, that lies buried through recoil, in a very thin layer below a glass surface (Johansson et al., 1992).

8.3.4 Proportional counters

Gas proportional detectors are widely used in low-level counting. Their volume ranges from a few ml to about 5 liters. The type of gas used depends on the application. The sample can be: (1) the counting gas itself or a part of it *(internal sample counter)* or (2) outside the counter, in solid form *(external sample counter)*. In the latter case the detector has a thin window allowing alpha and beta particles from the sample to enter the counter with minimum energy loss.

A variety of gases can be measured in internal sample counters, e.g., argon, methane, carbon dioxide and propane. When a noble gas or hydrogen is measured, a few percent of a hydrocarbon gas are added in order to improve the counting characteristics. External gas proportional counters are practically always of the flow type, where the counting gas is usually argon with 2–10% methane or pure propane.

The voltage of gas proportional counters must be set high enough to bring the smallest pulses well above the noise level. When a large dynamic range in pulse height is to be measured the large pulses may be in the region of limited proportionality. It seems probable that many counters referred to as proportional counters are actually working deep in the region of limited proportionality. Usually, this is of little consequence, as one is primarily interested in determining the quantity of radioactivity, not in the spectrum of the deposited energy.

In order to study the performance of gas proportional counters in a simple way and to determine the most suitable high voltage, their characteristic curve is usually measured, i.e., a curve showing the count rate, for a fixed source and fixed discriminator voltage (threshold), versus the high voltage. Figure 8.8 depicts a curve of this type for a sample emitting both alpha and beta rays. As the pulses of alpha particles are much larger than those of beta

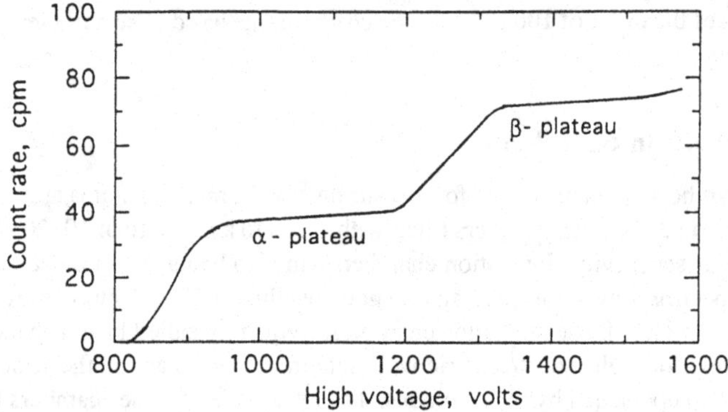

Fig. 8.8. A characteristic curve for a gas proportional counter exposed to a source emitting both alpha and beta particles.

particles, there is a high voltage range where practically only alpha pulses pass the discriminator of the counting channel. For an internal sample, or a thin external source, where all the alpha particles give a long track in the gas, this part of the curve rises slowly, giving an alpha plateau. For thick alpha sources, the plateau will be much shorter. At higher voltage, the counting rate begins to rise again when the largest beta pulses trigger the discriminator and a new plateau is gradually reached where practically all alpha and beta pulses are counted. Above this plateau the count rate starts to rise again, now due to after-pulses, and the counter will no longer function reliably. This voltage can even damage it and should be avoided. If the gas used is also suitable for Geiger counting, these after-pulses are suppressed and the detector enters the Geiger region.

8.3.5 Gas proportional counters for internal samples

Internal gas proportional counters are used extensively in low-level work for the measurement of tritium, ^{14}C, ^{39}Ar, and other radioactive gases or radioisotopes that can be synthesized as or converted to a gas. The wall of these counters is usually made of metal, most often copper, with an anode of stainless steel or tungsten. The anode diameter is typically 0.02 to 0.05 mm. These counters come in sizes from a few ml to up to about 5 liters. For some gases they can be operated at high pressure in order to increase sensitivity. A pressure of 3 atm CO_2 is common in radiocarbon dating, and argon has been measured at a pressure of 20 atm.

These counters are relatively simple to make, it takes about a day to assemble one. However, they are not available commercially any more, so gas proportional counting has, in the past 20 years, been losing ground to liquid scintillation counting in various type of low-level work, even where gas counting seems to have a distinct advantage.

8.3.6 Gas proportional counters for external sources

These counters are used extensively in low level work for measuring pure beta emitters, e.g., $^{90}Sr/^{90}Y$, ^{99}Tc and $^{32}Si/^{32}P$. Sometimes they are also used for the measurement of ^{137}Cs, a gamma emitter, because of their much higher counting efficiency compared to gamma counting with Ge-diodes.

In low-level work, these counters consist today of an array of 4 or more detector elements made of a single plate of acryl or copper (Theodorsson 1988). They are always of the flow type, i.e., with a constant flow through the detector of the counting gas from a steel cylinder at a rate of a few ml/min. Experience at the author's laboratory shows that these counters can work just as well at a greatly reduced flow, down to about 0.02 ml/min, if there is no leakage. The working voltage at this low flow rate will be about 100 volts higher than at high flow rates due to continuous outgasing of the acryl mass, but this does not affect the proper functioning of the detectors.

Finally, it should be noted that these counters may, with a suitable counting gas, work just as well in the Geiger region. The size of the pulses in gas proportional counters gives, anyway, little information of interest as the beta particles are not stopped in the thin active volume (about 1 cm thick) and they therefore deposit only a small part of their energy in the counting gas. Operating them in the Geiger region simplifies the electronics.

8.4 Scintillation counters

8.4.1 Scintillators

Luminescence played a decisive role in Röntgen's discovery of X-rays in November 1895 and scintillations excited by single alpha particles impinging on a screen coated with zinc sulfite (ZnS) were discovered a few years later.

During the 1930s, photomultiplier tubes were developed, primarily for the new television technique. It was, however, not until after World War II that they were applied to the detection of scintillations. The technique of modern scintillation counting was born in 1947–48 when (1) Coltman and Marshall reported the successful phototube detection of light scintillation produced by alpha, beta and gamma radiation, (2) Kallman discoverd the scintillation of organic crystals and (3) Hofstadter discovered scintillation properties of inorganic crystals and singled NaI out as the most promising one (Hofstadter 1975). The new techniques increased the detection efficiency of gamma rays up to two orders of magnitude compared to Geiger counters and, equally important, they allowed determination of the deposited energy.

The scintillation process begins when an energetic charged particle ionizes and excites atoms in the detecting medium. Most of the energy deposited in scintillators is degraded to heat, but a few percent are trapped in higher energy levels for a very short time, emitting fluorescent light when the atoms de-excite. The light flashes (scintillations) are detected by a photomultiplier tube.

Various factors influence the quality of a scintillator in a specific application. The following properties are desirable:

1. High conversion of absorbed energy to light.
2. Transparency to own fluorescence light.
3. Linear response.
4. Short decay time.
5. High probability for photon absorption, when used for gamma detection.

NaI crystals, activated with a small concentration of thallium, give the highest light conversion efficiency, about 13%. It is difficult to measure accurately the absolute value of this conversion. Therefore, relative efficiency is usually given, where the light emitted by anthracene, an organic crystal, is used as a reference. A standard phototube is used in this comparison, and it is implicitly assumed to have the same detection efficiency for the light from all scintillators, in spite of some variations in the emission spectrum.

Figure 8.9 shows the emission spectrum of three scintillators and the sensitivity curve for the most frequently used photocathode, of the bialkali type.

There are two main types of scintillators, organic and inorganic, with different scintillating mechanisms. Organic scintillators can be in the form of crystals, plastic materials or liquids. They have much shorter decay times but a lower light yield than inorganic crystals. Some properties of important scintillators are listed in Table 8.1

Organic scintillators are used for two different purposes in low-level counting: (1) for the assay of samples that are either mixed with a scintillation liquid or the sample itself is an scintillator, and (2) as a detecting medium in large anticosmic counters used for suppressing the cosmic-ray background contribution. This is discussed in more detail in Section 11.8.1.

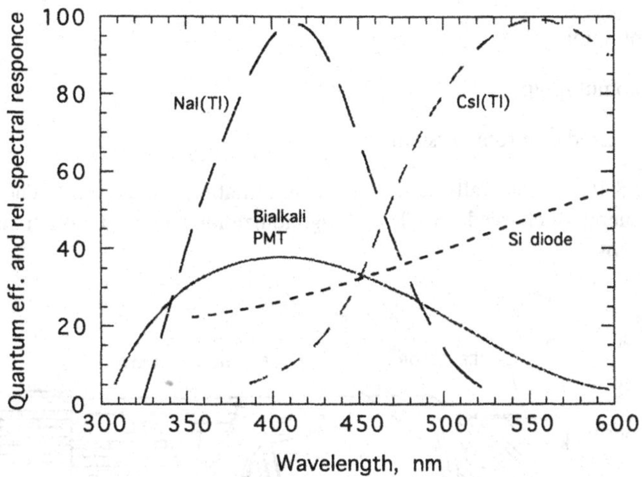

Fig. 8.9. Emission spectra of scintillators and quantum efficiency of a bialkali cathode.

Table 8.1. Parameters of important scintillators

Scintillator	Density g/cm^3	Wavelength at max. sens. μm	Rel. pulse height	Photons per keV	Decay time, ns
Anthracene	1.25	440	100	1.9	30
Toluene	0.86	285	70	1.3	2.5
Benzene	0.86	280	56	1.1	2.5
Acryl	1.06	350-450	35	1.0	2.5
NaI(Tl)	3.67	420	200	38	250
CsI(Tl)	4.51	540	100	52	1100
ZnS	4.10	450	260 (α)	50 (α)	>1000

8.4.2 Scintillation detector units

Scintillation counter systems consist of three basic constituents:

1. Scintillator.

2. Photomultiplying tube.

3. Associated electronic system.

Figure 8.10 schematically depicts a NaI scintillation detector unit. The NaI crystal absorbs the energy of charged particles through ionization and excitation, transforming a part of it into light.

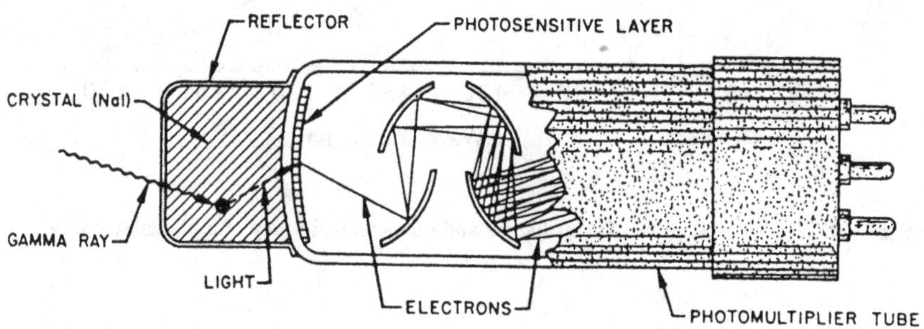

Fig. 8.10. A scintillation unit with a NaI crystal.

The surface of the scintillator, or the inside of the vessel holding it, is covered with a reflecting material, except the side that faces the phototube. A significant fraction of the photons hit, often after repeated reflections from the surface of the scintillator, the photocathode where a number of electrons are released. The fractional number of electrons emitted per photon is called the *quantum efficiency*. It depends on the wavelength of the light and has a maximum value of 0.2–0.3.

There is a potential difference of 100–200 volts between the cathode and the first collecting electrode (dynode). When the released photoelectrons impinge the first dynode with an energy corresponding to this potential difference, they will release 5–8 electrons, which are then drawn to the next dynode where the multiplying process is repeated. There are 9–12 dynodes, giving a total internal amplification factor of the order of 106–108, generating a relatively large output pulse at the anode, which is nearly proportional to the energy deposited in the scintillator. The following details are of importance in scintillation systems:

1. Absorption of energy from a charged particle or a photon in the scintillator.

2. Conversion of dissipated energy to light through fluorescence.
3. Loss of photons in the scintillator.
4. Collection of photons by the photocathode.
5. Electron emission at the photocathode.
6. Electron multiplication in the PMT.
7. Analysis of anode current pulse.

The number n_e of electrons released at the photocathode per absorbed energy (keV) can be expressed as

$$n_e = E_a\, S\, T_p\, G\, C \tag{8.4}$$

where:
E_a = energy absorbed in the scintillator,
S = number of photons converted to light per keV (scintillation efficiency),
T_p = fraction of photons not absorbed in scintillator,
G = fraction of photons that fall on the photocathode (light collection efficiency),
C = number of electrons released per photon falling on the cathode (quantum efficiency).

This shows that many factors affect the performance of scintillation detectors. Some of these are primarily of concern to the designer and of the system, but others the user must face.

8.4.3 Organic scintillators

Units with plastic scintillators and organic crystals are generally delivered ready to use by the manufacturer. The user, therefore, does not need to know much about the scintillation mechanism of these units. In the case of internal sample liquid scintillation systems, the user mixes the counting samples with some scintillation cocktail, which depends on the type of the sample to be measured. The scintillation processes of these samples will therefore be discussed briefly.

Organic liquid scintillators consist of a *solvent* (e.g., toluene, xylene) with one or more additives (*solutes*) in small concentrations in order to enhance light emission in the visible spectral region where the maximum sensitivity of the phototube lies (Figure 8.9). The energy of the charged particles is absorbed through ionization and excitation by the solvent molecules and then transferred very rapidly to solute molecules. These then emit a small part of the absorbed energy as light through flourescence, but most of it is lost as heat. The intensity of the fluorescence decays exponentially with a decay constant that is characteristic for the organic material used. In some cases a small part of the energy may be trapped in an energy level having a longer decay time.

The light emitted by the solute sometimes has a wavelength that is a little too short to secure maximum photon detection (good matching). A secondary solute is then added, a *wave shifter*, that effectively absorbs the light of the primary solute and reemits it at longer wavelengths.

The light output from a liquid scintillator may decrease when a sample is dissolved in the scintillation cocktail; we say that the counting sample is quenched. This effect can occur through two different processes:

1. *Chemical quenching.* The sample molecules interfere with the transfer of excitation energy from solvent to solute.

2. *Color quenching* occurs where the sample colors the liquid, causing light absorption in the path of the flourescence photons.

The first process does not depend on the location of the scintillation event in the vial, all pulses suffer the same amount of quenching. Color quenching depends on the path length of the photons in the vial, and therefore depends somewhat on the location of the event.

8.4.4 Inorganic scintillators

There are various types of inorganic scintillating crystals, mainly used for the detection of gamma radiation. The most widely used scintillator is sodium iodide to which a small amount of thallium has been added, designated by NaI(Tl), in order to improve the light yield. As NaI crystals are practically only doped with thallium, we will in the following simply speak of NaI scintillators.

The scintillation process is different in inorganic crystals from that in organic materials. The ionizing particles lift electrons from the ground state in the crystal lattice, the valence band, into the conduction band and a hole is formed in the valence band. The electrons as well as the holes wander freely around the crystal. Light is emitted by the thallium activation centers after collecting holes and electrons. Because of the relatively high atomic number of iodine ($Z = 53$), NaI has high gamma detection efficiency. The gamma absorption coefficient of NaI as a function of energy is shown in Figure 8.11.

NaI crystals can be made in large dimensions and different shapes. They are very hygroscopic and must be hermetically encapsulated. With the production of ever larger germanium crystals, NaI detectors have lost ground to these, whose energy resolution is about 35 times better and the background correspondingly lower. CsI(Tl) crystal units are now being produced, where silicon light-detecting diodes replace the photomultipler tubes. These may offer attractive possibilities for specialized types of low-level work, discussed in Section 15.5.

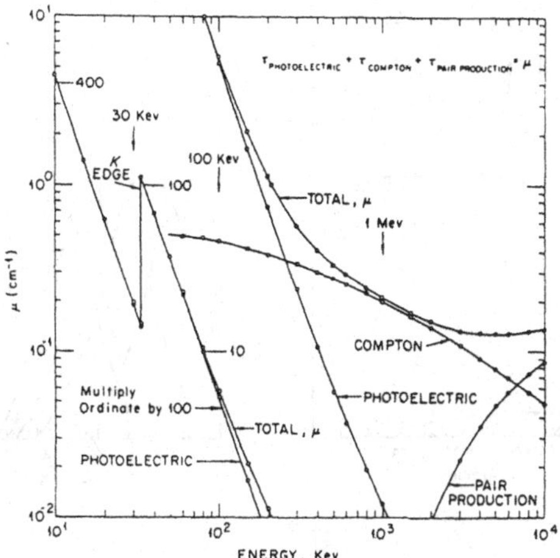

Fig. 8.11. Mass absorption coefficients of NaI.

8.4.5 ZnS(Ag) scintillation alpha counters

Zinc sulfide crystals, activated by a small concentration of silver, designated by ZnS(Ag), scintillate strongly when hit by alpha particles. This phenomenon has been used since the beginning of the century for the detection and measurement of alpha activity. Photomultiplier tubes were first applied in the recording of alpha scintillations in the early 1940s, initially by measuring the current of the phototubes. Some years later the alpha pulse counting technique was developed.

ZnS alpha counters come in two different forms: (1) as scintillation cells for the measurement of radon, their inside being coated with a thin layer of ZnS, and (2) as a flat ZnS screen, coupled to a photomultiplier tube, for the measurement of solid alpha samples.

As these detectors, important in various types of low-level work, are usually described only briefly in general texts on radiation detectors, they will be discussed below in some detail. The alpha scintillation yield of ZnS crystals is about 1.3 times higher than that of electrons in NaI crystals. This is, however, difficult to measure because of the very small size of the ZnS crystals and their partial absorption of the emitted photons. We can expect the scintillation yield per energy absorbed to be less for alpha particles, due to their high ionization density, than for electrons and muons. As these effects are of secondary importance in the use of the detectors, little attention seems to have been given to them. Knowing this particle dependence of the scintillation yield better will, however, help in determining the optimum lower pulse size discrimination level.

ZnS is only available as a powder in polycrystalline form. The size of the crystals is in the range of 7–20 μm. As the specific density of ZnS is 4.1 g/cm^3 a monocrystalline layer

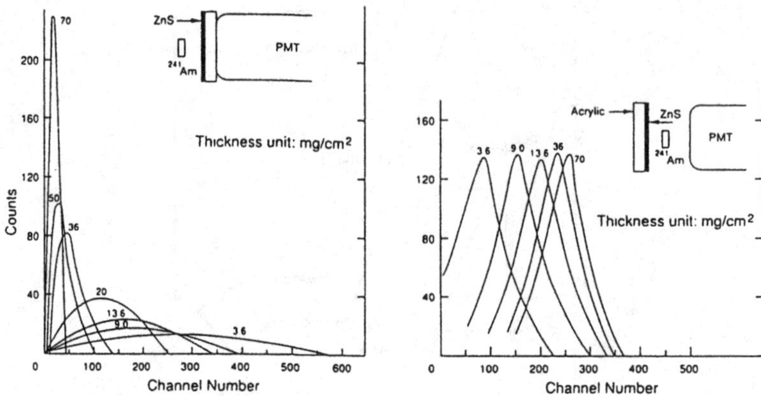

Fig. 8.12. Pulse height spectrum for alpha particles impinging on a layer of ZnS, both for transmission and reflection geometry. Liu et al., 1993.

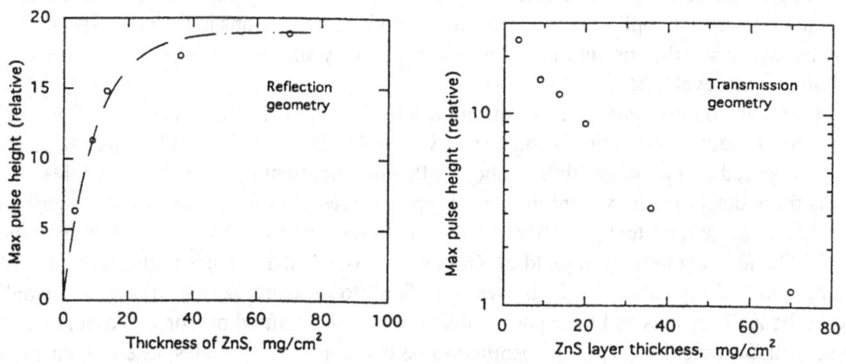

Fig. 8.13. Maximum pulse height for alpha particles impinging on a layer of ZnS at varying thickness, both for transmission and reflection geometry. Liu et al., 1993.

will have a thickness of 3–10 mg/cm^2. A layer of the polycrystalline ZnS will scatter the light and absorb a part of it. Liu et al. (1993) studied the size of the scintillation pulses for varying thickness of the ZnS, both in transmission and reflection geometries. Figure 8.12 illustrates these geometries and shows the results of the measurements. Figure 8.13 shows the size of the peak as a function of the ZnS thickness for both geometries. The results show, that for reflection geometry used in the scintillation cells described below, 95% of the maximum light pulse is reflected. In the transmission geometry, maximum pulse height is achieved with the thinnest layer, 3.6 mg/cm^2, even though the range of the ^{241}Am alpha particles is 9.2 g/cm^2 in ZnS. Both measurements give about 15 mg/cm^2 for the attenuation thickness of the scintillation light in ZnS. Alpha detection units based on scintillation in ZnS are described in section 15.3.

8.5 Semiconductor radiation detectors

8.5.1 Properties of semiconductors

Semiconductor detectors are crystals made of materials having very low conductivity, close to that of isolators. The electrons are immobile in isolators, they are all bound to atoms, i.e., bound in the valence band, that is separated from the conduction band by a broad (of the order of 10 eV) forbidden energy gap. No mechanism can lift the electrons across this gap, and isolators therefore carry no current when an electric field is applied.

In semiconductors, the energy gap is of the order of 0.5 V (Figure 8.14). Although the

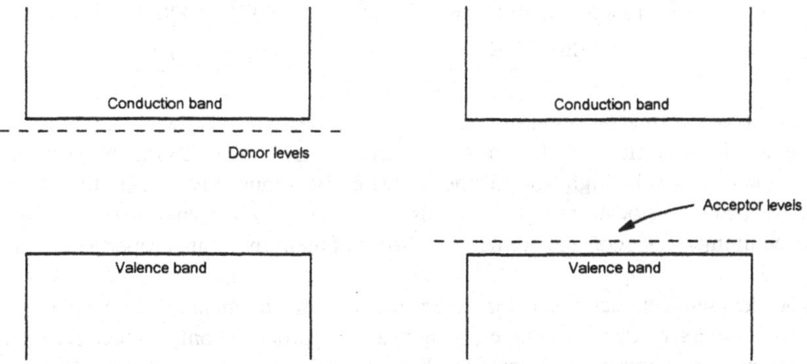

Fig. 8.14. Band structure of n- and p-type semiconductors.

average thermal energy of the electrons is only 25 meV at room temperature, electrons are continuously transferred to the conduction band through thermal agitation, giving rise to a leakage current when an electric field is applied. This current increases exponentially with temperature.

Small additives (dopants), sometimes added to control the influence of inevitable impurities, modify the conductivity of the semiconductor used in radiation detectors, forming a region where the number of free holes and electrons is almost the same. The interstitial dopant atoms have a much lower energy band gap (Figure 8.14). When a group V element is added to silicon or germanium in a small controlled concentration, they replace Si or Ge atoms in the crystal lattice. They have five electrons in the outer shell, the fifth being very loosely bound. This extra electron is practically free to move in the crystal due to its thermal energy, leaving behind an ionized atom. Such atoms are called donors. The free electrons dominate the conductivity of the crystal. When, on the other hand, atoms of elements that only have three electrons in the outer shell are added, they have a strong tendency to bind an electron from a neighboring Ge or Si atom, creating a positive atom, a hole. Such atoms are called acceptors. The hole can be filled with an electron from an adjacent atom, and the hole can thus wander around the crystal like a positive charge.

Table 8.2. Properties of silicon and germanium.

Property	Silicon	Germanium
Atomic number	14	32
Density, g/cm^2	2.33	5.33
Band gap at 300 K, eV	1.12	0.66
Energy per electron-hole pair, 77 K, eV	3.7	3.0
Hole mobility, 300 K, m^2 V^{-1} s^{-1}	480	1900
Hole mobility, 70 K, m^2 V^{-1} s^{-1}	11000	42000
Electron mobility, 300 K, m^2 V^{-1} s^{-1}	1350	3900
Electron mobility, 77 K, m^2 V^{-1} s^{-1}	21000	36000

For the successful operation of a semiconductor detector, (1) the crystal must be able to tolerate high electric fields for collecting electrons and holes, (2) the mobility of these charge carriers must be high, and (3) the charge carriers must have a long lifetime before being trapped by dislocations in the crystals. There are two principal materials that fulfill these requirements, silicon and germanium. Some of their important properties are given in Table 8.2.

The ideal semiconductor detector is a crystal made of pure material in which the number of charge carriers, electrons and holes, is equal and are produced only by thermal excitation of electrons from atoms of diode material. Earlier, neither germanium nor silicium could be purified sufficiently to fulfill these conditions. The conductivity of the impurities, usually of n-type, was always dominant. The crystals used for detectors were therefore doped, usually with lithium, which was made to drift at high electric fields into the crystal at elevated temperature, gradually giving the same spatial concentration as the impurities.

Semiconductor radiation detectors have a PIN diode structure, where I stands for the intrinsic carrier free region created by depletion of charge carriers when a reverse bias is applied to the diode. The higher the reverse bias is, the thicker the radiation sensitive depletion layer will become.

When an energetic particle passes through the carrier free layer of a semiconductor diode, it ionizes atoms along its track, creating equal number of free electrons and holes. The electrons are collected by the anode and the holes by the cathode, leading to a short electric pulse with a peak current of the order of about a microampere. The diode leakage current of the reverse biased diode must be much lower, otherwise noise pulses, caused by fluctuations in the leakage current, will interfere severely.

The detection process is similar to that occurring in ionization chambers. Normally the gas is nonconducting, but when ionized the electrons move towards the anode, and the positive ions to the cathode. These detectors are therefore sometimes referred to as solid state ionization chambers.

8.5.2 Germanium diodes

Earlier types of germanium diodes used as nuclear detectors were always doped with lithium, designated Ge(Li). The correct geometrical distribution of the lithium atoms in these diodes was seriously disturbed if the crystal was brought for a few hours to room temperature. This permanently deteriorated the energy resolution of the crystal. These Ge(Li) diodes therefore had to be kept constantly at 77 K.

In the mid 1970s, germanium refining methods had been improved to the point where the conductivity of impurities was brought to an insignificant level. Since the early 1980s, all new germanium diode detectors have been made of high purity germanium, generally referred to as HiPGe. In the following they will simply be referred to as germanium diodes. These new crystals can be kept at room temperature when not in use. The leakage current of these germanium diodes comes predominantly from thermally excited electrons of germanium atoms; the inner, or *intrinsic conductivity*, dominates.

The conductivity of pure germanium is so high at room temperature that the weak current pulses, produced by ionizing particles, are drowned in small current fluctuations. In order to suppress the current to an acceptable level, the diode is cooled to 77 K with liquid nitrogen.

Germanium detectors are made of a large high purity crystal with metal layers forming the electrodes. They come in three main types, planar, coaxial and well type diodes, shown schematically in Figure 8.15.

The first is a short germanium cylinder with the electrodes on the plane ends. The advantage of this configuration is that the positive electrode can be a very thin metallic layer, practically eliminating absorption of low energy photons in the dead surface layer. They are therefore used for measuring low energy gamma radiation and X-rays, with energy below about 50 keV. The coaxial diodes are the most common type. They are made of a cylinder where a hole is drilled along the axis of the diode and a metal contact, which forms the anode, is made inside the hole. The well type diodes have a well with a diameter of 10–16 mm drilled axially into the crystal, into which the sample can be put in order to increase the

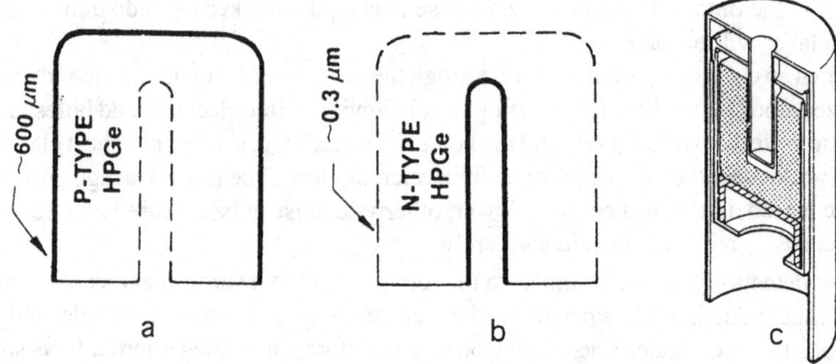

Fig. 8.15. Main types of germanium diodes: (a) coaxial, (b) low energy and (c) well type.

geometrical efficiency. This can increase the small sample counting efficiency considerably at low energies.

8.5.3 Silicone junction detectors

It is not yet possible to produce intrinsic silicone because of its high melting point. Silicone detectors must therefore be doped with lithium, which hardly migrates at room temperature. Because of the low atomic number of silicone, it is only used for the detection of charged particles, mainly alpha particles.

If the surface of an n-type silicone crystal doped with Li is oxidized, a thin p-type surface layer rich in holes is created. When a reverse bias is applied to this junction, it will cause the holes to diffuse still deeper into the n-type region and the electrons into the p-type, widening the depleted region, which can reach a thickness of 100–1000 mm. The price of the diodes is approximately proportional to the maximum depletion depth.

With a reverse bias voltage, only a weak leakage current, produced by thermal excitations of the electrons in the silicone base material, will flow through the diode. As the energy gap of Si is quite high, 0.7 eV, this current is very low, even at room temperature, so these detectors need no cooling.

Modern versions of these silicone diodes use implanted rather than surface barrier layers. They are more rugged and reliable than the older surface barrier detectors, but more expensive. They are produced in different sizes, with an area from about 30 to 3000 mm^2 (6.0 cm diameter). The FWHM alpha resolution depends mainly on the diode size, being about 12 keV for the smallest and 35 keV for the largest.

8.6 Neutron counters

Neutrons frequently give a significant contribution to the background of beta/gamma detectors. Their effect is seen most clearly in various peaks in the pulse height spectra of low-level germanium spectrometers in surface laboratories (Section 12.6.3) and in the background in gas proportional counters. Counters filled with propane give higher background then when filled with the same mass of CO_2, as propane is rich in hydrogen which has a high neutron cross section. It must be assumed that neutrons also give a significant background contribution in liquid scintillation counters.

In the present context we are primarily interested in neutron measurements for the determination of the neutron flux in our laboratories. The flux consists both of fast and slow neutrons. It is most convenient to use a gas proportional counter filled with boron trifluoride gas, BF_3, to detect. The ^{10}B isotope has a very high cross section for thermal neutrons (3800 barns), leading to the reaction:

$$^{10}\text{B} + n \longrightarrow {}^{7}\text{Li} + \alpha \tag{8.5}$$

96% of the events lead to an excited state where the energy of the alpha particle is 2.31 MeV. The large alpha pulses are easily separated from background gamma pulses, leading to a very low background. The ^{10}B isotope has a natural abundance of 19.6%. Boron in neutron counters is therefore usually enriched to about 95% in ^{10}B. This counter is practically only sensitive to slow neutrons. When fast neutrons are to be measured, the counter is surrounded with a hydrogen rich thermalizing material, usually a layer of either paraffin or polyethylene, about 5 cm thick. The background of the counter is determined by surrounding it with a thin foil of cadmium. A study of this kind was discussed in section 7.13.

Recommended reading

- Handbuch der Physik, volume XLV, Nuclear instrumentation, ed. S. Flügge, Springer-Verlag, 1958.
- Methods of experimental physics, Vol. 5, eds. L.C.L. Yuan and C. Wu, Academic Press, 1961.
- Radiation Detection and Measurement, Glenn F. Knoll, John Wiley and Sons, second edition, 1989.
- A Handbook of Radioactivity Measurements Procedures, 2nd ed., NCRP Report 58, National Council on Radiation Protection and Measurements.
- Measurement and Detection of Radiation, N. Tsoulfanidis, McGraw-Hill, New York, 1983.
- Twenty five years of scintillation counting, R. Hofstadter, IEEE Trans. Nucl. Science, 22, 13-18, 1975.

- Radoszewski, T., Application of liquid scintillation technique to low-level measurements, Huelva 1988.
- Nuclear Instruments and Methods, volume 167, 1979. A special edition with a good general review of nuclear detectors.
- A handbook of radioactivity measurements procedures, NCRP Report no. 58, issued by National Council on Radiation Protection and Measurements, 1985.
- Detektoren für Teilchenstrahlung, K. Kleinknecht, B.G. Teubner, Stuttgart.
- Atoms, radiation, and radiation protection, J.E. Turner, 2nd ed., John Wiley & Sons, 1995.

Chapter 9

Background: sources and components

9.1 Introduction

The background of a radiation detector is its count rate, or in rare cases its current, in the absence of a radioactive sample or, better still, with the same kind of sample as that to be measured, but one that is free of radioactivity. When the activity of weak samples is determined, various measures can be taken to reduce the background, but some count rate always remains. Its sources will be discussed in this chapter.

In spite of almost half a century of low-level counting, we still have a good deal to learn about the background of the different types of detectors. After the first 10–15 years of the technique, the methods used had reached a state that was generally considered acceptable and less time was allotted to background studies. The detectors used during this early period included proportional and Geiger counters (both for gaseous and solid samples), NaI scintillation detectors and liquid scintillation counters. Reading reports from this early period, it is easy to understand the moderate interest in background studies. The rewards seldom matched the large effort they required and the experimental technique used was primitive compared to what we have today.

A very thorough background study carried out at the Pacific Northwest Laboratory at the beginning of the 1970s (Kaye et al., 1972) is a good example of the difficulties and limitations of early studies. The background of a 3"×3" NaI detector was investigated: (a) with different shieldings (up to a 10 cm thick inner mercury layer), (b) with different types of photomultiplier tubes, (c) with background measurements made both in a surface laboratory and underground where the cosmic ray flux was reduced by a factor of about 80 and (d) with a 5 cm thick cadmium fluoride light pipe between the crystal and the phototube, for attenuating gamma radiation from the glass. This work showed that primordial radioactivity in the glass of the phototubes severely limited the potential use of NaI detectors in contamination studies.

A new era in low-level counting started in the mid 1980s spurred by large projects to verify the theoretically predicted rare events of double beta decay. Systems with backgrounds by orders of magnitude lower than earlier types were needed for these studies. At this time, new techniques were becoming available which greatly improved detectors. Further, the electronic technique had been revolutionized by the introduction of large scale integrated circuits and microprocessors. Larger resources were allotted to these projects than low-level counting had ever seen before. The work gave rich rewards, as will be described in Chapter 12. The massive effort invested in the development of these systems has immensely increased our knowledge of the background components of germanium spectrometers. The information collected in this work can now help us to better understand the background of other types of detectors and guide us in improving the systems.

Before the era of ultralow-level germanium spectrometers, our knowledge of background components was mostly qualitative, and construction of all types of low-level systems was more of an art than a scientific technique. This can no longer be said about germanium and Si(Li) spectrometers, although the former still needs some further studies. For other types of systems a good deal of work have to be done before we can design new ones based on quantitative knowledge.

In order to design an optimum system with the lowest possible background, in each case within the limits of available resources, we must have a thorough understanding of the sources of background, and be able to give a quantitative estimate for each significant component.

The background of beta and gamma detectors is a complex phenomenon, whereas for alpha detectors it generally only has one source, contamination of alpha emitting isotopes in the detector itself or close to it. The main part of this chapter therefore deals with background sources that are common to all types of beta/gamma detectors.

Methods to measure radiocontamination in materials are described in the next chapter. Chapter 11 then deals with means to reduce the background, and in Chapters 12–15 the backgrounds of individual types of systems are discussed.

9.2 Background of beta and gamma detectors

9.2.1 Background components

Beta and gamma detectors have common background sources:

1. Radioactive contamination.

2. Cosmic rays.

3. Electric disturbances.

The background count rate B can be divided into the following components:

$$B = \{B_\gamma(Ex) + B_\gamma(Ct) + B_\gamma(Rn) + B_\beta\} + \{B_\gamma(Sr) + B_N + B_\mu\} + B(El) \quad (9.1)$$

where the source of the components is:

$B_\gamma(Ex)$: External radioactivity in material outside the main shield.

$B_\gamma(Ct)$: Gamma-active contamination in the shield and materials inside it.

$B_\gamma(Rn)$: Radon, which diffuses into cavities in the shield, and its progeny.

B_β: Beta-active contamination in the wall of the sample detector.

$B_\gamma(Sr)$: Secondary photons (and electrons), excited by muons and protons in the shield.

B_N: Neutrons, induced by muons and protons, mainly in the shielding material.

B_μ: Muons, or muon leakage in systems with anticosmic counters.

$B(El)$: Electric disturbances.

These components will be discussed in the following sections.

9.2.2 External gamma radiation, $B_\gamma(Ex)$

All building materials contain primordial radioactivity: potassium, thorium and uranium (frequently written simply K/Th/U in the following). In most cases their concentrations lie within a factor of two from their mean values in the crust of earth (Table 6.2). In order to reduce the background contribution from this radioactivity, low-level beta and gamma detectors are enclosed by a thick shield of lead or iron. This component can thus be reduced to an insignificant level, i.e., made small compared to the residual count rate due to other sources. Usually 10 cm of lead or 20 cm of iron is used, following tradition rather than basing it on quantitative considerations. The reduction of this component is discussed in detail in Section 11.3.2.

9.2.3 Gamma contamination in shield and system components, $B_\gamma(Ct)$

Concern for radioactive contamination in detector materials is older than the low-level counting technique. Sensitive alpha contamination measurements were made as early as 1933. The alpha pulses from an ionization chamber made of the material under study were recorded on a photographic paper. The detection limit was about 5 Bq/kg (Bearden, 1933).

Libby was concerned with the purity of materials in his system. He tested different metals for his counter and found that copper, stainless steel, and lead-free brass all gave the same background, but that of aluminium was considerably higher. Systematic measurements of the radiopurity of materials started in early 1950s, and are still being conducted. In earlier studies, the emission of beta and alpha particles from the surface of plates was measured with thin window gas proportional counters. Later, NaI scintillation counters were used to measure the gamma contamination, but the primordial radioactivity in the glass of the photomultiplier tube limited their sensitivity. Furthermore, their potential was rarely exploited fully, as the use of expensive multichannel analyzers could seldom be afforded in the long counting runs of the low-level studies. One of the most extensive studies of this kind was reported by Rodriguez-Pasqués et al. (1972). They monitored alpha, beta and gamma contamination in a large number of batches of aluminium, steel and copper.

A major part of the development of ultralow-level germanium spectrometers consisted of identifying all contaminated system components, assaying the contamination in a large variety of materials used in the systems, and finding radiopure replacements when necessary. This work has brought us invaluable information as the large germanium diode spectrometers could measure much lower contamination levels than had been possible earlier, and each gamma contaminant could now be identified by the energy of its gamma peaks. The main results of these measurements are given below. In Chapter 10, the methods used are described, as is their sensitivity and reliability.

Lead. Lead is the most important shielding material because of its high atomic number, high density and favorable mechanical properties. However, it has the disadvantage that newly produced lead almost always contains substantial amounts of ^{210}Pb. Figure 9.1 shows an early demonstration of the nature of the contamination in lead, where the emitted X-rays and bremsstrahlung were measured with a well shielded four-by-four-inch NaI scintillation counter. These measurements are discussed in section 10.4. It has been confirmed with

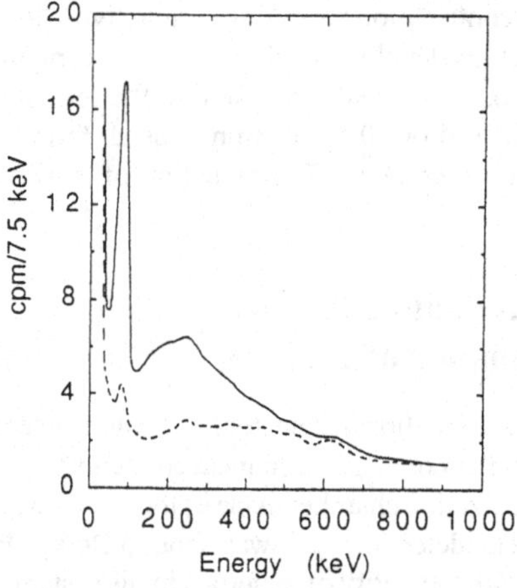

Fig. 9.1. Pulse height spectra of ^{210}Pb in lead with a four-by-four-inch NaI crystal. The lead samples were in the form of a 4 mm thick, 20 cm high cylinder. Curve (a) lead with 40 Bq ^{210}Pb/kg lead, (b) 400 Bq/kg. (From Kolb, 1966)

modern methods that lead is only contaminated with one radioisotope, ^{210}Pb. Thorium and uranium (usually in secular equilibrium with radium, from which ^{210}Pb is derived) are generally found in small concentrations in lead ores, but are separated very effectively in the refining process through the large difference in chemical properties. ^{210}Pb, which is in secular equilibrium with ^{226}Ra in the ore, will inevitably accompany lead through the refining process, giving the same $^{210}Pb/Pb$ concentration ratio in pig lead as in the ore. Actually, the concentration ratio in the final product may be somewhat higher as the coal used in the refining process may add some ^{210}Pb. This contaminant in lead decays with a half-life of 22 years after the production of lead. After 100 years it will have decayed to 4.3% of its initial value and to about 0.2% after 200 years.

^{210}Pb and its radioactive decay products, ^{210}Bi and ^{210}Po, emit very little gamma radiation, only 0.04 photons (45 keV) per disintegration in the decay of ^{210}Pb and this radiation is of no significance for the background. The ^{210}Pb radiation that affects our low-level detectors comes from characteristic X-rays of lead and bremsstrahlung, both produced by the relatively high energy beta particles of ^{210}Bi (maximum energy 1.2 MeV), the daughter nucleus of ^{210}Pb.

The $^{210}Pb/Pb$ ratio in newly produced lead varies strongly from one mine to another. Kolb (1968) measured ^{210}Pb in lead from 24 different mines in many parts of the world. Three of these gave a contamination level above 1000 ^{210}Pb per kg of lead (Bq/kg), the highest about 3000. Half of the samples gave a value below 300 Bq/kg. The ^{210}Pb concentration in lead from the same mine can show large fluctuations. Kolb (1988) measured samples from the same mine over a period of 3 months and found that the ^{210}Pb content varied by more than a factor of two over this period and in an irregular way. Figure 9.2 shows the concentration distribution in the samples measured.

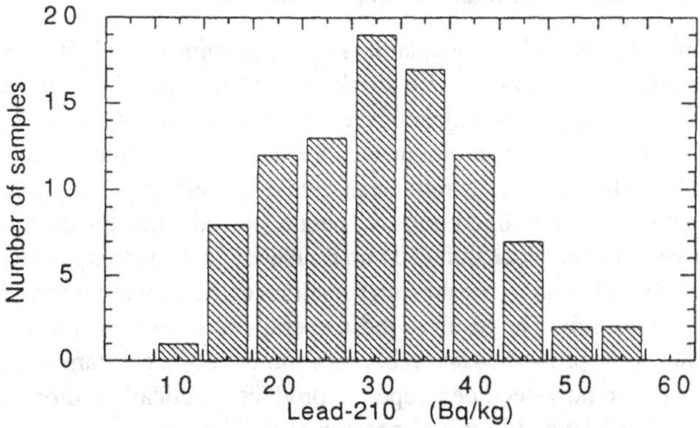

Fig. 9.2. Distribution of ^{210}Pb concentration in samples from the same mine, taken over a period of three months (from Kolb, 1968).

Newly produced good lead (except for special lead, discussed below) contains 20–50 Bq/kg. A type of modern lead frequently used, with 20–30 Bq/kg, comes from a Swedish firm

(Boliden). This is an acceptable concentration for most low-level counting systems, but in some cases the concentration must be below 1 Bq/kg.

By regularly monitoring the ^{210}Pb concentration of the ore, selection of the best batches, eventually separating the best minerals, and careful refining, lead of greatly improved quality can be produced, but at higher prices. This will be described in Section 10.4.

In low-level counting systems, an inner layer of old lead, which has been salvaged from ship wrecks or taken from old roofs, is frequently added. But "old lead" is an ambiguous term. It should be noted that the ^{210}Pb concentration in old lead depends both on the time from its production and its initial concentration. It takes 150 years for ^{210}Pb in lead with an initial concentration of 2000 Bq/kg to decay to 20 Bq/kg. It is therefore possible that 150-year-old lead is no better than good modern lead. Only direct measurement can prove the quality of the lead. Methods for assaying the concentration of ^{210}Pb in lead are described in Section 10.4.1. Section 11.4.2 discusses at what concentration the ^{210}Pb background contribution becomes unacceptable.

Iron. Earlier, the shields of low-level counting systems were generally of iron. Scientists were naturally worried about its purity, as iron and steel produced after 1952 could contain significant amounts of man-made radioactivity, especially ^{60}Co, which has been used widely in blast furnace crucible liners to monitor wear. Since iron is heavily recycled, the danger of ^{60}Co contamination is always present. But even iron produced before 1952 is not always completely radiopure, although it is well within the range acceptable for most shieldings. Arthur et al. (1988) found 13 mBq/kg of ^{226}Ra and 8 mBq/kg of ^{232}Th in such iron, but Heusser (1991) found no traces of radioimpurities in iron from the same period. Now that lead has replaced iron as the main shield in new systems, its radiopurity is of less concern. Iron is, however, still used in various system components.

Aluminium. Libby discovered radiocontamination in aluminium and all later studies have confirmed his results. Measurements with ultralow-level germanium spectrometers showed that the contamination comes mainly from the uranium series, having a concentration of 1–10 Bq/kg. The uranium is, however, not in secular equilibrium in aluminium, as it has lost the main part of its radium. An illustrative example of the effect of radiocontamination in aluminium is seen in the background improvement achieved when the original aluminium parts of a planar germanium diode unit (50 mm diameter, 15 mm thick) were replaced with copper (Kolb, 1988). In the final step the thin beryllium entrance window was replaced with one of more pure beryllium. The result of these operations is given in Table 9.1.

Aluminium has some properties favorable for components in germanium spectrometers. Methods have therefore now been developed to produce practically radiopure aluminium, with about 2 mBq/kg of Th and U, but it is rather expensive.

Copper. No material has been better assessed for radiopurity than copper. It is an important material for ultralow-level germanium spectrometers, both for system components and for inner shield. Its radiocontamination has been assayed in some of the best ultralow-level germanium spectrometers with counting times of up to weeks. These studies have shown that electrolytically refined copper, OFHC (oxygen free high conductivity) copper, is com-

Table 9.1. Background reduction of a planar Ge-diode when contaminated materials were replaced by pure ones. The 46 keV line is from ^{210}Pb and the 63 keV line from ^{234}Th in the ^{238}U series.

Mounting	46 keV cpm	63 keV cpm	5-150 keV cpm	Counting time
Al	0.41	1.02	29	15 h
Cu	0.015	0.18	14	13 h
Cu, new window	n.d.	0.032	9	19.6 d

pletely free of primordial and man-made radioimpurities, because of the refining processes used.

Mercury. Its high density (13.6 g/cm^2) and high atomic number ($Z = 80$) make mercury an attractive gamma absorbing material. It is completely free of radiocontamination as it does not occur with uranium or thorium in nature and it is easily purified by distillation. Until about 1985, it was practically the only shielding material that scientists felt sure to be absolutely free of radioactive contamination. A 2–4 cm thick layer of mercury, enclosed in a steel or acryl holder, was therefore frequently used for inner shields, attenuating gamma rays from system components outside this layer (Kulp and Tryon, 1952). Measurements with ultralow-level germanium spectrometers have verified the purity of mercury. Now that we know that both electrolytically refined copper and old lead are completely free of radiocontamination, mercury is no longer used in new systems.

Plastic materials. Various plastic materials, such as acryl, teflon and delrin, have been tested thoroughly in ultralow-level Ge-diode spectrometers and found to be extremely pure. Jagam and Simpson (1993) improved the sensitivity by using neutron activation, followed by activity measurement in a low-level germanium spectrometer. They carefully measured the radiocontamination in acryl and found about 15 mBq/kg of both thorium and uranium. This sensitivity has been improved by Milton et al. (1994). They measured alpha activity in the ash of 10 kg of acrylic with a 2.5 cm^2 surface barrier diode. A background measurement gave no pulse in the region of interest with a counting time of one week, and a chemical blank sample gave 5 counts per week. The sensitivity limit was about 2 mBq/kg for both uranium and thorium with a counting time of one week, probably the highest reported sensitivity in radiocontamination measurements. They found 5–10 mBq/kg of thorium and uranium.

Glass. Glass is important in low-level counting through its use in the photomultiplier tubes of scintillation detectors. It is always contaminated with K/Th/U. The producers cannot freely select glass for their tubes with lowest contamination, as there are constraints on the nature of the glass. It must be of high optical quality, high mechanical strength and stability, and the linear expansion coefficient must match that of the metal lead-in wires. The radiocontamination concentrations vary much from one type of glass to another, even between batches. Typically they contain 5–10 Bq/kg of all three elements (Jagam and Simpson, 1993). For the detection of solar neutrinos, thousands of large phototubes are needed. Work has been carried out to produce glass with greatly decreased contamination (Barton 1995). This development work has been successful. We will hopefully see better phototubes, but at higher prices, in the coming years.

Quartz. Quartz was earlier considered to be practically free of radiocontamination. Measurements have now shown that although quartz generally has very little radiocontamination, concentrations up to 1 Bq/kg may be found.

Electronic components. Printed circuit boards, solder and capacitors containing aluminium, can have traces of radioimpurities. It is therefore safer, when very low background is sought, to place the preamplifier outside the shield.

9.2.4 Radon background, $B_\gamma(Rn)$

^{222}Rn, the decay product of ^{226}Ra, has a half-life of 3.8 days. It belongs to the group of inert gases. It is present in varying concentrations in all buildings as well as in free air, as discussed in Section 5.4. It will, unless careful precautions are taken, diffuse into the shields of low-level counting systems where its mean concentration will be similar to that in the laboratory room, because its half-life is long compared to the exchange time of air in the shield with laboratory air.

In the early period of low level counting, little attention was given to the possibility of radon background disturbances. Scattered reports in the 1960s and early 1970s describe sporadic background fluctuations due to radon in NaI scintillation systems (Kloke et al., 1965) and in gas proportional counters. The development of ultralow-level germanium spectrometers brought serious attention to this background component (Heusser 1991; Heusser and Wojcik, 1994).

As discussed in Section 5.4, we can expect a mean radon concentration of 50–100 Bq/m^3 in our laboratories, about ten times of that in outside air. The radon concentration in basement rooms can rise to quite high values, 1000 Bq/m^3 is not uncommon, if ventilation is poor. The contribution of radon to the background of a detector is:

1. Proportional the radon concentration in the air.
2. Nearly proportional to the volume of free space surrounding the detector.
3. Dependent on the detector's gamma sensitivity.

The radon background response of different types of detectors is discussed in Chapters 12–15.

9.2.5 Contamination in detector wall, B_β

Low-level beta and gamma detectors are very sensitive to beta contamination in the wall of the container holding the detecting material, because the beta particles, escaping out of the wall into the detecting material, are counted with 100% efficiency, but gamma photons only with 1–10% efficiency. The container can be the metal or quartz tube of gas proportional counters, the mounting cup around a Ge-diode, the metal cap around a NaI crystal and the vial of a liquid scintillation counter. The materials for these should therefore be selected with utmost care.

Today, it is possible to measure the purity of materials of system components with high sensitivity. Earlier, the only reliable test of materials for gas proportional counters was to construct a counter of a selected material and measure its background. This was, of course, both time consuming and expensive. Based on rather limited information, quartz was generally assumed to be the purest available material for gas proportional counters. It was frequently used, in spite of its unfavorable mechanical properties, and the fact that a thin conducting layer had to be vacuum evaporated on the inner side of the tube. A comparison of the background of a large number of gas proportional detectors (Section 13.6) has, however, shown that quartz counters do not have lower background than counters of copper and stainless steel. Later measurements, made with ultralow-level germanium spectrometers, support this conclusion. This is a good example of earlier difficulties in low-level work, where scientists frequently had to rely on experience insufficiently supported by experiments.

What level of beta purity is needed for the wall material? Little seems to have been done to quantify this. An estimate will be presented here. Let us look at a wall that is contaminated with a beta-active nuclide with maximum energy E_{max} and a concentration a Bq/g. Let us consider a thin square element parallel to the surface of the wall with a mass density ρ, thickness dx g/cm^2 and an area dA cm^2. Figure 9.3 shows the paths of five beta particles. Particle A gives a pulse, but B goes away from the wall, showing the influence of geometry. Particles C and D come from atoms close to the inner surface of wall, but the first goes deeper into the wall, and the latter has not enough energy to penetrate out of it. Particle E gives a pulse, despite the large depth in the wall that it comes from.

The activity inside the volume element contains ($a\,\rho\,dx\,dA$) Bequerel. Half of the beta particles move in a direction towards the surface. If this layer is at a depth x (g/cm^2) from the inner surface, the number dN of beta particles, that penetrate from the volume element through the thin wall layer into the detecting medium, is according to Equation 4.8, given by

$$dN = e^{-\mu\rho x}\,\rho\,a\,dx\,dA/2 \tag{9.2}$$

where the factor 1/2 comes from the geometry. The total number N of beta particles escaping per cm^2 from a thin inner surface wall layer into the sensitive volume can be found by integrating (9.2) from zero to infinity (mathematically, in reality rather to the range R of the beta particles). This gives the total number N of beta particles escaping into the detecting medium

$$N = B_\beta = a \cdot A/2\mu \text{ beta particles s}^{-1}\text{ cm}^2 \tag{9.3}$$

where μ is the absorption coefficient, here given in cm^2/g.

The maximum tolerable concentration a for different detectors is discussed in Chapter 12. For a typical gas proportional counter in a surface laboratory the concentration should

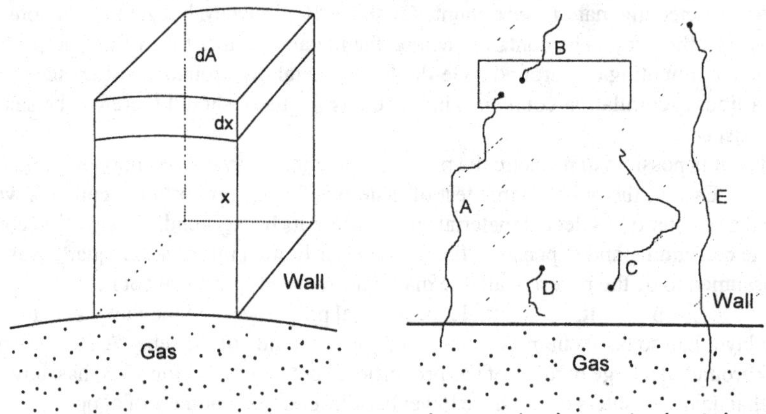

Fig. 9.3. A thin element in a counter wall, and the path of four typical beta particles (discussed in the text) emitted by radioactive atoms indicated by points.

be less than 0.7 Bq/kg (Section 12.3.5) and somewhat less in the mounting cup of germanium diodes. These values are about 20 times the gamma detection limit of good low-level germanium spectrometers and are therefore relatively easy to determine.

We now define the *effective virtual wall layer*, t_{vwl}, as the thickness where the number of disintegrations is equal to the number of beta particles that escape out of the wall, and it is according to Equations (9.2) and (5.9) given by

$$t_{vwl} = 1/2\mu = E_{\max}^{1.43}/34 \quad \text{g/cm}^2 \tag{9.4}$$

For $E_{\max}$ = 1.0 MeV this thickness is 0.029 g/cm^2, but the maximum range of the beta particles is 0.41 g/cm^2. As the maximum range of electrons R_β is a more common parameter it is convenient to express t_{vwl} in terms of R_β. The following expression gives a good approximation

$$t_{vwl} = R_\beta/14 \tag{9.5}$$

9.2.6 Secondary electrons and gamma rays, B(Sr)

Energetic charged cosmic-ray particles produce secondary radiation in the shield. Here we will look at the energetic secondary photons and electrons that usually give the main contribution to the background of low-level beta and gamma counting systems. This radiation is predominantly electromagnetic showers produced in the shield (a) by neutral pions, created by protons, or (b) by muons, mainly through the knock-on process. The energy spectrum of the electrons and photons ranges from a fraction of one MeV to hundreds of MeV.

The energy spectrum of this secondary radiation can be studied by gamma radiation detectors. Here we use a small CsI scintillation unit (described in Section 15.5) because of the ease of use and as it is free of radiocontamination. The response was measured: (a) without a shield, (b) under 5 cm of lead and (c) under a 10-cm-thick lead shield (Figure 9.5).

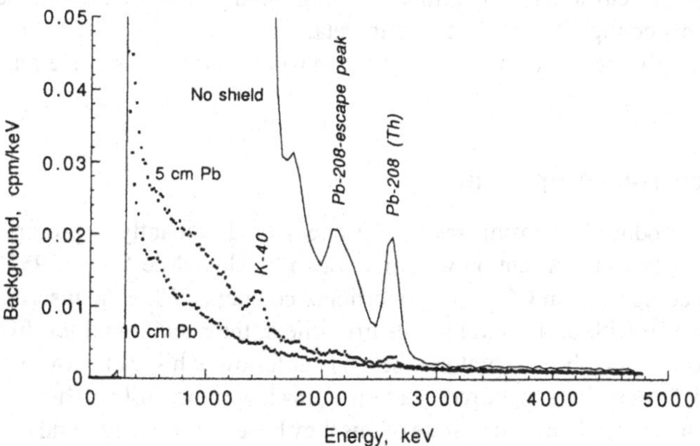

Fig. 9.4. The background spectrum of a CsI diode (diameter 4 cm, thickness 4 cm) under different shielding conditions.

With no shield, the secondary cosmic-ray electrons and photons are those produced in the surrounding building materials. These have relatively low atomic numbers and a critical energy of about 50 MeV (Section 7.10). The energy spectrum extends well above the 5 MeV limit seen in the figure. With no shield and inside 5 cm of lead, there is a sharp increase in the spectrum below 2.6 MeV, coming from gamma radiation emitted by primordial radioactivity in surrounding laboratory building materials, and when there is no shield it dominates over the weaker contribution from the secondary radiation. Above 2.0 MeV the spectrum only comes from secondary cosmic radiation. With a 10-cm-thick lead shield, more than 99% of the external secondary radiation is absorbed and it is replaced by new radiation produced in the shield. At this thickness, the gamma radiation from primordial radioactivity in surrounding building materials is suppressed to an insignificant level. All the background spectrum, also that below 2.6 MeV, now comes from secondary cosmic radiation.

The cosmic ray origin of the background spectrum is also clearly demonstrated in the difference in the background spectrum of a low-level germanium diode at the Joint Research Center IRMM in Belgium (Rainer Wordel, personal communication, Wordel et al., 1994). Figure 9.5 shows the background spectrum of a 100 cm^3 Ge-diode in the energy range from 580 to 750 keV, measured both in a surface laboratory (counting time 24 days) and in an underground laboratory with an overburden of about 500 mwe (counting time 40 days). This overburden reduces the cosmic ray flux by a factor of about 900, that is, to a negligible level. The removal of the secondary radiation reduces the background count rate by a factor of about 40.

Finally, anti-cosmic counters (guard counters) on all outer surfaces (external guard) of the lead shield of low-level germanium spectrometers, reduce the background by a factor of 10–20. In this arrangement the charged particles that induce the secondary radiation are detected and give a veto signal that eliminates background pulses due to coincident secondary radiation. This configuration, discussed in detail in Section 11.8.3, is the only way to reduce significantly the background due to secondary radiation in surface and near-surface laboratories.

9.2.7 Neutron component, B_N

Neutrons are produced by cosmic rays in the shield, predominantly by protons (or rather the nucleonic component). Attention was first drawn by Hassel de Vries (1957) to a neutron background component in CO_2 gas proportional counters. When he inserted 10 cm thick plates of paraffin with boric acid into his iron shield, the background fell from 5.5 cpm to 3.0 cpm. Most radiocarbon laboratories set up later followed his lead. However, as shown in Section 13.6, five counters in the radiocarbon dating laboratory in Trondheim, Norway, have no neutron absorbing layer, and in spite of this they have lower background than all counters in surface laboratories of similar size. This demonstrates that we do not know enough about this component.

The neutron contribution is seen most clearly in the background spectra of low-level germanium systems at small overburden. This is shown in Figure 9.5. The spectrum shows two recoil broadened Ge neutron peaks and one from the copper that surrounded the diode. All three peaks are due to fast neutrons. Spectra of this type are discussed in more detail in Section 12.6.3.

9.2.8 Muon background, B_L

Cosmic ray muons have a velocity close to that of light and their energy loss is therefore 1.8 MeV per g/cm^2 (see Chapter 5). When the detecting material is dense, solid or liquid, as in liquid scintillation detectors, germanium diodes and NaI crystals, most of the muon pulses are large compared to those we are counting, i.e., corresponding to less than 1–2 MeV. The pulses in the low energy tail of the broad muon peak, where muons pass through the edge of the detector, are generally of little significance in the normal energy window being counted. This background component will be discussed further in Sections 12.5, 13.7 and 14.4.

In gas proportional counters the situation is different. The muon pulses are of a size similar as those from beta particles of the sample and they must therefore be eliminated. This is achieved by the use of anticosmic counters (Section 11.8.1), which practically have a 100% muon detection efficiency. If, however, some muons, in spite of the guard counter, pass through the sample counter, but fail to activate the guard counter, we speak of a muon leakage. A typical CO_2 radiocarbon counter has a muon count rate of about 200 cpm. We expect undetected muons to contribute only insignificantly to its background, which is 0.5 to 1.0 cpm. The guard counter system must evidently be very effective. Actually, it is difficult to measure this small muon leakage.

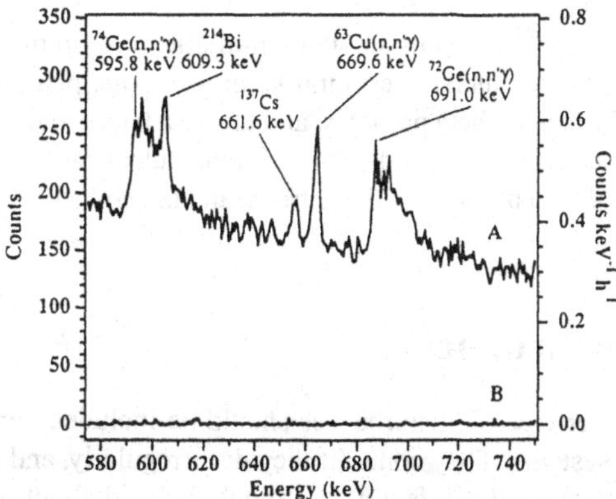

Fig. 9.5. The background of a low-level germanium spectrometer in a surface laboratory and underground (500 mwe)

9.3 Background of alpha detectors

Because of the short range of alpha particles, the background source always lies close to the detector, in the materials of the detector unit or very close to it. We look at four different types of alpha detectors: (1) Si(Li) diodes, (2) gas proportional counters with thin windows, (3) ZnS coated scintillating screens, where the light is detected by a photomultiplier tube, and (4) liquid scintillation counters.

Si(Li) alpha diodes give relatively sharp alpha peaks, which greatly reduces the background for individual alpha emitters. The diodes are usually operated inside a vacuum chamber where they see only the sample holding plate. The background therefore can only come from contamination in these two system components. New diodes will have a background of about 1 cpd/cm^2 in the energy range of 3–8 MeV, and a factor of about 10 lower under individual alpha peaks. Care must be exercised not to contaminate the diode and chamber through atoms recoiling from the sample. A low residual gas pressure is therefore maintained in the vacuum chamber in order to backscatter subliming and recoiling atoms or the sample is covered with a very thin vinyl film (Acena et al., 1994).

Gas proportional counters with thin windows. With proper care, the alpha background can be very low, a few cpd/cm^2, in spite of the fact that all alpha particles are counted in a broad energy window. For the lowest background the counter should be enclosed by a container that is flushed with radon-free air in order to eliminate contamination from radon.

ZnS alpha scintillation counters. Their background varies widely, from 0.1 to 1.0 cpd/cm^2, being presumably mainly due to the contamination in the scintillator. Their background is discussed in Section 15.3.

Liquid scintillation counters. Their vials and the scintillation coctail must be free of radiocontamination. Because of the much lower scintillation yield of alpha particles, their pulse size is similar to that of energetic beta particles, and the beta background can interfere seriously. As described in Sections 11.9.2 and 14.7, the alpha pulses can be discriminated very effectively from pulses of beta particles and electrons, and the alpha background reduced to a few pulses per day.

9.4 Electric disturbances

It can be disputed whether electric disturbances should be included here, as they are not found in well designed systems. They come (if they do) irregularly, and go (hopefully forever) after a short bad period. But a few words should be said about this annoying phenomenon, which many scientists have met, but few write anything about. A well known expert in the field of low-level counting once told the author of this book that it was really not the background that limited how weak samples his gas proportional counters could measure, but spurious pulses.

The electric disturbances may come through mains surges, radiofrequency emissions of nearby electric appliances, mechanical vibrations, and other sources. To minimize these the system should have a solid, loop-free ground connection, a filter inserted in the power line, the system may be enclosed in a Faraday cage, and it may stand on some form of antivibration platform.

It is a good practice in low-level systems to try to have some check on spurious pulses. Some commercial systems have a sensitive blind input to simultaneously pick up electromagnetic disturbing signals. Another quite common method is to count the number of pulses in subintervals, for example 10 minutes, and, at the end of the counting period, look at the distribution of the number of pulses in the subperiods. They should have a Poisson distribution. Before the PC-computer era, this was the best we could do. Today we have better and more varied possibilities.

False pulses frequently come in bursts that may only last for a short period. A system designed by C. Andersen at Risö National Laboratory, Denmark (personal communication), can find disturbances of this kind. The number of pulses in 20-second periods are stored in the memory of a computer. Spurious pulses are now not masked by the larger number occurring in 10 minute periods, but the number of periods is about 30 times higher. When a sample has been counted a program scans the periods and marks the ones where the number is above some given value, which depends on the mean count rate. For every subperiod there is either an OK mark or the number of pulses is stored in a new file, when it has exceeded the allowed number. These abnormal periods are scanned manually in order to get information about the nature of the disturbance. This system has proved to be very efficient in the measurement of weak radon samples. A variation of this method is to look in short subperiods at the length of the time interval from one pulse to a subsequent one. The length of these intervals should have a Poisson distribution. Storing the data is no problem now.

A powerful criterion for a genuine pulse is its shape. It seems probable that false pulses have a shape that is quite different from that of genuine pulses. The computer could for example measure the time from full pulse height to half pulse height and store the time distribution in subperiods.

The aim of this brief discussion is primarily to point out that some kind of a check should be built into low-level systems. Its type will, however, depend on the kind of system and the work that is being done. No general guidelines can be given.

A scientist struggling with spurious pulses frequently feels lost, he/she hardly finds any description in the technical literature of similar problems. The scientist may think that his troubles are caused by his incompetence. It would help many scientists, if reports of problems of this kind were published more frequently, where the nature of the trouble is described and methods used to identify and eliminate it.

Recommended reading

- G. Heusser, Low-Radioactivity Background Technique, Ann. Rev. Nucl. Part. Sci., 1995, 45, 543-590.

Chapter 10

Measurement of contamination

10.1 Radioactive contamination in low-level systems

In 1953 deVries reported that a small glass insulator in his low-level CO_2 counter had increased the background by a factor of three. This convincingly demonstrated the importance of the radiopurity of system components, even small ones. Over the next 30 years considerable effort was devoted to the study of contamination in materials, using different methods. Because of the limited sensitivity of available methods, the results were of modest value. Only the most contaminated materials could be identified and their impurity measured with sufficient accuracy. Primordial radioactivity was found in lead (^{210}Pb), aluminium and glass, and ^{60}Co sometimes in iron.

A new era in this field began with the development of ultralow-level germanium spectrometers in the mid 1980s. This work is described in Section 12.3. The most extensive program was that carried out at the Pacific Northwest Laboratory, Richland, Washington (Brodzinski et al., 1992). The background of the earlier type of germanium spectrometers was reduced by an order of magnitude by adding external anticosmic guard counters (Reeves and Arthur, 1988).

The main effort in the first phase of this work was to identify, through gamma peaks in the background spectra, the contaminants in the early germanium spectrometers and to locate them by the relative intensity of peaks compared to the abundance of the lines. Concurrently, an extensive study was carried out on radiocontamination in all prospective component materials (Reeves and Arthur, 1984, Arthur et al., 1988). When possible, they were measured in quantities of ten to a hundred times the amounts used in the spectrometers. Contaminated materials were gradually replaced with pure ones and in some cases better production processes have been developed (lead, aluminium and glass). Improved systems were operated underground and a part of their operating time was used to assay radioimpurities to still lower levels.

This work is important for the whole field of low level counting, as pointed out by the group at the Pacific Northwest Laboratory in a 1987 report (Brodzinski et al., 1987):

> Although the technological advancement discussed here was made with germa-

nium diode spectrometry, the developments are equally applicable to any other radiation detection technology and can be used to improve the sensitivity and cost effectiveness of any system used in low-level analyses.

The authors conclude their report with the following words, applicable to all development work in low-level counting: In summary, to construct ultralow-background detectors for laboratory use it is necessary to:

1. Fabricate the detector from selected materials that have a minimum of primordial radioactivity and are not prone to copious quantities of cosmogenic activity.
2. Avoid materials that have naturally radioactive isotopes.
3. Avoid the use of lead or solder containing ^{210}Pb close to the detector.
4. Shield the detector from the lead castle, if used, using very old lead, mercury, or some other dense material that does not contain high-equilibrium concentrations of cosmogenic activity.
5. Pressurize the detector space with dry oxygen or other gas to prevent the infusion of radon.
6. Use an electronic anticosmic shield to block background events due to primary cosmic interactions.

In the following a general review of the methods used to measure radiocontamination in materials is given, methods that meet the more modest needs of ordinary low-level counting. The sensitivity limit given below is the activity that gives a count rate that is equal to three times the standard deviation of a 24 hour background measurement. However, longer counting times are frequently used.

10.2 Alpha and beta surface activity

The assay of radioactive contamination in materials used in low-level counting systems was started in the 1950s and some 10 years later it was extended to a large variety of materials. Alpha and beta rays, emitted from the surfaces of different materials, were measured with flat, thin-window gas proportional counters with a window diameter of 2.5 to 5.0 cm. Later Si(Li) diodes were preferred for the measurement of low-level alpha activity because of their high energy resolution. Let us look briefly at the sensitivity of these methods. We can assume that the detection limits are approximately the same as those calculated for lead in Section 10.3. These limits are shown in Table 10.1.

When samples other than lead, glass or aluminium were measured, the activity was usually close to the detection limit. Earlier measurements must therefore be taken with reservations.

Table 10.1. Sensitivity in counting alpha and beta particles emitted from the surface of plates.

Particle	Detector	Window cm^2	Sensitivity Bq/kg
β	GPC	20	6
β	GPC	2×65	0.2
α	Si(Li)	30	1.2

One of these studies will be described here. It warrants special attention because of its very high sensitivity, its somewhat unexpected results and its implications (Al-Bataina and Jäecke, 1987). A cylindrical proportional counter with a diameter of 10 cm was used, ensuring that almost all emitted alpha particles were stopped in the gas. The materials to be tested were pressed to the cylindrical wall. The advantage of this system lies both in the very large area of the sample (1550 cm^2) and that it gives good alpha resolution. With the geometry used, practically nothing can disturb the result, the alpha sensing medium is only exposed to alpha particles from the sample. Figure 10.1 shows two of the resulting spectra, for cathodes of aluminium and brass. Both materials give spectra with peaks, which should not be seen from a contamination evenly distributed in the matrix of the sample. As pointed out by

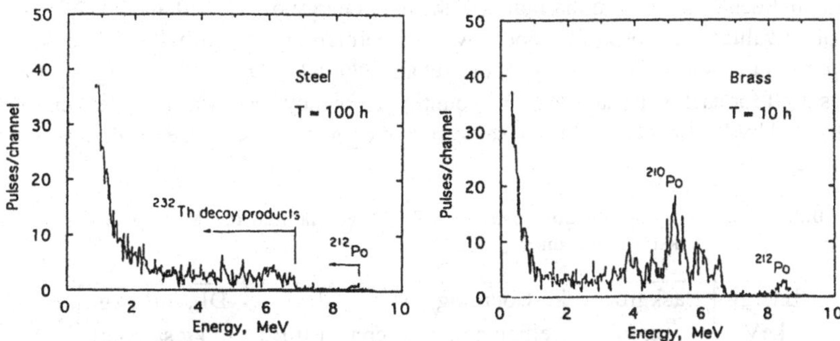

Fig. 10.1. Smoothed background spectrum of a gas proportional counter with an anode area of 1550 cm^2 (Al-Bataina and Jäecke)

the authors, the shapes of the spectra indicates that there is an increased concentration of the alpha decaying contaminant present in a very thin surface layer. We would expect it to be evenly distributed in the wall material, and a spectrum with a rather sharp drop in count rate above the energy of a given alpha line. Enhanced activity in a surface layer is well known in lead where ^{210}Po migrates towards a surface that has been cleaned mechanically, probably due to low energy surface traps (Sastawny et al., 1989). This phenomenon may occur for other atoms and in other materials. This is worth a closer study.

This method can be used to study the alpha emission from the cathode of ordinary large gas proportional counters, for example those used in radiocarbon dating. Measurements of this kind could give quite useful information about the background contribution of contamination in the counter wall material, as we know that for each alpha pulse there should come about 5 beta pulses because of the greater penetration power of beta particles. This would give us the value of B_β. These measurements can be made when normal samples are being measured. Long counting times are therefore possible, even weeks, which will give very good counting statistics. It should be noted that the alpha pulses are probably attenuated by space charge, compared to beta pulses. The pulses should be taken from the preamplifier in order to avoid overloading.

10.3 Gamma contamination

Contamination of man-made radionuclides in system components is rare. Iron, which frequently contains small amounts of ^{60}Co, is an exception. We are therefore primarily concerned with the contamination of potassium, thorium and uranium and the progeny of the latter two. Their concentrations can be determined by gamma counting. Today, gamma contamination is nearly always measured with low-level germanium spectrometers. In order to give some idea of the sensitivity of such contamination measurements, we estimate the detection limit (DL) of a system with a 100 cm^3 germanium diode (relative efficiency 20%) for radionuclides emitting gamma radiation with an energy of 0.5, 1.0 and 1.5 MeV. The background values are those of a good low-level germanium system with a diode of this size, operated in a surface laboratory (Mouchel and Wordel, 1992). A counting window of 8 keV is used for each gamma peak. The counting efficiency for a Marinelli beaker geometry, reported by Barton (1996) for the same size of crystal, is used. The result is shown in Table 10.2.

Table 10.2. Gamma detection limit using a 100 cm^3 germanium low-level spectrometer with sample in a Marinelli baker and a counting time of 24h.

Energy keV	Background cpm	Counting efficiency, %	DL cpm	DL mBq	DL, mBq/kg (5 kg sample)
500	0.20	1.7	0.030	34	7
1000	0.06	1.0	0.020	32	7
1500	0.036	0.77	0.015	32	7

Considerably higher detection sensitivity can be obtained by using ultralow-level germanium spectrometers with larger diodes and with longer counting times. Gamma counting with a low-level germanium spectrometer is evidently far superior to the older methods of alpha and beta counting, discussed above, as a comparison of Table 10.1 and 10.2 shows.

The detection limits in Table 10.2 also give approximately the limit of contamination that can be tolerated close to the detector.

The sensitivity of measuring uranium and thorium can be increased further by neutron activation followed by low-level germanium counting (Jagam and Simpson, 1993). This method can only be used when no interfering activities are produced. The sensitivity of this methods depends on transmuting isotopes with very long half-lives to short lived ones, giving higher count rates (Section 16.4).

10.4 Radioactive contamination in lead

Contamination in lead deserves a special section, because of the importance of this metal as a shielding material. In the early days of radiocarbon dating, contamination in lead was assayed by direct counting of the emitted beta particles from ^{210}Bi. As a preparation for setting up the first ^{14}C low-level gas proportional counter for ^{14}C dating, the radiocontamination in 15 lead samples of different origin was measured with a Geiger counter with a thin glass wall (de Vries and Barendsen, 1953). It can now be estimated that their measured contamination was from about 10 (detection limit) to 300 Bq/kg. Their beta absorption measurements indicated that ^{210}Bi, the decay product of ^{210}Pb, was the source of these beta particles. Later this was confirmed by the gamma spectrometric measurements of Weller et al. (1965) and Kolb (1968).

Table 10.3. The ^{238}U series from ^{210}Pb to ^{206}Pb

Nuclide	Particle	Half-life	Energy, MeV	γ/dis
^{210}Pb	β	22 y	0.063	
	γ	-	0.046	0.042
^{210}Bi	β	5.0 d	1.16	
^{210}Po	α	138 d	5.30	
^{206}Pb		Stable		

Table 10.3 shows the lowest part of the ^{238}U series, from ^{210}Pb to the stable isotope ^{206}Pb. Any of the three radioactive isotopes in the series can be used for the determination of the concentration of ^{210}Pb in lead, where one of the following five types of radiations is measured:

1. Energetic beta particles of ^{210}Bi, $E_{max} = 1.2$ MeV.
2. Alpha particles of ^{210}Po.
3. 46.5 keV, yield 4.5%, gamma radiation of ^{210}Pb.
4. X-rays produced by ^{210}Bi beta particles in the lead.

5. Bremsstrahlung produced by ^{210}Bi beta particles.

Two recent reports on ^{210}Pb in lead deserve special attention: (i) A report on the methods that have been used at the Physikalisch-Technische Bundesanstalt in Braunschweig, Germany, by Kromphorn (1993). This institute has been leading in the assay of ^{210}Pb in lead since the mid 1960s (Kolb, 1968). (ii) The most thorough study on lead of the highest radiopurity was that on ancient lead, recovered from a Roman ship sunken near Sardinia (Alessandrello et al., 1991). The paper gives a good description of the methods used.

There are five main methods for the measurement of ^{210}Pb in lead. These will now be described and their sensitivity estimated, assuming a detection limit at 3 σ of the background count rate with a counting time of 24 hours.

1. *NaI scintillation gamma detector.* This method gave the first direct evidence that lead was only contaminated with ^{210}Pb (Weller et al. 1965). Kolb (1968) also measured ^{210}Pb in lead with a well shielded NaI crystal (diameter 10 cm, thickness 10 cm). The lead was in the form of a 4 mm thick annular cylinder, annular around the NaI crystal and 20 cm high. The spectra of two of his samples containing about 40 and 400 Bq/kg respectively were shown in Figure 9.1. It can be estimated roughly that the detection limit of this method, using a well shielded 4×4 in^2 NaI crystal, is about 10 Bq/kg. This method is relatively simple and can be used in laboratories that have a suitable NaI spectrometer. If the NaI crystal is shielded with standard lead bricks (usually $5\times10\times20$ cm^3 or $2\times3\times4$ in^3) and contamination in the same type of bricks is to be determined, the inner shielding layer can be replaced by the bricks to be measured. Bricks of known ^{210}Pb concentration are needed for reference.

2. *Gas proportional beta counting.* Commercial gas proportional counters with thin windows have been used to measure the 1.2 MeV beta particles of ^{210}Bi emitted from plates of lead. When a plate lies close to the window, 1.0 Bq/kg will give 0.11 cpm/cm^2 (Kromphorn, 1993). A counter with a window diameter of 25 mm will have a background of about 0.2 cpm and a detection limit of 6 Bq/kg. A counter with a 65 cm^2 window on each side, sandwiched between two similar, but larger, flat guard counters, has a background of 0.7 cpm and a ^{210}Pb sensitivity of 0.2 Bq/kg (Theodorsson 1991).

3. *Si(Li) diode alpha counting.* Let us assume that we have a lead plate close to the window of a large (30 cm^2) Si(Li) diode having a background of 2 counts per day and detecting 80% of the alpha particles escaping from the surface of the plate. It will further be assumed that the surface emission rate is equivalent to all alpha disintegrations in a layer of 4 mg/cm^2 (about 1/4 of the range of the 5.3 MeV alpha particles in lead). This gives a minimum detectable alpha activity of about 1.0 Bq/kg.

4. *Gamma counting with germanium spectrometers.* Spectrometers with planar diodes are most suitable for measuring the 46 keV photons from ^{210}Pb and the X-rays produced by the energetic beta particles of ^{210}Bi. Coaxial germanium diodes are, however, more often used to measure the ^{210}Pb contamination. Kromphorn (1993) reports the use of a well shielded (10 cm lead) planar germanium diode, with a diameter of 50 mm, a thickness of 15 mm, and

with a thin beryllium entrance window. The 45 keV gamma line and the 84.7 $K\alpha_1$ X-ray line (Figure 10.2) both gave a detection limit of about 20 Bq/kg.

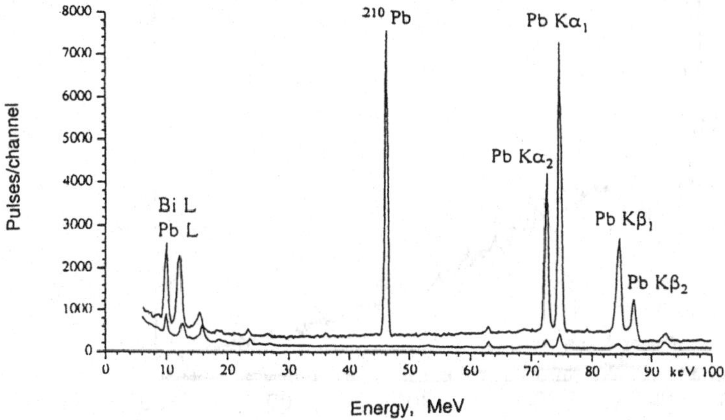

Fig. 10.2. The spectrum of the background (lower) and of ^{210}Pb in lead, measured with a planar germanium diode. The counting time was 41 days. Kromphorn, 1968.

For the measurement of the bremsstrahlung produced by the beta particles of ^{210}Pb, a coaxial germanium diode is most suitable. Figure 10.3 shows the anticoincidence spectra of recent Boliden lead (with about 20 Bq/kg) and old lead measured with an ultralow-level germanium spectrometer (Heusser et al., 1989).

The spectrometer is operated underground (overburden of 15 mwe) where the cosmic ray flux is reduced by a factor of 3 from its surface value. The detector, a 170 cm^3 diode (relative efficiency 32%), is behind a heavy shield and external guard counters reduce the background by a factor of about 12. The lead tested formed the inner 5 cm layer of the shield. This geometry is inconvenient for the measurement of a series of samples. Having a 2–3 mm thick plate of the lead to be measured on the top of the diode, as well as surrounding the diode, would be preferable, but would reduce the sensitivity somewhat.

When the lead surrounds the diode, each Bq/kg of ^{210}Pb in the lead will give 0.015 cpm in the X-ray window (70–92 keV) and 0.081 cpm in the bremsstrahlung window (92–500 keV) (G. Heusser, personal communication). The background and detection limits (3 σ in 24 hours) are given in Table 10.4, both for Heusser's system and for a low-level germanium spectrometer in a surface laboratory with the same size of a germanium diode (Preusse, 1993).

5. *Chemical separation.* The most sensitive, reliable and accurate method is to measure the alpha activity of ^{210}Po after chemical separation from lead. This method, which takes a few hours, is described in the report by Kromphorn mentioned above. 15 g of lead were dissolved in acid, a known amount of a calibrated solution of ^{209}Po tracer was added and the polonium then separated chemically and deposited on a silver plate. The chemical

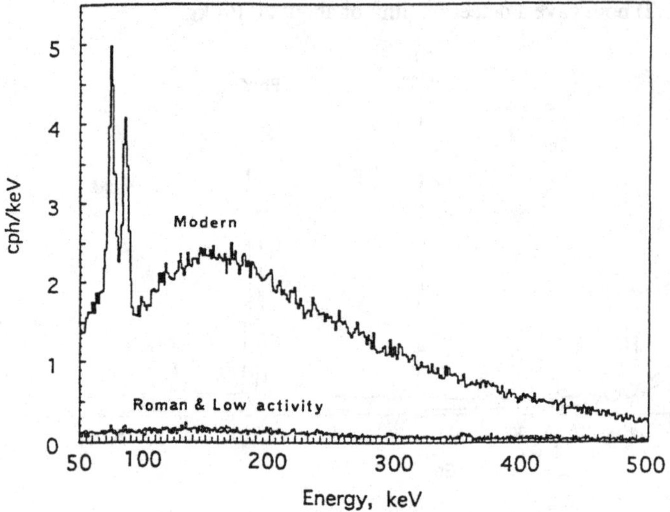

Fig. 10.3. Spectra of recent Boliden (with about 20 Bq/kg) and old lead mesured with an ultralow-level germanium spectrometer (Heusser et al., 1989).

Table 10.4. Detection limit of ^{210}Pb in lead measured with a germanium spectrometer in a surface laboratory (Preusse, 1993) and in an ultralow-level system with an overburden of 15 mwe (Heusser, 1991).

Radiation	Range keV	cpm per Bq/kg	Surface lab. Backg. cpm	Surface lab. DL Bq/kg	Lab. under 15 mwe Backg. cpm	Lab. under 15 mwe DL Bq/kg
X-rays	70–92	0.015	1.6	7	0.06	1.3
Bremsstr.	92–500	0.081	22	5	0.71	0.8

yield was 80–90%. The alpha activity was measured with a Si(Li) diode with an absolute counting efficiency of 30% and a background of about one count per day under the ^{210}Po alpha line. The detection limit was 0.02 Bq/kg. Considering the high sensitivity, reliability, and apparent simplicity of this method, it is surprising that it has not been used more often, rather than spending long, valuable counting time in germanium spectrometers.

Table 10.5. Comparison of methods to measure ^{210}Pb in lead

	Detector	Particle	Sens. Bq/kg
1	NaI (10cm×10cm)	γ	10
2a	GPC, 5 cm^2 window	β	6
2b	GPC, 2×65 cm^2 window	β	0.2
3	Si(Li), window 25 cm^2	α	1.2
4	Ge, planar	γ	20
4b	Ultralow-level Ge	γ	2
5	Chem. separation, 15 g Pb	α	0.02

The concentration of ^{210}Pb in lead used in shields of systems described in the literature is rarely given, but frequently stated that "radiopure lead" or "old lead" has been used. This is not satisfactory in view of the methods available to measure the ^{210}Pb concentration and the wide availability of the equipment needed. It is sufficient to know the ^{210}Pb concentration with about 20% accuracy for samples that lie in most cases from 20 to 100 Bq/kg. This can be measured with sufficient accuracy with all the methods compared in Table 10.3. It should, however, be pointed out that care must be exercised in direct counting of alpha or beta activity from lead plates. When the alpha activity is measured, the surface should be cleaned by etching, not mechanically. When the beta activity is measured in a surface laboratory, it should be remembered that a plate of copper, or any other clean material, cannot be used for background measurements as it will modify the secondary radiation. Therefore, a radiopure lead plate of the same form as the samples should be used for background determination. Furthermore, the method must be calibrated with plates of known ^{210}Pb concentrations. If these measurements are made in an underground laboratory, where the flux of cosmic rays has been reduced to an insignificant level, we get lower background and a radiopure copper plate can be used as a background reference.

Chapter 11

Background reduction

11.1 Optimum background reduction

Now that the background sources of low-level detectors have been described (Chapter 9) and methods to measure radioactive contamination in system materials and components (Chapter 10), we discuss means to reduce the background. Factors that are common to the main detectors of interest will be addressed in some detail in this chapter. Effects of concern primarily for one detector type only will be treated in Chapters 12–15. Some overlapping with these chapters cannot be avoided.

We can frequently reduce the background of our detector, but this usually costs time and money. How far shall we extend our efforts, where shall we stop? This is a matter of compromise. We should have a fixed general tolerance level at which we have obtained acceptable background reduction. This will be illustrated by an example. We look at the selection of an optimum lead shielding thickness in a hypothetical system. Its background comprises two components, $B(Ex)$ due to external gamma radiation that has penetrated through the shield, and B_0, which is practically independent of the shielding mass. $B(Ex)$ should be reduced by the shield to a level where it will not significantly affect the accuracy or sensitivity of our measurement of feebly radioactive samples. Figure 11.1 shows how the total background count rate B (curve T) decreases with the thickness of the lead layer.

We assume that the total background B is 7.0 cpm, 6.0 cpm coming from $B(Ex)$ and 0.4 cpm from B_0. $B(Ex)$ will fall exponentially with the thickness of the lead, by a factor of 11 for 5 cm of lead, which reduces $B(Ex)$ to 0.6 cpm. It is evident that we should increase the shielding thickness. By adding a further 5 cm of lead, $B(Ex)$ is again reduced by a factor of 11, to 0.054 cpm, which is then 12% of the total background. Few scientists would find it worthwhile to reduce B(Ex) further, as this will neither significantly improve the accuracy nor the sensitivity.

Whatever we do to suppress a background component requires an effort, sometimes a large one, in other cases only a small one. When the effort needed is small it is natural to include a safety factor, i.e., to suppress the background contribution well beyond what might be of practical value. When the effort is large, we must weigh carefully the benefits of a

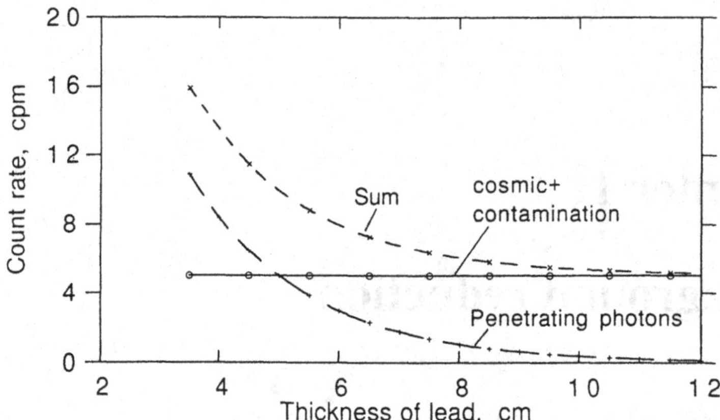

Fig. 11.1. Background of a hypothetical system at varying lead shield thickness.

further reduction against the expenses involved. We have only looked at the benefits (lower background); let us also look at the cost. Addition of further 3 cm to the 10 cm of lead will reduce $B(Ex)$ further by a factor of 4 and bring its background fraction to 3% of the total background, from 0.054 cpm to 0.41 cpm. How much would this add to the mass of the shield? Let us assume that the inner detector space is $30\times30\times30$ cm^3. The total weight of a 10 cm thick lead shield is then 1310 kg. By adding 3 cm, its weight will be increased from 1310 kg to 1880 kg! A large number of systems can be found where excessive shielding is evidently used. The thickness is generally chosen according to tradition, and some extra thickness is frequently added to be well on the safe side.

If an increased thickness is chosen in cases of this type, it is more economic to fine tune the shielding thickness by an inner layer of the required thickness, rather than add another layer of the standardized 5 cm thick lead bricks.

One further remark should be added. For a given lead thickness, $B(Ex)$ is proportional to the concentration of potassium, thorium and uranium in the materials of the laboratory building. These concentrations can vary by a factor of about 5 from a building made of materials with low concentrations to one where they are high. The shield in the latter laboratory would need a 3 cm thicker shield than in the former, in order to reduce $B(Ex)$ to the tolerance level. In spite of the large additional shielding mass that may be in question, the natural gamma radiation in the laboratory hardly ever seems to be taken into consideration.

With the information we have today, the design of new systems should be based on quantitative arguments of the type presented above. For this analysis we must be able to quantify, with reasonable accuracy, each background component, know the value of B_0 and have a fixed tolerance level. It seems reasonable to accept a background contribution ΔB in the range 10% to 20% of B_0 from a component that is difficult to reduce. Going below 10% is an extravagance, except where it costs very little. 10% would in most cases be considered fully acceptable, while 20% would be reasonable when a further reduction would require considerable effort. In the following we will generally use 10% as a tolerance level.

11.2 Background and overburden

11.2.1 Overburden

Overburden M_{ob} is the sum of the mean mass per cm^2 in building materials — floor plates and walls — through which the cosmic ray particles must pass in order to penetrate into the laboratory or into the shield. It is either given in g/cm^2 or in meters of water equivalent (mwe). A typical concrete floor plate has a thickness of 0.50 mwe. When the overburden of a system is specified, the shielding mass is usually excluded.

The background of low-level beta/gamma systems comes mainly from secondary radiation induced by cosmic-ray muons and protons in the shield. The overburden (Section 7.11) may attenuate their flux significantly. It is therefore an important parameter. If we want to analyze the background of a system, or compare the background of similar systems, we must know the overburden.

Figure 11.2 shows the overburden in three different laboratories, two at surface level and one in a shallow underground location. The overburden in buildings may absorb a large part of the cosmic ray protons (producing neutrons) and it can attenuate the muon flux by 10–20%. It is surprising how little attention this receives in descriptions of low-level counting systems.

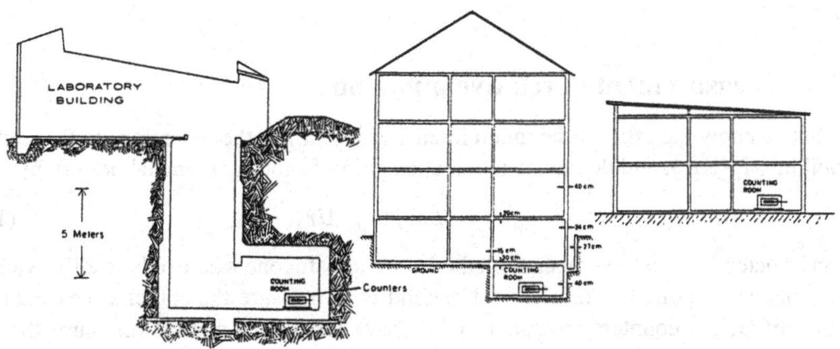

Fig. 11.2. Laboratories with different overburden.

If the laboratory is in a multi-storey building, it is decidedly best to locate low-level systems in the basement. But the advantage of lower background may be outweighed by the inconvenience of always having to go downstairs in order to attend the measurements. In order to be able to select the best location for our measurements, the overburden should be estimated and we must know the background reduction it will bring us.

Most laboratories have only the overburden which the building gives. A number of shallow underground laboratories have been specially constructed, below the laboratory buildings, under a layer of a few meters of soil or rocks, where the cosmic proton flux has been reduced to an insignificant level. The deepest of these is below the basement of the University of Bern, having an overburden of 70 mwe (Oeschger et al., 1975, 1981), where the muon flux is reduced by a factor of 11. Furthermore, a number of institutes have access to working space still deeper, in tunnels or mines, with an overburden up to 4000 mwe. Povinec (1994) gives a list of deep underground laboratories. These facilities are outside the coverage of this book, but they have played an important role in the development of better low-level germanium spectrometers, in studies of radioimpurities of materials and in the analysis of background components (described in Chapter 12). Here we are mainly concerned with the intensity of cosmic rays in laboratories with an overburden less than 20 mwe.

The number and thickness of floor plates above a laboratory gives a reasonably good estimate of the overburden, but for a more accurate value we must take into consideration that a part of the cosmic particles come through the walls at a declining angle and may escape one or more of the floor plates.

For a detailed background analysis we need to know the relative intensity of both of these fluxes. The muons are very penetrating, and produce energetic electrons/photons that are the main source of background in low-level beta/gamma systems. They also produce neutrons, but at a rate that is usually an insignificant background source. The background component due to protons comes both from the neutrons they produce, and from the neutral pions that induce electromagnetic cascades. Direct, but only relative, determination of the flux of muons and protons is therefore desirable. This will be discussed in the next two sections.

11.2.2 Measurement of relative muon flux

We want to know the ratio of the muon intensity $I_\mu(Lab)$ in the laboratory to that outside the building $I_\mu(Air)$, and define the muon attenuation factor A_μ in the laboratory by

$$A_\mu = I_\mu(Lab)/\,I_\mu(Air) \tag{11.1}$$

A_μ can be determined by a number of methods. We look for one where only readily available instruments are required. The classical method is to measure the coincidence count rate of a pair of Geiger counters separated by 10 cm of lead. This method measures the hard component, which at sea level consists 99% of muons and 1% of protons. This is a simple method and it gives the most unambiguous result. However, suitable Geiger counters are probably rarely found in our laboratories today, except possibly in the attic.

A more convenient method is either to use a NaI scintillation unit or a germanium spectrometer. The NaI unit is better as it is relatively inexpensive, available in most laboratories, and is easy to transport. 2"×2" NaI crystals are the most common size and they are quite convenient for the purpose discussed here. The spectrum, as well as the count rate, depends somewhat on the vertical orientation of the crystal axis, and all measurements should therefore be taken with the same orientation. As a horizontal axis is most convenient inside shields, it should be chosen.

The muons give quite large pulses, about 4 MeV per cm of their straight track in NaI, or 20 MeV for 5 cm in the 2"×2" NaI crystals. We must take into consideration that some of the large pulses do not come from muons, but from secondary electrons and photons, produced both by muons and protons. These secondary particles are mainly produced in electromagnetic cascades, initiated by neutral pions, induced by protons, and muon knock-on electrons.

Preusse (1993) studied the high energy part of the spectrum of a 3"×3" NaI crystal in a surface laboratory. He had a large flat gas proportional counter (60×65 cm^2) 5 cm above the NaI crystal and registered both the coincidence and anticoincidence spectra of the NaI crystal. He also calculated the expected pulse spectrum of muons hitting the crystal using the Monte Carlo method. The result of his work is shown in the three spectra in Figure 11.3.

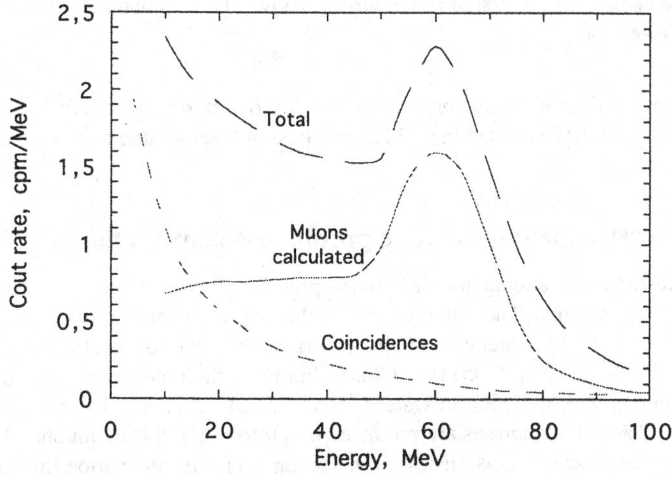

Fig. 11.3. Cosmic ray spectra of a 3"×3" NaI scintillation detector: a) spectrum in coincidence with pulses from a large flat gas proportional counter, (b) the anticoincidence spectrum, (c) total spectrum and (d) calculated spectrum for muons hitting the crystal. From Preusse, 1993.

Practically all muons traversing the NaI crystal also give a pulse in the large proportional counter, and nearly all these muon NaI pulses are registered in the coincidence channel. A comparison of the measured coinincidence spectrum and the Monte Carlo calculated muon spectrum clearly shows that secondary electrons and photons contribute significantly to the coincidence channel, especially at lower energies. We see from the figure, that, by setting a rather narrow muon counting window, the electron/photon component can be made small, so that we will mostly be counting muons.

The relative muon intensity can also be found from the high energy pulse height spectrum of germanium spectrometers, if the muon count rate is known for a germanium diode of similar size with insignificant overburden.

If N(Lab) and N(air) are the count rates in this muon channel of a NaI unit, in our laboratory (eventually inside the shield) and in free air, the muon attenuation factor A_μ, which is a function the overburden M_{ob}, is

$$A_\mu(M_{ob}) = N(Lab)/N(Air) \tag{11.2}$$

This attenuation factor, describing the penetration of muons, was discussed in section 7.11, where it was shown that it can be expressed by the following empirical equation (7.14)

$$A_\mu(m_{ob}) = 10^{(-1.32 \log d - 0.26 (\log d)^2)}$$

where $d = 1 + (m_{ob}/10)$. A similar equation, which only takes into account the first power of $\log d$, has been described by Skoro et al. (1995). They used it to determine the overburden from the value of A_μ, measured as described above. Their equation is equivalent to the following expression:

$$A_\mu(m_{ob}) = 10^{(-1.47 \log d)} \tag{11.3}$$

This equation gives a satisfactory value when the overburden is less than 40 mwe. For the depths we are most interested in (<20 mwe), it is satisfactory and more convenient than Equation 11.2.

11.2.3 Measurement of relative proton and neutron fluxes

We are interested in the attenuation of both the proton and neutron fluxes inside the shield of our low-level systems. The relative neutron flux can be measured inside a lead shield by using a BF_3 neutron counter covered by a thermalizing layer of paraffin as described in Section 7.13. Coccioni et al. (1951) used this technique in their measurement of the neutron production rate at varying depth in water (Section 7.12). Their results show that at zero overburden, 92% of the neutrons are produced by protons and 8% by muons. The proton flux decreases exponentially, as discussed in section 7.11. Its attenuation thickness is not known with satisfactory accuracy, but 1.6 mwe is a widely accepted value, so

$$A_p = e^{-m_{ob}/1.6} \tag{11.4}$$

The neutron attenuation factor A_n is defined in the same way as that for muons:

$$A_n(m_{ob}) = N(Lab)/N(Air) \tag{11.5}$$

The attenuation factor of neutrons can therefore be written

$$A_n = 0.92\, A_\mu + 0.08\, e^{-m_{ob}/1.6} \tag{11.6}$$

where A_μ is given by equation (7.19). Instead of measuring the proton flux, the relative production rate of neutrons and muons is measured as a function of overburden. The results will be similar to those shown in Figure 11.4, which are based on Equation (7.19) and Equation (11.6).

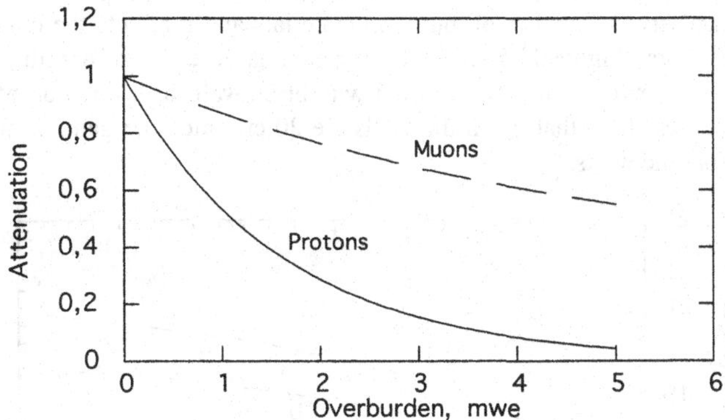

Fig. 11.4. The attenuation factor of cosmic-ray muons and neutrons at varying overburden.

11.3 Reducing external gamma radiation

11.3.1 Laboratories with reduced gamma flux

In some laboratories the gamma flux from natural radioactivity can be a factor of 5 to 20 lower than in normal rooms. This brings the additional advantage that it is easier to maintain low radon concentrations in the laboratory, because of lower exhalation rates from floor plates and walls. Low gamma flux can be achieved in the following ways:

Shielded laboratories. The simplest way is to cover the walls of the laboratory with iron or lead. This is practically always done in rooms with whole body counters, but rarely in normal low-level laboratories.

Walls of clean materials. Pure materials (sand and cement) can be selected when the laboratory is constructed. In an underground laboratory at the University of Bern the gamma flux is only 1/15 of that in ordinary buildings as a result of careful material selection (Oeschger and Loosli, 1977). The primordial gamma radiation was reduced by a factor of 25 by using pure concrete and a 3 cm Pb lining in a low-level laboratory in Vienna (Aiginger et al., 1986).

Low-activity geological formations. Advantage has, in some cases, been taken of the low concentrations of natural radioactive elements in special geological formations. Some salt mines have very low concentration of K/Th/U. The Asse mine in Germany has, for example, a gamma flux that is only about 1% of that in normal environment.

11.3.2 Absorption of external gamma radiation in the main shield

Primordial radioactivity is found in all building materials and it affects the count rate of our low-level detectors. Figure 11.5 shows the air gamma dose rate of uranium, thorium and potassium at the center of a standard room with brick walls as a function of the wall thickness. The curves show that when the walls are 20 cm thick the gamma flux comes predominantly from the walls.

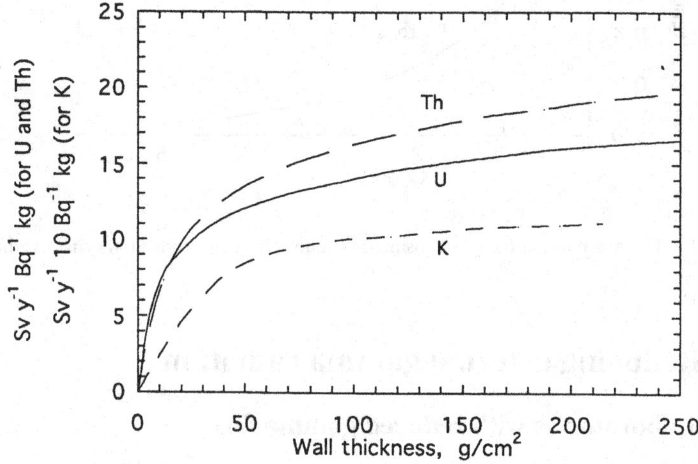

Fig. 11.5. Absorbed air dose rate from K/Th/U in a room with varying wall thickness. From Bruzzi et al., 1993.

The content of natural radioactivity in building materials varies widely, just as in bedrock and soil. Concentration differences of a factor of about 5 are observed, going from a building made of materials with low concentrations to one where they are high.

In a low-level counting system the background component $B(Ex)$ due to external gamma radiation can in principle be reduced practically to zero by enclosing the detector with a sufficiently thick shield. However, we should strive for an optimum thickness and avoid making the system too bulky.

We first discuss, in a general way, the penetration of external gamma radiation, with a wide energy spectrum, through lead. Let us denote the energy dependent gamma flux in the laboratory, outside the shield, by $\Phi_{Lb}(E)$ and the flux inside the shield by $\Phi_{Sh}(E)$. If we (incorrectly) treat this as attenuation of gamma radiation under good geometry (Section 4.4) we have

$$\Phi_{Sh}(E) = e^{-\mu x}\,\Phi_{Lb}(E) \tag{11.7}$$

Figure 11.6 shows the factor $e^{-\mu x}$ as a function of energy for 5, 10 and 15 cm of lead. According to this graph there is a broad maximum at 2 MeV, but the penetration falls rather rapidly below 1 MeV. If this describes the attenuation correctly, there would be practically no low energy photons inside a shield of 10 cm of lead. This is, however, not the case. A large part of the attenuation is through Compton scattering where the photons do not vanish, as good geometry assumes, but are transformed into lower energy photons. The low energy part of the spectrum is thus continuously replenished from the higher energy part.

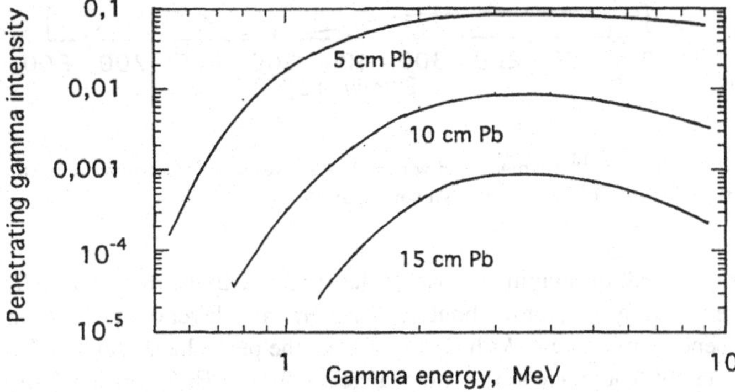

Fig. 11.6. Penetration of gamma radiation in 5, 10 and 15 cm of lead at varying energies, assuming good geometry.

This important energy spectrum shift is well illustrated by the attenuation of the 0.66 MeV gamma rays from ^{137}Cs in water. Figure 11.7 shows the energy spectrum measured with a 2"×2" NaI scintillation counter: (a) in air, and (b) in water with a distance of 30 cm, which is equal to one attenuation length, between source and detector. The undisturbed spectrum shows a 0.66 MeV photopeak rising high above a rather flat Compton plateau, but in water low energy photons are dominating. This is similar for lead; Compton scattered photons are produced in the shield with reduced energy and they dominate the spectrum inside the shield, not the original high energy photons. Under normal shielding thickness, the energy spectrum will be dominated by this Compton scattering, with maximum intensity at about 150 keV.

We now look at the selection of optimum shielding thickness for a system where a CsI scintillation detector replaces the unspecified hypothetical system in Section 11.1. This detector unit, described in Section 15.5, has the advantage, compared to conventional NaI scintillation counters, that it is practically radiopure and very compact. Its gamma background spectrum was recorded under different shielding conditions: (a) without a shield, (b) behind 5 cm of lead and (c) behind 10 cm of lead. The resulting spectra are shown in Figure

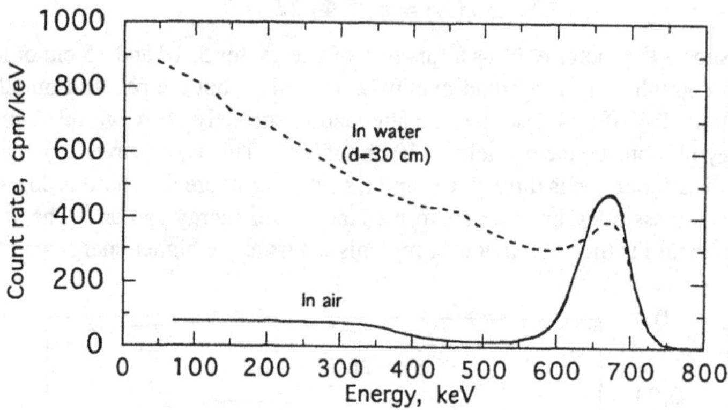

Fig. 11.7. The spectrum of ^{137}Cs measured with a 2" ×2" NaI scintillation counter: (a) in air and (b) in water with a distance of 30 cm between source and detector.

11.8. As to be expected, the continuum part of the spectrum in the region from 0.5 MeV to 2.6 MeV is attenuated by a factor of about 11 by a 5 cm thick layer of lead. The attenuation is almost independent of energy. With 10 cm of lead, the peaks have completely vanished, and a further 5 cm of lead made almost no change. At 10 cm $B(Ex)$ is evidently insignificant, and the residual background comes practically from secondary radiation alone. At 5 cm of lead, $B(Ex)$ is evidently about half of the residual background B_0. As $B(Ex)$ is reduced again by a factor of 11 going from 5 cm to 10 cm of lead, $B(Ex)$ is reduced to about 4% of B_0. There is thus very little to gain by a thicker shield, we are actually well below the 10% tolerance limit, actually a total thickness of about 8.5 cm would have been sufficient. In underground laboratories and when external guard counters are used, B_0 may be decreased by a factor of 10 or more. In this case a thicker shield is required.

The result of this study applies equally well to other detectors: NaI, gas proportional and liquid scintillation counters. The result is also valid for the continuum of germanium diode spectra, but the suppression of high energy gamma peaks to an insignificant value may require a thicker shield, as discussed in Section 12.7.

11.4 Reduction of radioactive contamination

Reducing the background contribution of radioactive contamination in materials used in the systems is a major task in the technique of low-level counting. We should know (or have a good estimate of) the contamination in the materials used, and at what level it can give a background contribution that exceeds what we are ready to accept, the tolerance level. Extensive measurements on the radiopurity of a large number of materials, made during the

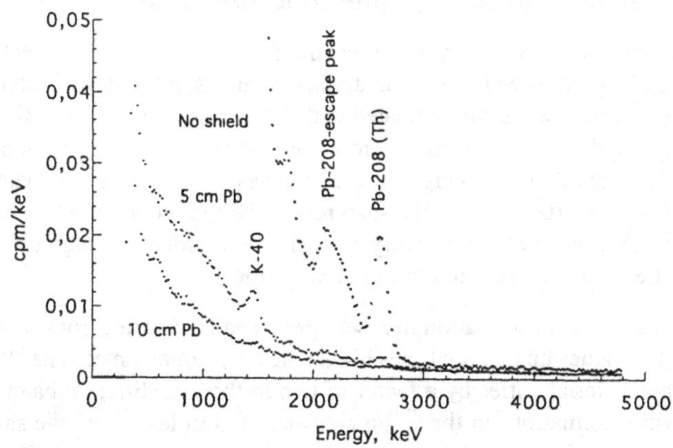

Fig. 11.8. Background of a CsI scintillation detector with different shielding (Theodorsson 1996).

last 10 years with ultralow-level germanium spectrometers, have now brought us detailed information on which we can base our selection of radiopure materials.

11.4.1 Measurement of contamination inside the shield

Contamination in materials inside and in the shield can be detected and identified in systems with germanium detectors. This is possible only to a limited degree in systems with NaI scintillation detectors, both because of their low energy resolution and primordial radioactivity in the glass of the phototubes. Gas proportional and liquid scintillation counters are only affected through their total count rate.

The gamma flux inside the shield of the latter systems can probably be quantified and crudely analyzed using the small CsI scintillation unit mentioned above, as it is compact and has no inner contamination. Under 10 cm of lead the CsI crystal only gives a continuum background spectrum of 0.82 cpm per 100 keV at 1 MeV due to secondary radiation, as seen in Figure 11.8. Its count rate in 100 keV energy intervals can be determined to an accuracy of about 1% with a counting time of one week. We should therefore be able to detect a deviation of a few percent from the count rate in a clean shield. We might also see faint peaks of the contaminating radioisotopes. It should therefore be possible to study the gamma flux from primordial radioactivity inside the shields of these systems, below the level that gives an appreciable background increase in the respective detectors. The CsI unit would replace a gas proportional counter during the measurement, and in a liquid scintillation system it would replace one of the two phototubes. If a second measurement is made with the other phototube also removed, the difference spectrum will give the background contribution of the phototube.

11.4.2 ^{210}Pb in lead and acceptable concentrations

Due to good mechanical properties, high atomic number, low neutron cross section and reasonable cost, lead is the most widely used material for shields in low-level systems. Modern lead is, however, nearly always contaminated with ^{210}Pb, as discussed in sections 9.2.3 and 10.4.1. Its background contribution comes from bremsstrahlung, and from characteristic Pb X-rays, induced by the relatively energetic beta particles of ^{210}Bi. Approximately only one photon comes for every 100 beta particles, so relatively high contamination levels can be tolerated. The ^{210}Pb concentration can vary by almost two orders of magnitude in recently produced lead. Lead can be classified into four categories:

Modern lead. The ^{210}Pb concentration in newly produced lead varies from about 20 Bq/kg to 1000 Bq/kg, but values up to three times higher are occasionally met. The ^{210}Pb concentration in the raw material varies by a factor of two to three in different batches from the same mine, giving fluctuations in the ^{210}Pb concentration in lead from the same supplier (Section 10.4). Certified low-activity lead, with a ^{210}Pb concentration of 20–50 Bq/kg, is commercially available at a price about twice that for regular lead. This will be referred to as good modern lead.

Selected modern lead. Some producers offer better lead, containing about 5 Bq/kg. The lead is presumably a selected, carefully processed, part of the raw material from a mine that generally has a low ^{226}Ra concentration in the ore. This lead is, however, more expensive still.

Specially produced modern lead. Special lead is produced for the electronic industry that makes severe demands on the alpha purity. The ^{210}Pb content of this lead is below the detection limit, less than about 0.3 Bq/kg (Alessandrello et al., 1991). This is supposedly obtained by separating ^{226}Ra free Pb minerals in the ore and careful processing. A very thorough study of this lead and ancient Roman lead revealed that the ^{210}Pb contamination was below the detection limit in both types (Section 10.4.1). The contamination from ^{238}U and ^{232}Th was less than 0.09 Bq/kg, measured with a counting time of ten days using an ultralow-level germanium system in an underground laboratory.

Old lead. Centuries-old lead has been recovered from the wrecks of several sunken ships, from roofs of old buildings, etc. If the lead is 300 years old its ^{210}Pb has decayed to an insignificant level. No other activity is found in old lead, demonstrating the high separation of uranium and thorium from lead in the production process.

In order to keep the cost of the main shield low, the following should be observed. Any lead can be used for the outer 5 cm layer. For the inner 5 cm layer, good modern lead should be used. In the literature this lead is often referred to as low activity lead, even radiopure lead. But when is it sufficiently pure? The ^{210}Pb concentration that can be tolerated in different systems is discussed in Chapters 12–15. The main results are listed in Table 11.1. The values, except those for germanium spectrometers, should be considered as educated guesses, calling for more reliable figures. It is evident that lead available at normal prices contains, in some cases, too much ^{210}Pb. An inner layer of some pure material must then be added.

Table 11.1. Tolerance levels (TL) of ^{210}Pb in the inner lead layer of a shield.

System	TL, Bq ^{210}Pb per kg
Proportional counters for solid samples	200
Gas proportional counters	20
3"×3" NaI crystals	20
Ge-diodes, 200 cm^3, surface laboratory	5
Ge-diodes, with external guard	1.0
Ge, external guard, 15 mwe overburden	0.20
LSC in a surface laboratory	200
LSC with a guard	40

11.4.3 Inner layer of main shield

Earlier, mercury was frequently used for the inner shield, but today essentially only tested lead or copper are used. Due to more intense secondary radiation, copper will increase the background by 30–50% in the energy range of 100–500 keV, compared to an inner layer of pure lead as discussed in Section 12.8. The innermost layer should therefore also be of lead, but of sufficient purity. In systems operating in underground laboratories, where the cosmic ray flux and its associated secondary radiation has been suppressed to an insignificant value, copper is usually preferred for the innermost layer, as it is readily available and absolutely radiopure.

It should be noted that an inner shielding layer of lead need not be of ^{210}Pb free lead, but its concentration must not be higher than the tolerance level for the particular system (Table 11.1). How thick must this inner lead layer be? It is assumed here that the concentration in the main shield is 20–40 Bq/kg. We may then need to attenuate the bremsstrahlung radiation by a factor of 10, never by more than a factor of 50. If we use nearly radiopure lead (containing less than 0.5 Bq/kg), we are dealing with an expensive material that we should not waste by having an inner layer thicker than necessary. Nevertheless this layer is often 5 cm thick as determined by the dimensions of standard lead bricks.

Let us look at the absorption of the bremsstrahlung, which is more penetrating and more intense than the accompanying ^{210}Pb X-rays. If we use (conservatively) the gamma absorption coefficient at 400 keV to calculate the absorption, the attenuation thickness is 0.34 cm. A 5 cm thick lead layer reduces the bremsstrahlung to 4×10^{-7}, which is excessive. One cm of lead reduces it to about 10% and two cm to 1%. If the main shield is made of good modern lead, a 1.0 cm thick layer of the old lead would probably be enough in most cases and a thickness of more than 2 cm is never needed. A thicker inner layer is a waste of valuable material.

Graded shielding has been popular for many years in low-level beta- and gamma-counting systems, especially in germanium spectrometers, for the suppression of the X-rays induced in the lead shield. This shielding consists of two layers on the inner side of the main lead shield. The outer layer is usually 2 mm of cadmium, for the absorption the X-rays from the lead, and the inner 2 mm of copper for the absorption of X-rays from the cadmium. This will effectively suppress the X-rays of lead, but will add considerably to the secondary radiation in the energy range 100–500 keV. Modern germanium spectrometer therefore do not have a graded inner shield of this type.

11.4.4 Innermost shielding layer

In low-level gas proportional counting systems, an extra inner absorbing layer of radiopure material, either of mercury or lead, is usually inserted between the sample counter and the guard counter system. This technique was introduced by Kulp and Tyron (1953) in order to absorb gamma radiation from contamination in the iron shield and components inside it. They surrounded their sample counter by an annular vessel filled with mercury, giving a 2.5 cm thick shielding layer. This reduced their background from 4 to 2 cpm. H. de Vries (1953) used old lead for the same purpose. Most later gas proportional counting systems have an inner layer of lead, 2–4 cm thick. This layer is also quite effective in reducing the background component due to secondary radiation, as discussed in Section 12.8. It should be noted that the total lead thickness, the sum of the main and inner shields, usually needs to be only 10 cm.

11.4.5 Reducing contamination in system components

Germanium spectrometers. Earlier Ge-diode systems had a significant background component B(Ct) coming from radiocontamination inside the shield. This accounted for possibly about half of the count rate of early germanium spectrometers. Improvements obtained in a system described in Section 9.2.3 give a good indication of the contribution of common contaminants in earlier Ge-systems; when copper replaced aluminium in the diode holder, the background was reduced by a factor of two.

Today, new germanium systems are readily available that are practically without radiocontamination, except for traces of cosmogenic activity below 1 mBq/kg. This state of the art has been achieved through an extensive study of the radiopurity of all materials used and subsequent selection of pure ones. These systems are the standards against which other types of low-level systems should be compared.

Gas proportional detectors. The experience from the development work on modern germanium spectrometers should now be transferred to other low-level systems. Gas proportional counter systems have a number of components inside the shield that are a potential source of contamination: guard counters, paraffin/boron layers and various structural components.

The inner layer of lead or mercury reduces B(Ct) considerably in these systems, usually by a factor of about four.

There are two ways to decrease B(Ct): (a) to replace all system components inside the shield, which we suspect to be contaminated, with new components made of radiopure materials, or (b) to have as few system components inside the shield as possible. The latter can be achieved by using external guard counters, as discussed in section 11.8.

NaI scintillation detectors. A large number of background studies of NaI scintillation counters were carried out in the 1960s and early 1970s. They revealed a considerable internal K/Th/U contamination in the glass of the photomultipler tubes, contributing 30–50% of the total background count rate of a well shielded detector. The rest came from secondary radiation, B(Sr).

In order to determine the relative size of these two contributions, one must: (a) measure the background spectrum with a multichannel analyzer, (b) have long counting times, of the order of a week, to get good statistics, and (c) measure the spectrum up to about 4 MeV in order to see the part of the $B(Sr)$ spectrum that is undisturbed by natural radioactivity (above 2.6 MeV, highest energy natural gamma line). During the most active period of these NaI background studies, multichannel analyzers were expensive and rarely available for weeks, as needed in this work. Furthermore, the spectrum seldom extended beyond 2.6 MeV. Therefore, the detailed information we need is rarely found in the technical literature.

A detailed study, performed by Stenberg and Olsson (1968), is a rare exception. Their system is shown in Figure 11.9. Their data are of enhanced value as they had an array of 45 Geiger guard counters behind a 10 cm layer of lead, and an additional 8 cm thick lead layer between the guard and the NaI detector (3"×3" NaI crystal). The separation of the spectrum into coincidence and anticoincidence channels makes it possible to separate the two background components. Figure 11.10 shows three spectra with logarithmic X- and Y-scales. It shows clearly a large number of gamma peaks from K/Th/U. As the low energy peaks are apparently unattenuated, as seen in their relative intensity, the source of radiocontamination is evidently inside the shield; it all comes from contamination in the glass of the photomultiplier tube. An analysis of the spectra shows that B(Ct) is about 70% of the background component of the secondary radiation.

Miller et al. (1956) determined, in a very thorough early study, the contamination in a few photomultiplier tubes by placing them at the end of the NaI crystal of a scintillation unit. Their results showed that B(Ct) was 30–50% of B(Sr). The influence of phototube contamination can be reduced by inserting an absorbing layer between the tube and the NaI crystal. The most effective way is to use a 3–5 cm thick NaI crystal without a Tl activator, but quartz has also been used. This arrangement is, however, of little interest today, as germanium spectrometers have largely replaced the NaI units. In the near future we may, however, see NaI scintillation counters with phototubes where the radiocontamination has been reduced by an order of magnitude, as discussed in Section 9.2.3.

Liquid scintillation counters are sensitive to radiocontamination in the glass of the photo-

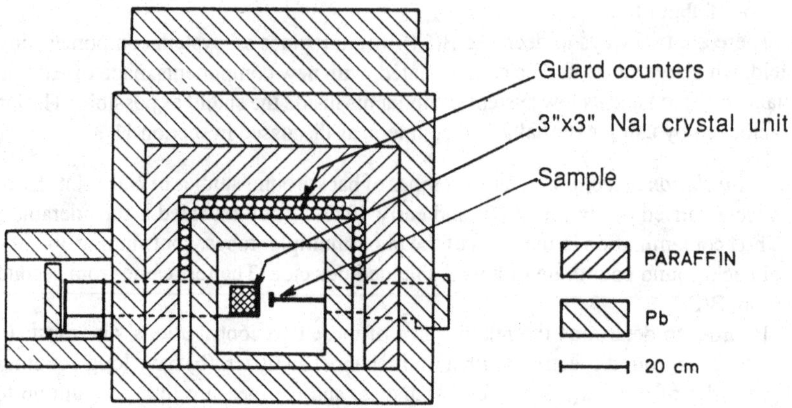

Fig. 11.9. System with a 3" ×3" NaI scintillation counter used for background studies. The system had both a guard counter system and an inner shield. From Stenberg and Olsson, 1968.

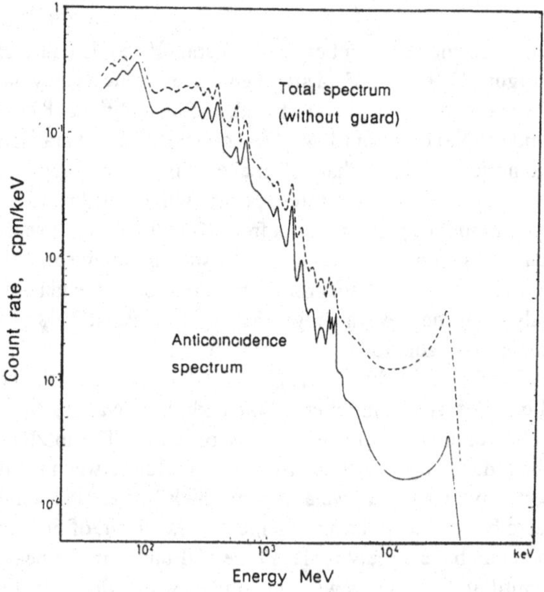

Fig. 11.10. Background spectra of a 3" ×3" NaI crystal. From Stenberg and Olsson, 1968.

tube, mainly because of the much smaller pulses that must be counted. Cerenkov pulses produced by beta particles from radiocontamination in the glass of the phototubes give a background contribution. This is discussed in more detail in section 14.4. A new generation of phototubes with very low radiocontamination may, in coming years, improve low-level liquid scintillation counters,

11.5 Radon

Radon is present in every house and every laboratory, where its air concentration is typically 50 Bq/m^3, about 10 times higher than in outside air (Section 6.5). Its concentration depends on ventilation. Radon can be a serious disturbing factor as its concentration varies in a sporadic way. The tolerance level, defined in Section 11.1, is not appropriate here, because it only applies to a background component that gives a constant contribution. If radon gives a count rate increase equal to the statistical standard deviation σ of a single measurement, it can be said to be disturbing. This value may be significantly less than 10% of the background, used earlier for the tolerance level of a constant background source. Let us look at a typical gas proportional counter having a background count rate of 2 cpm, and assume that we are measuring a sample that is close to the background, for example giving a net count rate of 1.0 cpm. If this sample is measured for two days, the standard deviation will be 0.032 cpm, but 10% of the background is 0.2 cpm, or 6 times higher than σ. For radon we will therefore set the limit of acceptable concentration where it increases the count rate by σ for the measurement of a weak sample and for a typical counting period.

From 1960 to 1985 there were a number of reports about radon disturbances in low-level counting systems of various kinds. During this period, little was known about radon in houses. In the mid 1980s, large efforts were initiated in many countries to study the concentration of radon in our environment. It was found that radon can build up to very high concentrations inside houses, frequently up to many hundreds of Bq/m^3, and in some cases to thousands of Bq/m^3. The concentration of radon in the laboratory can be minimized by: (a) good ventilation and (b) reducing emanation from floors and ceilings by a suitable paint. The effect of radon on different types of systems will be discussed below.

Ge-diode systems. In the mid 1980s it became clear that radon could be a serious disturbing factor in the new generation of low-level germanium spectrometers. Its presence in these systems is immediately seen in the gamma peaks in the background spectrum, where the 242, 259 and 352 keV lines of ^{214}Pb and the 609 keV line of ^{214}Bi are prominent. As the number of pulses in a few adjacent channels above and below a gamma peak are always used for determining the background, radon will only disturb when measuring gamma peaks that lie very close to those of radon. We can therefore use the tolerance level as defined in Section 11.1. Let us look at the strongest radon line, the 609 keV line, and assume that radon is filling a free space of 3 liters around a 30% relative efficiency germanium diode, and that the counting efficiency for the 609 keV line is 1%. For a radon concentration of 1.0 Bq/m^3, the total count rate in the peak will be 0.02 cpm and about 0.0006 cpm/keV in the mid peak channel. The tolerance level is then about 10 Bq Rn/m^3. Small radon peaks will, however, only rarely interfere with measurements of other radioisotopes.

Radon can be eliminated in a rather simple way in these systems by venting nitrogen boiling off the Dewar vessel into the inner detector space. Typically this gives about 1000 liters of gaseous nitrogen per day, which should be enough to keep the space surrounding the detector radon free.

Gas proportional counters. In these systems a major part of the background pulses comes from electrons released from the wall of the counter, a smaller part from direct interaction with gas molecules. The background is therefore nearly proportional to the surface area (cathode) of the counter for both the component of radon $B(Rn)$ and for the secondary radiation $B(Sr)$. $B(Rn)$ is approximately proportional to the free air space around the counter. Let us look at a counter with a volume of 1.0 liter. It will have an outer surface area of about 800 cm^2 and the thickness d of the free space surrounding the counter will be a few mm. Hedberg and Theodorsson (1996) have measured B(Rn) for a 1.0 liter counter where d was approximately 2 mm and found a tolerance limit of 130 Bq/m^3. This is a lower value than one could expect, considering the limited attention that scientists, working with these systems, have paid to radon. It seems likely that some radiocarbon laboratories are having trouble because of intermittent high radon concentrations.

The radon tolerance limit for each counter should be estimated and the radon concentrations in the laboratory air. This can be measured by a number of methods using readily available instruments (Sections 15.3 and 17.3). If there is any risk that the radon concentration may exceed the acceptable level, precaution should be taken either to decrease its concentration or its effect on the counter by either of the two methods mentioned above. Further, the free space around the sample counter can be filled with some radiopure material.

Liquid scintillation counters. Because of the small size of the vials (20 ml) the free space around them is very small, about 30 times smaller than around a 1 litre gas counter. The sensitive mass of a 20 ml sample (see section 12.3) is similar to that of typical gas proportional counters, but the effective volume of radon is smaller. We could therefore expect a tolerance level about 30 times higher, or 4000 Bq/m^3. This is, however, a very crude estimate. It is desirable to measure this parameter directly. The solid daughter products of radon may plate out on the vial. The energetic beta particles of ^{210}Bi may, if they penetrate the wall, give a significant background contribution because of their 100% detection efficiency. The absorption of beta particles is, on the other hand, high in the wall.

11.6 Secondary gamma radiation

Let us first look briefly at the direct background contribution of muons and protons, i.e., when these particles travel through the detectors. In well shielded gas proportional counters, these events induce more than 95% of the background pulses. In low-level systems, the sample gas counters are therefore always surrounded by a guard counter or an array of separate guard counters, the active shield. Detectors with a solid or liquid medium hardly need an active shield of this type, as most of the pulses of muons and protons are large and above the usual energy region of interest. These events, therefore, only give a relatively small background contribution.

Although direct interaction of muons and protons only gives a small, even insignificant, background contribution, this does not mean that they can be disregarded. On the contrary, the secondary radiation these particles produce in the shield — electrons, gamma radiation and neutrons — is the predominant source of the residual background. The neutron component will be discussed in the next section.

The secondary electrons and gamma radiation come primarily from electromagnetic showers produced in the shield and material inside it; (a) by muons, primarily through knock-on electrons, and (b) by protons through neutral pions. Although the proton flux at sea level is only about 1% of that of muons, they contribute significantly to the showers.

There are two ways to reduce this background contribution significantly; (1) to have the systems in a laboratory with thick overburden, preferably underground, or (2) by having a thick layer between a good guard counter system and the sample counter, preferably an external guard. The latter possibility is discussed in Section 11.8.2. The effect of secondary radiation is discussed in more detail in Section 12.5.

11.7 Neutrons

The nucleonic component (mainly protons) of cosmic rays produces neutrons in the atmosphere. This production continues in the laboratory building materials, and finally, but not least, in the lead shield of low-level systems and in the materials inside it (Sections 7.11–7.13). Most of the neutrons in the laboratory are thermal, having been slowed down by the relatively low mass number building materials. As discussed in Section 7.13, and well illustrated by Figure 7.21, lead is very efficient in producing neutrons, which have an evaporation type energy spectrum, i.e., having high energy. Usually there is little material inside the shield to slow down these neutrons.

The neutron flux inside the shield, thus, consists of two components: (a) a small one, resulting from the fast and slow cosmic ray neutrons in the laboratory diffusing into the shield, and (b) a larger fast component, produced in the lead. The fast neutrons make a few collisions with lead nuclei before they diffuse out of the shield. The neutrons give background pulses through:

1. Direct elastic and inelastic collisions of fast neutrons with detector atoms.
2. Photons, emitted in inelastic collisions of fast neutrons.
3. Photons, emitted after capture of a slow neutrons.

The relative importance of the three processes depends on the type of detector.

Pulses of recoiling atoms due to collisions of fast neutrons. The fast neutrons inside a shield give pulses when they collide with detector atoms as the recoiling atoms are ionized. These events are best know from low-level germanium spectrometers where the energy of the recoiling atoms is 0–30 keV. These collisions are discussed in more detail in Section 12.6.3. Here we show the recoil energy distribution when fast neutrons collide inelastically with ^{74}Ge atoms, where the nucleus emits a 596 keV photon (Figure 11.11).

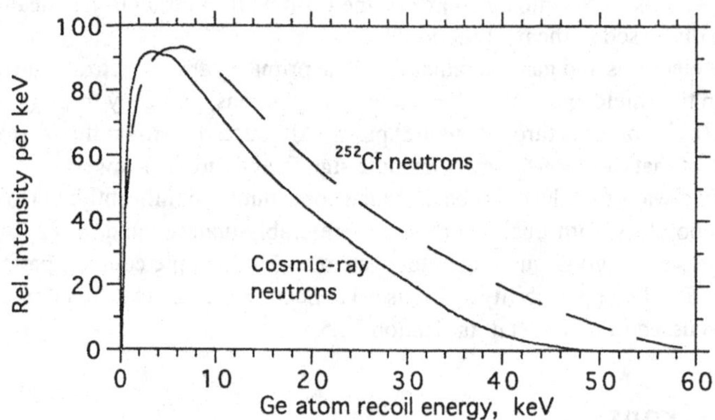

Fig. 11.11. The energy spectrum of recoiling ^{74}Ge nuclei caused by fast neutron collisions.

When cosmic ray neutrons collide with atoms in gas proportional and liquid scintillation counters similar recoil pulses appear, but the transferred energy in both detectors is larger as the atoms have a lower mass number. The distribution of the recoil energy in these detectors can be derived approximately by taking into account the difference in mass number A of the atoms. The maximum energy transferred to a recoiling nucleus in a head-on collision is given by

$$E_{rec}/E_n = 1 - \left[(A-1)/(A+1)\right]^2 \tag{11.8}$$

where E_n and E_{rec} are the energies of the neutron before collision and the energy of the recoiling nucleus. If we assume that the mean recoil energy is approximately half the value of the transmitted energy in a head-on collision, the mean recoil energy in a collision with a ^{74}Ge nucleus is 0.026 E_n , it is 0.142 E_n for ^{12}C and 0.5 E_n for ^{1}H (protons). The mean recoil energy is therefore 5.4 times larger for ^{12}C nuclei than for ^{74}Ge and 19 times larger for hydrogen nuclei.

Figure 11.12 shows the calculated recoil energy spectra of ^{12}C and ^{1}H. This figure shows that a large fraction of the recoil pulses falls into the counting window in most low level work with gas proportional and liquid scintillation counters.

Neutron induced gamma background. The photons can be induced either by fast neutrons, through inelastic collisions (mainly $(n, n'\gamma)$ reactions), or through capture of a slow neutron.

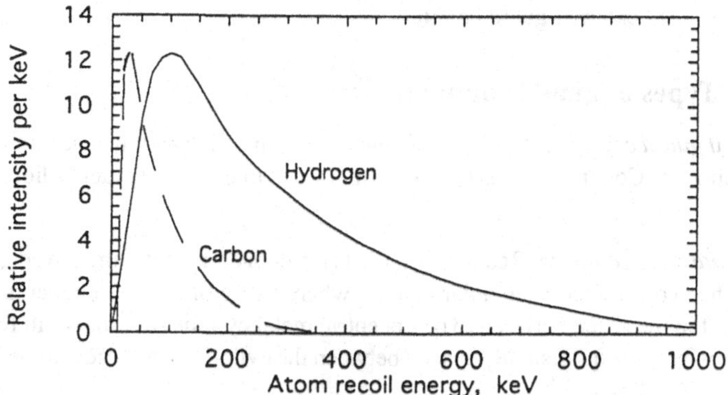

Fig. 11.12. Calculated recoil energy of hydrogen and carbon atoms after collision with cosmic ray neutrons.

A major portion of gamma pulses can be eliminated with external guard counters, but some of the pulses are delayed and are therefore not excluded by the veto signal.

11.8 Guard counters and their background reduction

The anticoincidence technique was introduced by Libby in 1947. His sample counter was almost surrounded by an array of 11 guard counters. Their function was to intercept all high energy charged cosmic-ray particles (99% muons) that could hit the sample detector. The guard counters have other names: veto, anticosmic and anticoincidence counters. In this book the shortest name is usually preferred, i.e., guard counters, and its blocking signal is called a veto signal. The guard counter system is often referred to as the active shield.

The guard counters have 100% detection efficiency for traversing muons and protons. With good geometrical coverage they will detect practically all muons hitting the sample detector. Most guard counter systems are of the three different types, shown in Figure 11.13: (a) an array of separate cylindrical counters, (b) a single annular detector unit or (c) a set of flat counters. The latter two types are either gas proportional counters, or scintillation counters with a liquid or acryl scintillator.

Originally, the only objective of the guard counters was to eliminate the direct background contribution of muons. When, somewhat later, an absorbing layer of radiopure material was added between the guard counter and sample detector, the scientists only aimed at attenuating gamma radiation from radiocontamination in the main shield and material inside it, including the guard counters. However, it soon emerged that the combined effect of the guard and inner shielding layer also attenuated the background from secondary radiation formed in the outer shield. The different types of guard counters will first be described, then studies of the background reduction they give, and finally the most effective guard counter arrangement, i.e., external guard counters.

11.8.1 Types of guard counters

Cylindrical guard counters. In his guard counter system, Libby applied separate cylindrical Geiger-counters. Counters of this type are still used in many of the older radiocarbon dating systems.

Annular guard gas counters. Reath et al. (1951) introduced a greatly improved guard counter system. They constructed an annular counter where a ring of counter elements is built into a single unit in the space between two concentric metal cylinders as shown in Figure 11.13. The anode wires are equidistantly spaced between the cylinders. A large number of counters

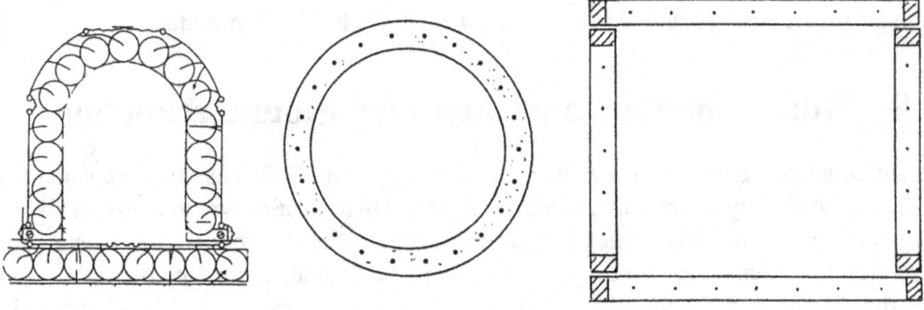

Fig. 11.13. Three main types of guard counter systems.

of this type, of varying size, were built at the radiocarbon dating laboratory in Trondheim, Norway, and used for a series of interesting background studies discussed in the next section, as well as for the counting systems of the Trondheim group (Nydal et al., 1980).

Large flat guard counters. These guard counters are made in the shape of large flat, thin boxes. They are of two different types: (1) gas proportional counters and (2) scintillation counters. Loosli et al. (1967) tested a single large external gas proportional guard counter (60×70 cm^2) on the top of a lead shield, but they seem not to have followed up this work. Large flat guard counters were first routinely used in a multicounter system developed for radiocarbon dating in Heidelberg. Four such counters were used to shield 9 tightly packed 4.0 litre gas proportional counters (Schoch et al., 1980). The main shield was outside these counters, but a 5 cm thick layer of lead inside.

When ultralow-level germanium spectrometers were developed in the mid 1980s, they were shielded on all outer sides when intended for use in surface or near surface laboratories. These guard counters were both of the gas proportional type (Heusser et al., 1989) and scintillation type, either with a liquid or acryl scintillator (Reeves and Arthur, 1984). The plastic scintillation counters are simpler to make than those with a scintillating liquid, but they are more expensive. The scintillator is 5–10 cm thick and the light flashes are viewed by 6–8 photomultiplier tubes at their edges. As muons lose about 1.7 MeV per cm of their track, their pulses can be separated from those of natural gamma radiation by pulse height discrimination. This significantly reduces the pulse rate of the guard counters and thus the dead time. Pulse height discrimination between muon and gamma pulses cannot be applied when gas proportional counters are used. However, their count rate is acceptable, as their gamma sensitivity is much lower.

Annular scintillation guard counters were introduced in the late 1970s. They came into routine use with the introduction of the low-level Quantulus scintillation system in 1984. The scintillation liquid in the Quantulus type guard counter is between two acentric tubes (Figure 14.8). A set of photomultiplier tubes at both ends detect the light pulses. In addition to their use in the Quantulus system, these guard detectors have also been used for gas proportional counters (Mäntynen et al., 1987). Their significant neutron and gamma efficiency enhances their suppressing effect on secondary cosmic radiation. Similar guard counters made of plastic, often in the form of a well counter, have also been used as guard counters. They have also been used extensively as anti-Compton counters as described in Section 15.8.

NaI guard counters. Although large NaI crystals are rather expensive, they have been used in a number of systems as guard counters, where they either have an annular form or have a large well in the crystal. They suppress not only the background pulses from muons, but also a large part of the pulses coming from the secondary radiation showers. They also reduce background pulses of Compton interactions in the sample detector, where the scattered photons are registered by the NaI crystal. The high efficiency of this arrangement was convincingly demonstrated in a study where the background of a set of small gas proportional counters was measured inside different shields and guard counter systems in three laboratories (Loosli et al., 1986). In the Harwell laboratory, with very little overburden, the background of the gas proportional counters inside a large annular NaI crystal guard counter was slightly lower than in the underground laboratory in Bern, where the cosmic ray flux is reduced by a factor of 11 compared to its surface value.

11.8.2 Guard counters and secondary radiation

We now discuss studies that have demonstrated that the main effect of the inner shield is to attenuate the component of secondary radiation, rather than to absorb gamma radiation from contamination outside this layer. One of the earliest indications of the combined effect of a guard counter system outside an inner shield was the study of Steinberg and Olsson (1968), discussed in Section 11.4.5. Figure 11.9 shows the system used. Both the passive and active shields of the 3"×3" NaI scintillation counter detector have already been described. The background spectrum (Figure 11.10) was registered with a multichannel analyzer, not only up to 3 MeV, as was usually done, but up to 40 MeV in order to capture also the broad peak (at 26 MeV) of direct hitting muons. The reduction of the muon peak in the anticoincidence spectrum shows that 90% of the muon pulses are intercepted by the guard system, reflecting the geometrical coverage of the guard. Below about 15 MeV the count rate rises with decreasing energy. This shows clearly the presence of secondary radiation. The coincidence pulses below 3 MeV come only from the secondary radiation. The attenuation of the 0.51 MeV annihilation peak in the anticoincidence spectrum shows that the guard counters eliminate about 75% of the secondary radiation. Less attention than deserved was paid to this important study. It demonstrated clearly the importance of the inner shield and paved the way for external guard counters, discussed in the next section.

Further evidence of the suppression of secondary radiation was found early in the 1970s in a comprehensive background study carried out at the ^{14}C dating laboratory in Trondheim, Norway (Gulliksen and Nydal,1979). The effect of an inner layer of varying thickness was studied, as well as the effect of adding a second and third guard counter. Figure 11.14 shows one of the system configurations used. A large annular gas proportional guard counter was constructed inside of which an up to 5.5 cm thick layer of lead could be inserted. Alternatively, additional guard counters of the same type could be placed inside it. The background was measured as a function of the thickness of the lead layer. The result is shown in Figure 11.15. With no lead, the background was 3.2 cpm but dropped to 0.69 cpm with a 5.3 cm thick inner lead layer. The background curve can be interpreted as showing: (1) the attenuation of gamma radiation from contamination outside the sample counter, (2) the attenuation of secondary radiation, or (3) a combination of both effects. A new interpretation of this curve is presented in the next section.

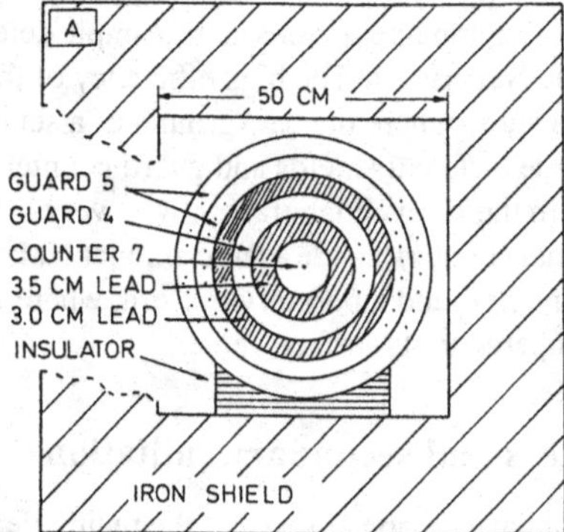

Fig. 11.14. A system with a large annular guard counter inside of which an up to 5.5 cm thick inner layer could be inserted (Nydal et al., 1980).

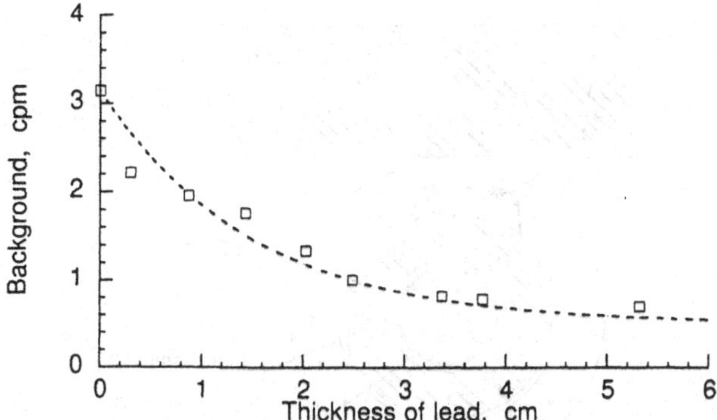

Fig. 11.15. Background of a gas proportional counter as a function of thickness of an inner lead layer (Nydal et al., 1980). The curve drawn is explained in the text.

Let us look at how the inner layer attenuates the background component of secondary radiation. Figure 11.16 illustrates this schematically. When the main shield is 10 cm of lead, the added inner layer insignificantly alters the gamma radiation the sample counter is exposed to. The layer will naturally attenuate the secondary radiation formed outside it, but this will be compensated for by new secondary radiation formed in the inner layer. This is illustrated by tracks A and B in Figure 11.16; the photon from muon A is absorbed in the inner layer, but muon B produces a compensating photon in the layer. When muon C passes through the guard counter, it gives a pulse there. It then creates a photon/electron shower in the inner lead layer, and one of its photons or electrons give a background pulse in the sample detector. The two pulses, in the guard and sample counter, are in coincidence, and this event will therefore be eliminated from the counting channel by the guard veto signal. If the secondary particle is a neutron (track D), the sample counter pulse may be delayed and therefore not eliminated.

As the inner shield serves mainly to reduce the secondary radiation, it should be made as thick as possible. Maximum reduction of the secondary radiation is then obtained when all the absorbing mass is placed inside the guard counter. This is achieved by covering all six sides of the shield with large, flat guard counters, here called *external guard counters*. Practically no charged particles can then enter the shield without being registered.

11.8.3 External guard counters

Nearly all background pulses of modern low-level germanium spectrometers in surface laboratories come from secondary radiation. The background drops to 1–2% when the systems are moved to deep underground laboratories, where the cosmic radiation is reduced to an insignificant level (10^{-3} or more). When these systems are shielded on all sides by flat guard counters (external guard), as discussed in section 11.8.1, the background is reduced

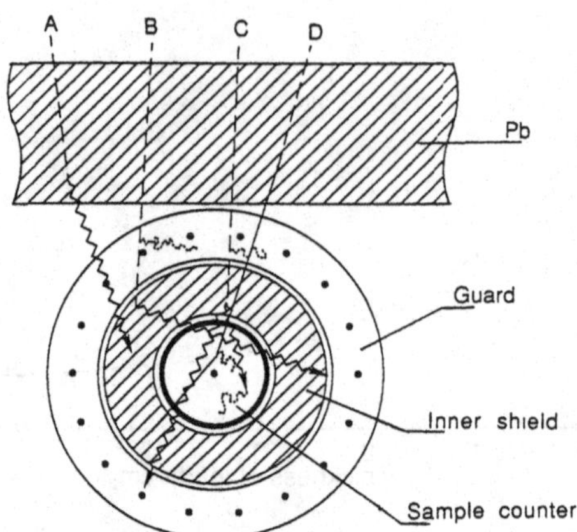

Fig. 11.16. Illustration of events in main shield, guard counter, inner shield and sample detector. The events are explained in the text.

by a factor of about 6 in systems with small overburden (to $0.17 \times B$), and a factor of about 15 in systems in laboratories with an overburden of 10–20 mwe. The ratio f_r of the residual background to total background thus depends on the overburden. What is the source of the residual fraction f_r of the background and why this difference with overburden? The guard counters detect practically all charged cosmic ray particles entering the shield, and all prompt background pulses are eliminated in the anticoincidence channel. Two explanations seem to be possible:

1. Some of the neutrons, produced in the shield, may give delayed pulses, that are not suppressed by the veto signal.

2. The pulses come from external neutrons, diffusing into the shield.

Neither explanation seems likely to be correct. The second is improbable as some studies have shown that covering the outer sides of the shield with boronated paraffin, which absorbs all external neutrons, seems to have no effect on the background. Further studies are needed here.

We can now calculate the background reduction as a function of the thickness of the inner shield if we know how effectively it absorbs the secondary radiation from the main shield. Monte Carlo calculations, discussed in Section 12.1.4, show that the attenuation thickness of secondary radiation in lead is 2.0 cm. If we assume that a fraction f_r of the sample counter pulses that originate from events in the inner shield are not eliminated by the veto signal in systems in surface laboratories, the background B is given by

$$B(x) = B(0)\,\{f_r + (1 - f_r\,e^{-x/2})\} \tag{11.9}$$

where $B(0)$ is the background without an inner layer. The results of the background measurements shown in Figure 11.15, discussed in the last section, can now be described quite well by this equation by assuming a value of 3.0 cpm for $B(0)$ and using the value 0.17, given above, for f_r. The equation thus describes the attenuation of the secondary radiation, as well as the residual background. This strongly indicates that practically all the background pulses of this counter come from secondary radiation, i.e., that the contribution of radiocontamination inside the shield is insignificant.

11.9 Background reduction through pulse shape analysis

Considerable background reduction can sometimes be obtained when particles, usually of a different type, give pulses of similar size but of different shape. Two such cases will be discussed briefly here, but will be described in more detail in later chapters.

11.9.1 Pulse rise analysis in proportional counters

When tritium is measured in a gas proportional counter filled with methane to 3 atm, the maximum range of the beta particles (E_{max} = 18 keV) is less than 2 mm. The electrons set free in the ionization process drift towards the positive anode and enter the gas multiplication volume close to the anode, almost simultaneously. This gives a pulse with a fast rise time. A relativistic electron or muon with a track length of about 5 cm in the counter gas will give a pulse of similar size as that from tritium, but its ionization will be spread over a much larger range of distance from the anode. The electrons will therefore reach the anode with a considerable spread in arrival time, giving a more slowly rising pulse. This can be effectively used to separate tritium pulses from background pulses.

The same technique can also be used with ^{14}C, although the difference in spread of drift time is much smaller. A typical gas proportional counter used for radiocarbon dating has a diameter of about 6 cm and a working pressure of 3 atm of CO_2. Most of the tracks of the beta particles will be less than 2 cm and this will give faster rising pulses than the majority of background pulses. This can be used to obtain considerable background reduction. However, classification of some slowly rising ^{14}C pulses as background pulses (Mäntynen et al., 1987) cannot be avoided. This will be discussed in more detail in Section 13.2.

11.9.2 Pulse decay analysis in scintillation detectors

In most organic scintillators, the fluorecence decay time for the main part of the excitation energy is very short, from 3 to 5 nanoseconds (prompt or fast decay time). Some of the molecules are, however, in an excitation state with a much longer decay time, from 200 to 400 nanoseconds. The intensity of the slow-decay component is much lower than that of the prompt one. This difference is important because the ratio of the two components depends on the type of exciting particle, or rather the density of ionization, as high density favours excitation of states with slow flouorescense. The difference in decay time can be used to separate pulses from different types of particles — electrons, protons, and alpha particles — both in solid and liquid organic scintillators. This technique is important for discriminating alpha and beta pulses in liquid scintillation counters, discussed further in Section 12.7.

11.10 Background of alpha detectors

The source of background in alpha particle detectors is always close to the detector because of the short range of the alpha particles. The background source is radiocontamination: (1) in the detector material, (2) in materials facing the detector, (3) radiocontamination deposited on the detector window, and (4) radon in the air surrounding the detector.

Alpha counting systems can be produced with very low background. The problem is to maintain this low background during years of use, as the samples that are measured in the system are a potential source of radioactive contamination. The background of alpha detecting systems will be discussed in Sections 12.5, 12.6 and 12.7.

Recommended reading

- Heusser, G., Low-radioactivity background technique, Ann. Rev. Nucl. Part. Sci., 1995, 45, 543-590.

Chapter 12

Germanium spectrometers

12.1 Low-level germanium spectrometers

The last 15 years of development of ultralow-level germanium spectrometers has brought us commercial systems that are completely free of primordial and man made radioactivity, at least in quantities of any significance, except in systems operating deep underground. In surface and near surface laboratories (in which we are mainly interested here), the background pulses come predominantly from cosmic radiation: its primary particles and the secondary radiation they induce in the shield and materials inside it.

The cosmic-ray pulses are either prompt or delayed. The prompt pulses are caused by (a) direct interaction of a muon or a proton, or (b) by their secondary particles, which give a pulse that is in coincidence with the muon/proton. These pulses can be eliminated by an external guard counting system, but the delayed pulses cannot.

The background pulse-height spectra of germanium detectors give detailed information because of their high energy resolution. Studies with low-level germanium spectrometers have provided information which can help us in better understanding the background of other types of low-level beta/gamma systems, and will in the future, hopefully, bring us better systems of various types, both of simpler construction and with lower background.

When structural details of a germanium system are given, it is, in principle, possible to account for all the main features of the spectrum we measure:

(a) the intensity of the various neutron peaks,

(b) the size of the 511 keV peak,

(c) the shape of the continuum,

(d) the total count rate in a wide energy window and

(e) the spectrum above 3 MeV (rarely studied).

We still have a good deal to learn before we can give a detailed quantitative analysis. When we can do this for germanium spectrometers, it will also be easier to analyze the spectra of other types of low-level beta/gamma systems. These problems will be discussed in this chapter.

12.2 Development of germanium detectors and systems

The first lithium ion drifted germanium diodes (denoted Ge(Li)), with a volume well below one cm^3, were made in 1963. This new detector represented a revolution in gamma spectroscopy as it offered an energy resolution up to 50 times better than NaI scintillation detectors. As the photoelectric efficiency of germanium is much lower than that of NaI, because of its lower atomic number, large germanium detectors were needed. With increasing experience, diodes over 100 cm^3 could in the 1970s be produced with the lithium drifting technique. They suffered from the disadvantage that they had to be constantly kept at low temperature in order to conserve the compensating spatial distribution of lithium.

The technique was greatly improved in the mid 1970s when the manufacturers succeeded in producing very pure germanium, where intrinsic conductivity dominates that of the remaining impurities, even at the temperature of liquid nitrogen. This made lithium doping superfluous. The new diodes did not need cooling during storage and they could be produced in larger sizes without impairing their energy resolution, which is typically 1.8 keV FWHM at a gamma energy of 1.33 MeV. Their size is indirectly given by their *relative efficiency*, which is the peak detection efficiency of the 1.33 MeV gamma line of ^{60}Co compared to that of a 3"×3" NaI crystal, both measured at a distance of 25 cm from the source. The mass or volume of the diodes is, however, usually more significant in background studies. The relative efficiency can easily be checked by the user, whereas it is more difficult to determine the size of the diode, which is incapsulated in a metal holder. The relative efficiency is approximately 5 times its volume in cm^3. Presently (early 1996), diodes with a relative efficiency of 35-45% (mass about 1.0 kg) are most common in new systems, but larger diodes are available, with efficiencies above 100%. The large crystals have practically the same energy resolution as the smaller ones.

The high purity germanium diodes are operated at a reverse high voltage that secures full depletion of the germanium crystal. Applying higher voltage increases the electric field, and the drift velocity of the electrons and holes formed by the ionizing particles. This minimizes drift time, carrier recombination and trapping. A high voltage about 3 times that required to fully saturate hole velocity is usually applied.

The germanium diodes come in different configurations as dicussed in Section 8.5.2 and shown in Figure 8.15. The most common is the *coaxial* type. It is made of p-type high purity germanium and the n+ contact is in the form of 0.5–1.0 mm thick diffused lithium layer on the outer surface. The axial core of the crystal is removed and a metallic contact made on the inner surface. The outer dead layer absorbs significant amounts of gamma radiation and X-rays with an energy below about 60 keV. The low energy range can be extended by starting with an n-type germanium crystal and adding a very thin, about 0.3 micrometers, outer contact layer by boron ion implantation. These diodes have a thin beryllium entrance window.

For maximum detection efficiency with small samples, germanium well detectors are sometimes used. The diameter of the well is 10–15 mm, sample volume of 1–2 cm^3. The detection efficiency is above 90% in the energy range of about 50 to 150 keV, and is still about 15% at 1 MeV.

Today, coaxial diodes with a relative efficiency of over 100% are produced, but they are expensive. Their advantage is twofold. Diodes of larger mass intercept and absorb a larger fraction of the emitted photons from the sample, and a larger fraction of the Compton scattered photons are absorbed in the crystal, increasing the height of the photopeak compared to the Compton continuum. This increases the detection efficiency and reduces the Compton background of peaks at energies above that being measured. The *peak to Compton ratio* of germanium detectors is therefore an important parameter. It is defined as the ratio of the number of counts in the highest photopeak channel to the mean number of counts per channel in the continuum, taken in the flat part of the spectrum below the Compton edge. This is usually specified for the 1.33 MeV line of ^{60}Co. The ratio depends on the relative efficiency of the diode as shown in Figure 12.1.

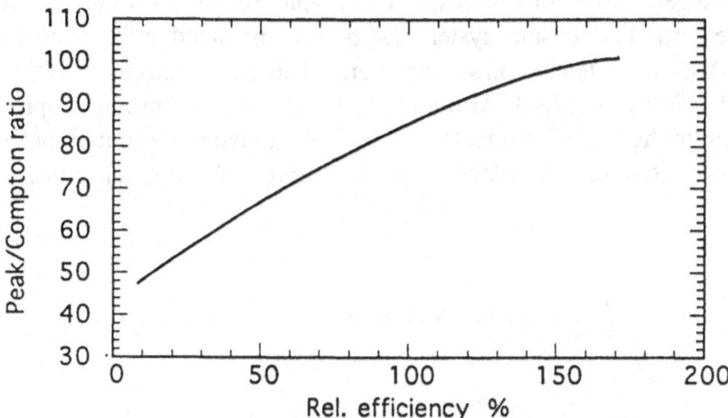

Fig. 12.1. The peak/Compton ratio versus diode volume.

12.3 Development of ultralow-level Ge-spectrometers

A new generation of low-level germanium spectrometers has emerged from the development of systems primarily intended for the study of double beta-decay, a process of considerable theoretical importance. Theory indicated that the isotope ^{76}Ge (relative abundance 7.8%) might decay through simultaneous emission of two beta particles, having a very long half-life. This presented a lucky coincidence as the same material could be used as a source and detector. It was evident that a large diode with extremely low background was needed for this study.

Interest in this problem intensified as ever larger, high purity germanium diodes became available in the 1980s. Large projects were initiated at a number of laboratories for studies of this kind and resources were invested in them on a scale never before seen in low-level counting. One of the largest projects was carried out by a group at the Pacific Northwest Laboratory (PNL) in the state of Washington, USA. This work has been well documented and the discussion below is largely based on its results. The main effort in the early phase of this project was to identify the contaminants in the existing systems and eliminate them in new ones. This work was described in Section 10.1.

Although the lowest background of the new spectrometers was obtained in underground laboratories, it was also important to have the best possible systems in normal laboratories. The anticosmic counter technique, which had been so successful in radiocarbon dating and various other kinds of low-level work, was therefore applied to a new generation of germanium spectrometers, but with a very important modification: the guard counters, in the form of large flat boxes, were placed on the outer surface of the lead shield. These counters have already been described in Section 11.8.1.

Figure 12.2 illustrates the rewards obtained, showing schematically the background continuum of three generations of systems: (1) of a typical assembly when the project started, (2) after an (external) anticosmic system had been constructed and contaminated parts removed and (3) of a third generation system operated deep underground (4000 mwe), with an inner layer of 400-year-old lead (Arthur et al., 1988). The background improvement from the first system to the second is a factor of about 70, a factor of about 30 probably coming from the anticosmic shield and a factor of up to 2 from removing impurities.

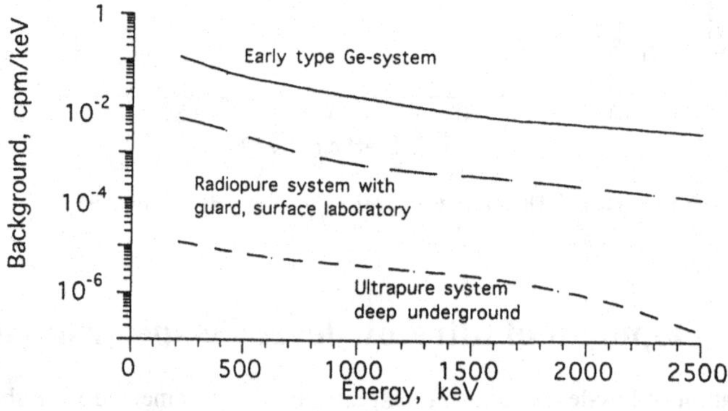

Fig. 12.2. The background continuum of three generations of low-level germanium spectrometers: (1) a typical 1980 system, (2) a radiopure system with external guard and (3) a third-generation system operated deep underground (4000 mwe).

12.4 General background features

The detailed information that the spectra of low-level germanium systems gives is the key to better understanding of the background of all low-level beta and gamma counting systems. The main features of the background spectra of the low-level germanium spectrometers in surface laboratories are:

1. A broad muon peak, at about 45 MeV in germanium diodes with a volume of 200 cm^3.
2. A continuum produced by electrons and photons which are secondary or tertiary radiation of the muons and protons, extending from the smallest pulse height to the largest.
3. An annihilation peak at 511 keV.
4. Gamma peaks produced by neutrons through a number of processes and in different materials.

The monoenergetic gamma radiation, which gives various peaks in the background spectrum, is induced by secondary neutrons, either fast of thermal, in the germanium diode itself or the shield or material inside it.

A general discussion of these background components is presented in the next three sections.

12.5 Muons and secondary radiation

The response of a 3"×3" NaI crystal to muons and their accompanying secondary electrons and photons was discussed in Section 11.2.2. The direct interaction of muons gives only a small contribution to the spectrum below 10 MeV, but above this energy the muon count rate increases slowly, forming a low, broad peak at 60 MeV as shown in Figure 11.3. Electrons and photons give pulses in all this range. Their contribution is small at the highest energies, but it begins to increase below 20 MeV and dominates at low energies. This secondary radiation is the main contribution to the energy region of the background spectrum we are most interested in, i.e., below 3 MeV.

Figure 12.3 depicts three high energy background spectra. Two are measured in a surface laboratory (overburden of 2 mwe), one without a shield (curve A) and the other behind 10 cm of lead (curve B). According to Equation (7.14) the muon flux in the surface laboratory is reduced by about 9% by the lead in the top layer of the shield. The count rate in the broad peak is reduced somewhat more, presumably as the secondary radiation also gives some contribution at these high energies, and this radiation is more affected by the lead than the muons. The direct hitting muons give a broad peak with a medium value at about 50 MeV. The third curve is measured in a shallow underground laboratory (overburden 15 mwe).

The response of germanium diodes is similar. The high energy pulses are from direct interacting muons and protons and their secondary radiation. Figure 12.3 depicts three background spectra, two taken in a surface laboratory with an overburden of 2 mwe (curves A and B) and the third curve (C) in a laboratory with an overburden of 15 mwe. The muon flux in the surface laboratory is reduced by about 9% by the lead at the top of the shield. The count rate in the broad peak is reduced somewhat more, as the secondary radiation still gives some contribution at the highest energies. This radiation is more affected by the lead than are the muons.

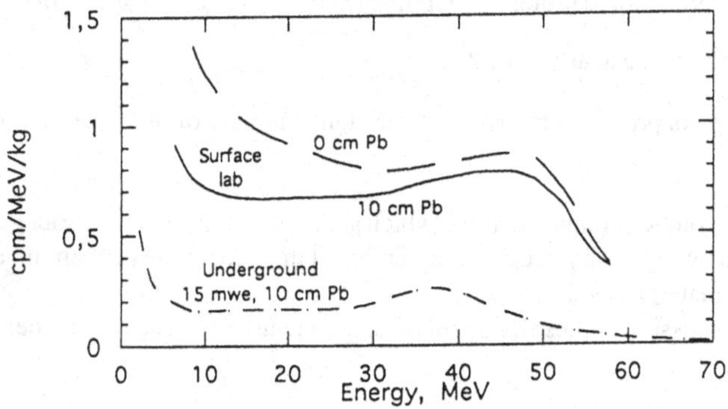

Fig. 12.3. High energy spectra of two 170 cm^3 diodes. Spectra A and B are taken in a surface laboratory, and spectrum C in a laboratory with an overburden of 15 mwe. A is without shield above, B and C inside shield. From Preusse 1993 and Heusser 1995.

Below about 30 MeV the difference between curves A and B increases with decreasing energy. Without a shielding layer above, the electrons and photons hitting the diode are secondary radiation induced in building materials with much higher critical energy and a lower absorption coefficient. The critical energy is 40 MeV for concrete, compared to 8 MeV for lead. Curve C is similar to curve B, but the count rate is lower as the flux there is only about 35% of that in the surface laboratory.

The secondary radiation we see clearly in these spectra comes mainly from electron/photon showers. Muons produce these through knock-on electrons as do protons through their secondary neutral pions. The relative contribution of the muon and proton components is not known, but that of muons is probably considerably larger in the surface laboratory. Overburden reduces the proton component much faster than the muon component.

As seen in Figure 12.3, the intensity of the secondary radiation depends markedly on the critical energy of the surrounding materials. If there is an inner shielding layer of material of low atomic number, such as copper or iron, the intensity of secondary radiation in the energy range 100–500 keV will be significantly increased. We will see examples of this in background spectra shown later in this chapter.

The 511 keV annihilation component comes from the recombination of positive electrons (from the electromagnetic showers) and negative electrons. The intensity of this peak therefore depends, like the secondary radiation, somewhat on the material of the inner shielding layers.

12.6 Gamma peaks

12.6.1 Sources and types of peaks

The main gamma peaks in the background spectra of modern low-level germanium spectrometers are induced by the cosmic radiation. Those from primordial and man-made radioactivity are usually very weak or entirely absent. In surface laboratories they are quite different from those measured in shallow underground laboratories, where the nucleonic component (mainly protons) has been fully absorbed. Generally, the peaks come from four different sources:

1. Neutrons interacting with germanium.
2. Neutrons interacting with the inner shielding layers.
3. Muons and protons producing secondary positrons giving the 511 keV annihilation peak.
4. Primordial and man-made radioactivity, absent in the best systems.

The neutron lines are relatively strong in surface systems but almost absent in systems having an overburden of 5 mew or more, where the nucleonic component (mainly protons) has been absorbed.

12.6.2 Energy and sources of gamma peaks

The cosmic ray gamma peaks seen in background spectra depend on the shield and the materials inside it. Table 12.1 lists the energies of the more prominent of these lines, their sources, half-lives and processes. The table shows also the sources of the frequent gamma ray peaks. Small peaks of primordial (mainly radon) and man-made radioactivity may be seen in the background spectra.

12.6.3 Neutron produced gamma peaks

Before discussing background spectra of low-level germanium detectors it is useful to look at their response to neutron irradiation, both to fast (Chasman et al., 1965, Lester and Smith, 1969, Bunting and Kraushaar, 1974) and slow neutrons (Chien and Chen, 1991, Chao and Chung, 1992). The neutrons react with all five stable isotopes of germanium as shown in Table 12.2.

Table 12.1. Cosmic ray produced gamma lines.

Energy, keV	Isotope source	Half-life	Process
53.4	^{73m}Ge	0.5 s	^{72}Ge(n,γ), ^{74}Ge(n,2n)
66.7	^{73m}Ge	0.5 s	^{72}Ge(n,γ), ^{74}Ge(n,2n)
72.8	Pb X-rays	Prompt	
74.8	-	Prompt	
84.2	-	Prompt	
87.4	-	Prompt	
109.9	^{19}F	Prompt	^{19}F(n,n)
139.5	^{75m}Ge	46 s	^{74}Ge(n,γ), ^{76}Ge(n,2n)
159.5	^{77m}Ge	53s	^{76}Ge(n,γ)
198.3	^{71m}Ge	21 ms	^{74}Ge(n,γ), ^{76}Ge(n,2n)
500.2		Prompt	^{72}Ge(n,g)^{73}Ge
511.0	Annihilation		
562.8	^{76}Ge	Prompt	^{76}Ge(n,n'γ)
595.8	^{74}Ge	Prompt	^{74}Ge(n,n'γ)
651.1	^{114}Cd	Prompt	^{113}Cd(n,γ)
669.6	^{63}Cu	Prompt	^{63}Cu(n,n'γ)
691.0	^{72}Ge	Prompt	^{72}Ge(n,n'γ)
770.6	^{65}Cu	Prompt	^{65}Cu(n,n'γ)
803.3	^{206}Pb	Prompt	^{206}Pb(n,n'γ)
846.8	^{56}Fe	78d	^{56}Fe(n,n'γ)
962.1	^{63}Cu	Prompt	^{63}Cu(n,n'γ)
1412.1	^{63}Cu	Prompt	^{63}Cu(n,n'γ)
2223.0	^{2}H	Prompt	^{1}H(n,γ)

Table 12.2. Isotopes of germanium and their excitation by neutrons.

Isotope	Abundance	Cross section for (n,γ), barn (a)	Daughter nucleus of (n,γ)	Half-life
^{70}Ge	20.5%	3.5	^{71m}Ge	20 ms
			^{71}Ge	11 d
^{72}Ge	27.4%	1	^{73m}Ge	0.53 s
			^{73}Ge	stable
^{73}Ge	7.8%	14	^{74}Ge	stable
^{74}Ge	36.5%	0.1	^{75m}Ge	49 s
		0.4	^{75}Ge	82 m
^{76}Ge	7.8%	0.1	^{77m}Ge	54 s
		0.4	^{77}Ge	11 h

(a) Approximate values, data from Chasman et al., 1965.

Fast neutron response. Each time a fast neutron collides inelastically with a germanium atom, the atom recoils with a kinetic energy which is distributed continuously from nearly zero energy, in grazing collision, to a maximum energy, E_{max}, in head-on collisions, given by

$$E_{\text{max}}/E_n = 1 - [(A-1)/(A+1)]^2 \tag{12.1}$$

where E_n is the energy of the neutron before the collision and the energy of the recoiling nucleus. The recoil energy spectrum will therefore give good information about the energy spectrum of the fast neutrons. This was discussed in Section 11.7.

Let us first look at the spectrum of a germanium diode exposed to fast neutrons produced by energetic deuterons in beryllium. The neutrons have somewhat higher energy than those from cosmic rays. The response of the germanium diode is shown in the pulse-height spectrum in Figure 12.4. The peaks at 563, 590, 691 and 868 MeV are recoil broadened. Although the neutrons from the ^{252}Cf source are fast, surrounding laboratory building materials partly thermalize the neutron flux, and weak, sharp thermal peaks at 500 and 596 keV (the latter superimposed on the recoil broadened peak).

Let us look more closely at the shape of these recoil broadened peaks, taking the 691 keV pack of ^{72}Ge as an example. When a neutron collides with the nucleus it can lose anywhere from a small fraction of its energy (in a grazing collision) up to a maximum of 5.2% in a head-on collision (see Section 11.7). Furthermore, the cosmic ray neutrons have a continuous energy distribution, that of evaporation neutrons, with a maximum at about 0.6 MeV and extending up to several MeV. The collisions therefore give a continuous energy spectrum. When neutrons collide inelastically with ^{72}Ge nuclei they are left in an excited state, but fall immediately to the ground state, emitting a photon with an energy of 691 keV. If this photon is absorbed photoelectrically in the germanium diode, the size of the output pulse is proportional to the sum of the energy of the photon and the ionized recoiling ^{72}Ge atom.

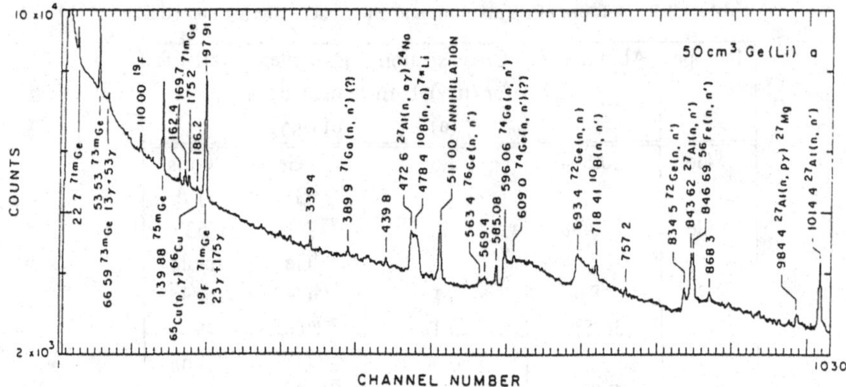

Fig. 12.4. Spectrum of a germanium diode irradiated by fast neutrons from a beryllium target bombarded by deuterons (Bunting et al., 1974).

The recoil energy spectrum can be found by subtracting 691 keV from the energy scale and subtracting its background value from the count rate. Figure 12.5 shows the energy spectrum of recoiling ^{72}Ge atoms after collisions with cosmic ray neutrons and fission neutrons from a ^{252}Cf source. The figure shows that the cosmic ray neutron energy spectrum is similar to that of fission neutrons from ^{252}Cf, but somewhat higher in energy.

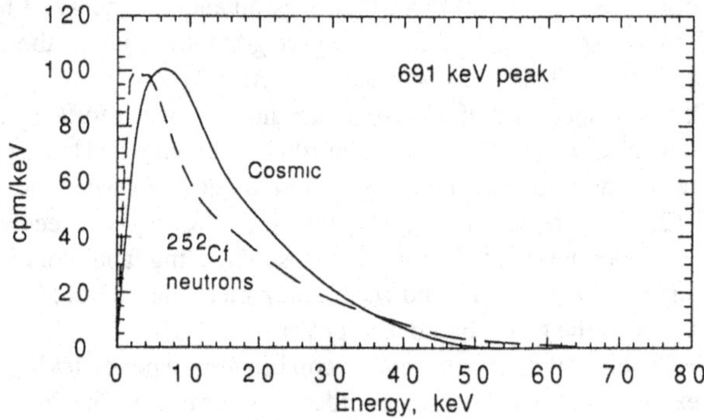

Fig. 12.5. Energy spectrum of recoiling ^{72}Ge atoms that have been hit by fission neutrons from a ^{252}Cf sources and cosmic ray neutrons.

Thermal neutron response. Chao (1992) and Chao and Chung (1992) studied the response of a germanium diode to thermal neutrons escaping from a graphite cylinder, 84 cm in diameter and 57 cm high, with a ^{252}Cf neutron source in its middle. Figure 12.6 shows the recorded spectrum. The 595.8 keV gamma line has the highest intensity. In this spectrum it has the shape of a sharp peak as it now comes from the capture of a slow neutron, or $^{73}Ge(n, \gamma)^{74}Ge$, but it is superimposed on a weak recoil broadened peak of fast neutrons, escaping through the graphite. Chao (1992) gives the following relative intensities for the main thermal peaks, compared to the 595.8 keV peak:

Peak, keV	139.5	198.3	500.2	595.8	868.0
Rel. Intensity, %	180	140	36	100	30

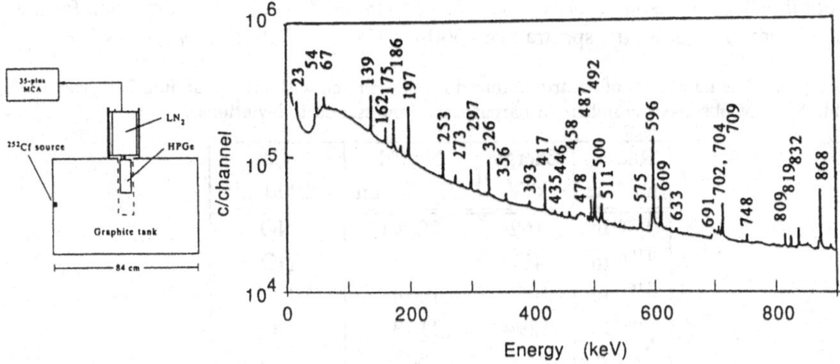

Fig. 12.6. Spectrum of a $60cm^3$ germanium detector exposed to thermal neutrons. (From Chao and Chung, 1992).

The peaks at 139.5 and 198.3 keV are usually ascribed to thermal neutron capture (Skoro et al., 1992), but an $(n, 2n)$ reaction is also sometimes given as a possible process (Heusser, 1993). The presence of these peaks in this spectrum of fast neutrons supports the latter explanation. This is further supported by the following three experiments:

1. If a thermal neutron component is a part of a predominating fast neutron flux, a sharp peak at 596 keV should be seen superimposed on the recoil broadened fast neutron peak. This is, however, not seen in the background spectra.

2. Roy et al. (1989) studied the influence of an outer cadmium layer on the gamma peaks of indium and germanium. A low-level system with a relatively small Ge-diode (73 cm^3) with a few grams of indium for making contact to the diode material was used in this study. Table 12.3 shows the main results. All the thermal indium lines vanish when the inner Cd layer is added. The 140 and 199 (or rather 139.5 and 198.3 keV) lines are little affected. This supports the hypothesis that the main part of these two peaks comes from fast neutrons when there is no thermalizing material inside the lead shield.

3. Neutrons produced in lead, both fast and thermal neutrons, were measured inside a shield of increasing thickness (Section 7.13). The relative values of both the fast and

thermal neutron fluxes were measured with a BF_3counter. This study showed that the thermal flux, as measured with a bare counter, was not influenced by the increasing flux of fast neutrons as the lead shield became thicker (Figure 7.19). For all thicknesses, the thermal flux was equal to that measured in the laboratory without a shield. The thermal neutrons evidently diffuse freely through the lead, as is to be expected, as there is little neutron absorption in the shield and the neutrons produced in the lead do not increase this flux.

For direct comparison of the response of germanium detectors to thermal and fast neutrons, Figure 12.7 shows the spectra from 500–900 keV for both energy groups.

Table 12.3. The intensities of neutron induced lines, with and without a cadmium foil covering the shield. NO: not observed. Numbers in parenthesis show standard deviations.

Radio-nuclide	Energy keV	No Cd c/1000 min	Cd c/1000 min
^{116m}In	162	20 (7)	NO
^{116m}In	417	17 (5)	NO
^{116m}In	1097	10 (4)	NO
^{116m}In	1294	11 (3)	NO
^{114}Cd	558	NO	
^{206}Pb	803	20 (4)	11 (4)
^{75m}Ge	140	36 (8)	26 (6)
^{71}Ge	199	37 (7)	37 (7)
^{74}Ge	596	68 (12)	112 (12)
^{74}Ge	691	111 (13)	100 (12)

12.7 Modern low-level Ge-spectrometers

Germanium spectrometers in surface laboratories. Based on the results of the aforementioned development work, modern low-level germanium spectrometers, which are practically free of primordial and man-made radiocontamination, are now produced. The system described below has been selected because its background spectrum has been measured both in a surface laboratory and deep underground, with an overburden of about 500 mwe, where the muon flux has been reduced by a factor of about 1000 (Mouchel et al. 1992, Wordel et al. 1994). The system is shown in Figure 12.8. The diode has a volume of 100 cm^3 (a relatively small diode by present standards), its relative efficiency is 20%, and the energy resolution is 1.9 keV FWMH at 1.33 MeV. Its counting efficiency for a cylindrical sample in a holder with an inner diameter of 32 mm and sample height of 10 mm (mass density of sample 1.0 g/cm^3) is shown in Figure 12.9.

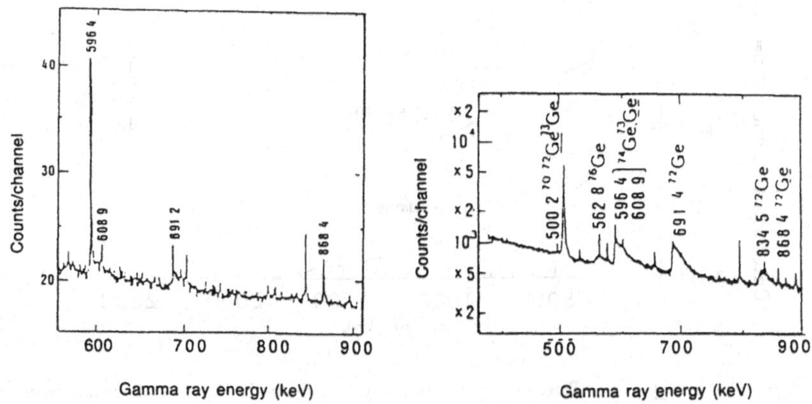

Fig. 12.7. The spectrum of germanium detectors irradiated by thermal neutrons (left) and fast neutrons (right).

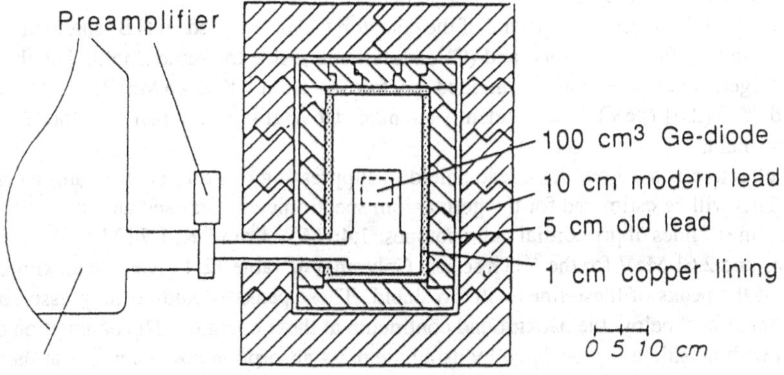

Fig. 12.8. Low-level germanium spectrometer with a 100 cm^3 germanium coaxial diode (relative efficiency 20%), energy resolution 1.9 keV at 1.33 MeV. From Mouchel and Wordel, 1991.

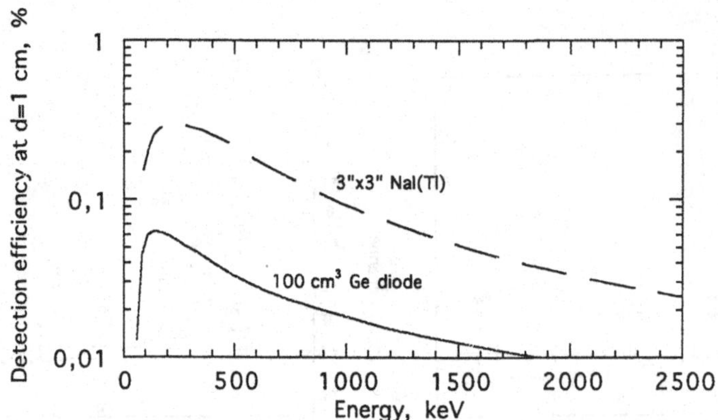

Fig. 12.9. Full energy peak efficiency of germanium system shown in Figure 12.8 with a cylindrical sample holder on the top of the diode (Wordel et al. 1994). For comparison the detection efficiency of a 3" × 3" NaI scintillation counter with a point source 1.0 cm above the crystal is also shown.

All parts of the system are made of carefully selected radiopure materials. The cryostat, detector holder, and end cap are made of oxygen-free high-conductivity copper. The detector is shielded by 15 cm of lead, the outer 10 centimeters having a ^{210}Pb concentration of 17 Bq/kg (Boliden lead), and the inner 5 centimeters about 1 Bq ^{210}Pb/kg (old lead). The preamplifier is placed outside the shield. It should be noted that any lead could have been used for the outer 10 cm layer, as discussed in Section 11.4.3.

Figure 12.10 shows the background of the detector, (A) unshielded and (B) shielded (15 cm of lead) in a surface laboratory, and (C) underground inside the same shield. The three intense energetic lines seen in the unshielded spectrum — from ^{40}K (1.46 MeV), ^{214}Bi (1.76 MeV) and ^{208}Tl (2.61 MeV) — are evidently attenuated to negligible intensities in the 15 cm thick lead shield.

How effective is a 10 cm thick lead shield in suppressing external natural gamma radiation? This will be estimated for the germanium spectrometer discussed above for three energetic gamma lines of primordial radioisotopes: 1.46 MeV from ^{40}K, 1.76 MeV from the ^{238}U series and 2.61 MeV for the ^{232}Th series. Column 3 in Table 12.4 gives the maximum intensity of the peaks of these lines with no shield. These peaks should be suppressed by the lead shield well below the background continuum at these energies. The attenuation of the peaks with no shield is calculated by using the mass absorption coefficient, μ, at these three energies (column 2) for photons impinging vertically on 10 and 15 cm thick layers of lead. The resulting count rates are given in columns 4 and 5. Finally, column 6 shows the measured continuum count rate at these energies as measured with a 15 cm shield, which is only slightly lower than behind 10 cm of lead. The table shows that all three lines would give a significant peak with 10 cm of lead, but their background contribution is insignificant at 15 cm.

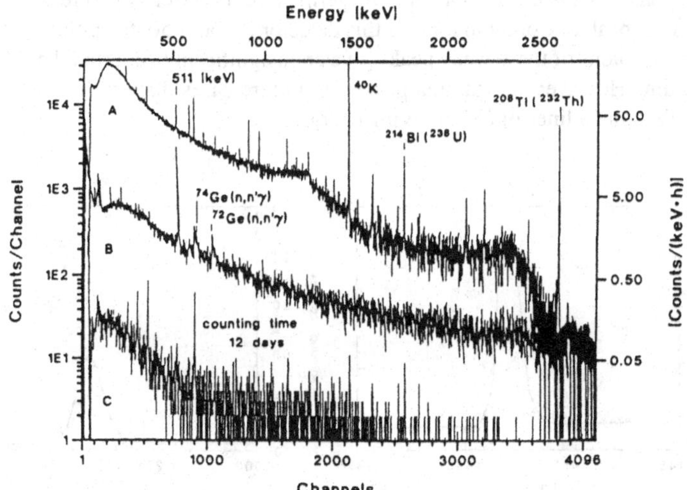

Fig. 12.10. Background of a low-level 100 cm^3 germanium spectrometer without shield (top), behind 15 cm thick shielding in a surface laboratory with counting time 24 days (middle), and behind 15 cm thick shielding with an overburden of about 500 mwe, counting time 12 days (bottom) (Wordel and Mouchel, 1992).

Table 12.4. Penetration of primordial gamma lines in a low-level germanium spectrometer.

Line, MeV	μ, cm^2/g	0 cm Pb cpd/keV	10 cm Pb cpd/keV	15 cm Pb cpd/keV	Continuum cpd/keV
1.46	0.0538	4940	11.3	0.5	6.7
1.76	0.0480	320	1.4	0.1	5.4
2.61	0.0422	810	6.9	0.6	2.4

Figure 12.11 shows this penetration through a 10 cm thick lead shield for 1.46 and 2.61 MeV gamma photons in a system with a 178 cm^3 germanium detector with a relative efficiency of 39%. As the peak to Compton ratio of this detector is about 50, the pulses of the Compton scattered photons of these weak peaks give an insignificant background contribution in the continuum. However, the gamma peaks can in rare cases disturb the measurement of isotopes with gamma lines of closely lying energies.

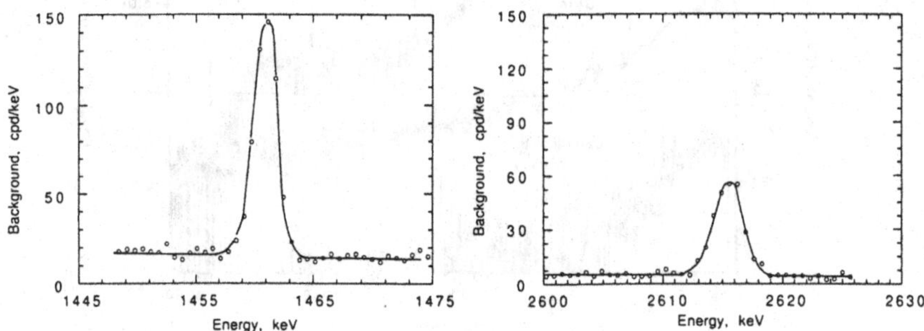

Fig. 12.11. The penetration of primordial gamma lines through 10 cm of lead in a germanium spectrometer.

Germanium spectrometers in a shallow underground laboratory. The background spectrum of a system in Heidelberg, operating underground at an overburden of 15 mwe, is shown in Figure 12.12.

The muon flux in this laboratory is about 35% of the flux above ground and the neutron flux about 50 times lower (Heusser 1995). Actually it does not take much overburden to reduce the neutron line intensities to insignificant values. The strongest neutron lines in systems in surface laboratories may rise 50–80% above the background continuum. If the neutron flux is reduced by a factor of about 10 the neutron peaks will vanish below the background continuum. An overburden of about 4 mwe, corresponding to an earth layer of about 2 meters, gives this reduction. This does not seem much, but such laboratories are rather scarce. There is no trace of neutron peaks in the background spectrum, only lines induced by muons: the 511 keV annihilation line and the lead X-ray lines. However, when the shield had a 2 cm thick inner layer of iron a weak fast neutron iron peak at 843 keV was seen. The large reduction in background obtained by using guard counters proves the cosmic-ray origin of a predominant part of the spectrum.

12.8 Effect of inner shielding layer

Let us look at how layers inside the lead shield, for example of copper or mercury, affect the background spectrum: (a) the continuum, (b) the 511 keV annihilation peak and (c) the

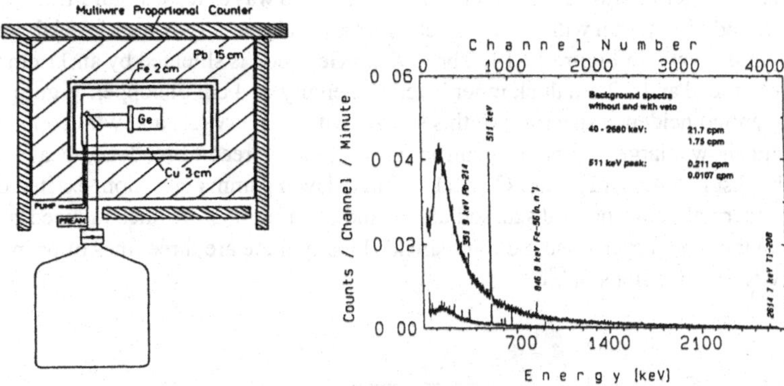

Fig. 12.12. A low-level germanium spectrometer shielded with six external guard counters and its background spectrum. The laboratory has an overburden of 15 mwe (Heusser 1995).

neutron induced gamma peaks. We look first at the background continuum and the 511 keV peak. These depend on the atomic number of the inner shield, as discussed in Section 12.6. Figure 12.13 depicts two spectra measured in the Heidelberg underground laboratory, one inside a 15 cm thick layer of lead and the other with a 2 cm thick inner layer of iron. The continuum peak at about 160 keV is 3.3 times higher with an iron inner layer (2 cm thick) compared to lead. The 511 keV is 1.5 higher for lead than iron because of higher pair production.

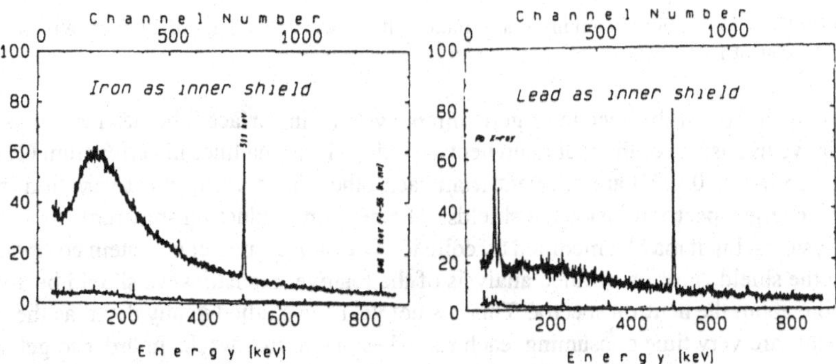

Fig. 12.13. Two spectra measured in the Heidelberg underground laboratory showing the effect of inner shielding layer on the continuum and the 511 keV peak.

We then look at neutron lines. Background curve B in Figure 12.10 shows only neutron peaks induced in germanium as the inner layer of the shield was of lead. Figure 12.15 shows the background of a system with a 2-cm-thick inner layer of mercury in a surface laboratory. This system is shown in Figure 12.14. The 0.93 kg Ge-diode is shielded by an 11 cm thick outer layer of lead and a 2 cm thick inner layer of mercury or 4 cm of copper. Figure 12.15 shows the anticoincidence spectrum of this system with mercury (A) and with copper (B). The spectra show a large number of neutron lines: (1) some recoil broadened peaks of Ge induced by fast neutrons, (2) sharp Ge peaks from delayed gamma emission, excited either by fast or thermal neutrons as discussed above and (3) a number of lines induced by fast neutrons in the inner copper and mercury shield. Finally there are three lines of primordial radioactivity and two lines of ^{60}Co.

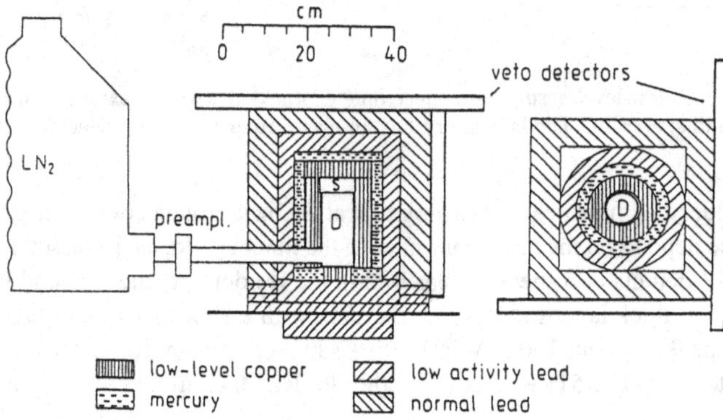

Fig. 12.14. A low-level germanium spectrometer with three external guard counters and an inner shielding layer of mercury.

A comparison of the spectra of germanium systems in surface laboratories shows that the relative intensities of the four main neutron induced gamma lines in germanium (139.5–198.3–595.8–691.0 keV) are different from each other. This must reflect variation in the neutron energy spectrum inside the shields. The neutron production spectrum is the same in all systems, but it may be modified by collision with atomic nuclei in system components and in the shield. From a detailed analysis of the relative intensities we should be able to say more about the neutron spectra. This has not yet been studied in any detail as the measurements are very time consuming, each run takes one to two weeks in order to get sufficient statistical accuracy. Table 12.5 gives interesting information about the background of a low-level germanium spectrometer in a surface laboratory with different inner shields.

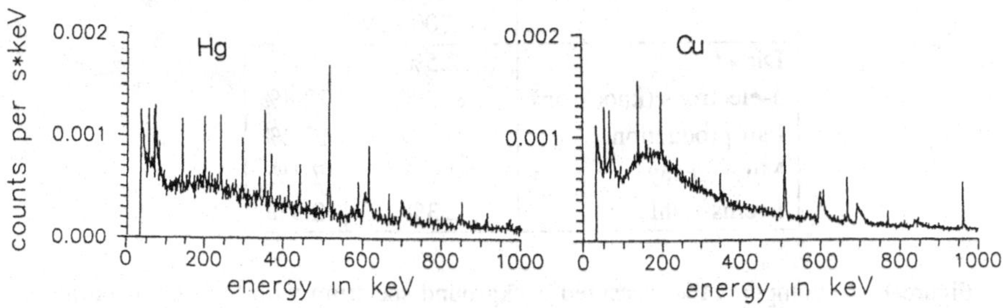

Fig. 12.15. The anticoincidence spectrum of a system with an inner shielding layer of mercury (A) and copper (B).

Table 12.5. Background components in cpm with different inner shielding layers (Lindstrom, personal communication).

	10 cm Pb	10 Pb+6 Cu Teflon	15 Pb+6 Cu Teflon
40-2.7 MeV	90	109	105
0.511 MeV	1.34	0.66	0.64
0.691 MeV	0.19	0.075	0.058
2.7-3.0 MeV	1.47	1.08	1.01
0.139 MeV	0.035	0.125	0.045

12.9 Monte Carlo calculated muon spectrum

Vojtyla (1995) has calculated the pulse height spectrum from muon reactions in a low-level Ge-system using the Monte Carlo method. He used a sophisticated program, GEANT, developed at CERN. His calculations give a deep insight into the processes that are the source of the muon background. According to Vojtyla the relative contribution of different muon interaction mechanisms in the induced background are as shown in Table 12.6.

Table 12.6. Muon processes contributing to the secondary radiation background component of a low-level germanium spectrometer, calculated by the Monte Carlo method (Vojtyla, 1995).

Mechanism	Continuum 50-2700 keV	511
Direct	2.5%	
δ-electrons (knock-on)	81.2%	72.8%
Pair production	10.0%	17.2%
Muon decay	4.0%	7.1%
Bremsstrahlung	2.3%	2.9%

Figure 12.16 compares the measured background spectrum with the calculated muon induced spectrum. The shape of the spectrum, the total background count rate as well as the 511 keV peak found from these calculations are in good agreement with measured values.

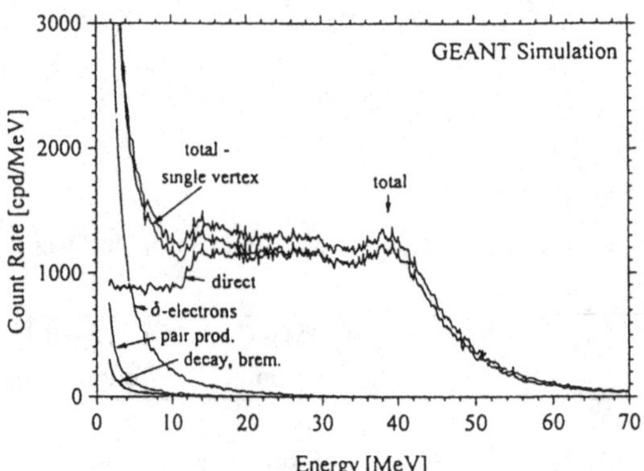

Fig. 12.16. Comparison of a measured background spectrum with a calculated muon induced spectrum.

These calculations also show where the individual events that give a background pulse occur in the lead. Figure 12.17 shows that the background contribution decreases exponentially with the depth from the inner surface of the lead, with an attenuation thickness

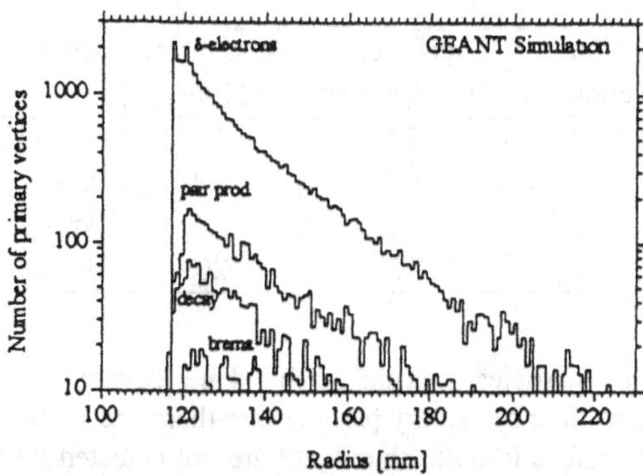

Fig. 12.17. Number of primary vertices (points of muon interaction) as a function of distance from the vertical axis of the diode. The inner diameter of the shield is 10 cm.

of 2.0 cm, corresponding to an attenuation factor of 0.044 cm^2/g. The radiation length in lead is 0.56 cm. The attenuation factor is the same as that for photons in lead in the broad minimum at about 3 MeV.

12.10 Anticoincidence background reduction

The anticosmic arrangement gives high background reduction and will probably be used increasingly in coming years. The guard counters are either large flat gas proportional or scintillation counters on the outer sides of the shield (external guard counters). The veto blocking time is usually kept short in order to minimize dead time, about 10 μs. Increasing the dead time gives only a small further background reduction (Heusser, 1994). It is best to have guard counters on all six sides of the shield, but for economic reasons the shield is only partly covered in some systems. The background reduction is characterized by the *veto reduction factor*, which is the ratio of the total count rate divided by the anticoincidence count rate in a broad window, from 100–1500 keV. In the Heidelberg system with its 6 guard counters, the top counter accounts for 76% of the total effect, the bottom counter for only 0.6%.

Systems with external guard counters are still rare, most of them recent, and information about them limited. The background reduction is considerably smaller in systems in surface laboratories. The veto reduction factor of four of these are compared in Table 12.7.

The background of good germanium spectrometers deep underground is only 1–2% of its surface value. It is therefore evident that the residual anticoincidence background count rate of a surface system, where the guard has reduced the background by a factor of 5–10, is still predominantly caused by the cosmic radiation. What is the source of this count rate and why is this difference in veto reduction factor between surface and shallow underground

Table 12.7. Comparison of veto factor of different systems.

Overburden mwe	Guard counters	Veto factor 50–2500 keV	Veto factor 511 keV	Reference
1.5	6	5.2	10	Pointurier, 1996
2.0	3	2.4	2.8	Preusse, 1993
3.2	1	2.2	2.4	Vojtyla et al., 1994
15	6	13	20	Heusser, 1995

laboratories? All charged particles entering the shield are registered by the guard counters, and eventual pulses from their secondary particles are therefore suppressed. Nut neutrons, both fast and thermal, diffuse into the shield and are not detected by the guard counters. Systems at NWL had a 10 cm thick external guard counter on all sides with a plastic scintillator, and inside these there was a cadmium foil. This arrangement strongly attenuates external neutrons, both fast and thermal. This does not seem to attenuate the neutron lines in the germanium detectors. Delayed pulses of secondary radiation seem most plausible, but even this explanation does not seem to fit facts well enough. Neutrons can give delayed secondary gamma radiation, but at 15 mwe the neutron flux is probably too low to make this hypothesis acceptable. Further studies are needed here.

12.11 Background versus mass of germanium diodes

Modern germanium diodes are practically free of radiocontamination and the background pulses come predominantly from cosmic rays. We would therefore expect that all low-level germanium spectrometers, operating under similar overburden, would give approximately the same background count rate per mass. Large diodes might, however, give somewhat lower values because of self-shielding effects. The background count rate in the energy window 50–1500 keV for a number of systems has been collected from the literature and through personal communication. The result is shown in Figure 12.18. Also shown are the 511 keV line and the 596 keV peak. All systems are in normal buildings, but there may be a small difference in the overburden. The figure shows that the wide window background is similar for all diodes, with a tendency of decreasing values for larger diodes. There is, as to be expected, some scattering in the 511 and 596 peaks, both of which are sensitive to differences in the inner shielding layers.

12.12 Future prospects

No sector of low-level counting has contributed so much to the technique of measuring weak radioactivity as germanium spectrometry. In spite of the major progress made in the last 10 years, various details of their background spectra are still not understood well enough and further studies are therefore needed. The most important contribution could probably come

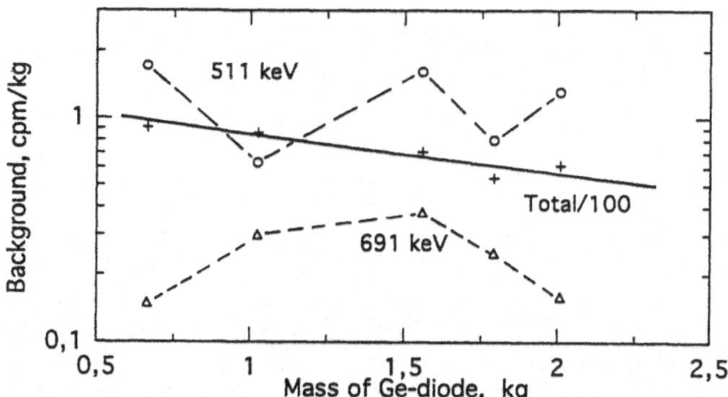

Fig. 12.18. The specific count rate (cpm/kg) of low-level germanium spectrometers with diodes of different mass: in the 50–1500 keV window and the 511 and 591 keV lines

from a portable system with external guard counters with which the background could be measured at different overburdens and various special studies made. This would in itself not bring us direct improvements in these systems, but would strengthen the basis of these systems and the low-level counting technique.

Low-level germanium spectrometers used for general work can be improved considerably by adding external guard counters. A trade-off between lower background and price and complexity will probably be made by having a guard counter only at the top of the shield. The attenuation thickness of secondary radiation in lead is 2.0 cm. A 5-cm layer of lead below a guard counter would therefore reduce the secondary radiation by a factor of 12. It would probably give the best results to have 5 cm of lead above the top guard counter and 5 cm below it, as this would increase the geometrical coverage leading to lower background. Samples would then be changed from the side.

Either a guard counter of scintillating acryl or a gas proportional counter of the flow type would be used. The latter type is simpler and less expensive and would present a better solution if a good design can be developed. Sandwiching the counter between two layers of lead would decrease its background, reducing the dead time. This arrangement would probably decrease the background by a factor of 3.0 to 3.5.

Recommended reading

- Low Level Gamma Spectroscopy, Canberra.
- Technique and problems of low-level gamma-ray spectrometry, W. Westmeier, Int. Radiat. Isot., 43, 1992, p. 305–322.

Chapter 13

Gas proportional counter systems

13.1 Gas proportional counters

Low-level gas proportional counters are of two types: (a) for internal gaseous samples (e.g., tritium, ^{14}C and ^{39}Ar) and (b) for external solid samples (e.g., ^{90}Y, ^{99}Tc and ^{32}P). The latter are often also called window counters, as they have a thin window so that alpha and beta particles from the sample can penetrate into the counter's active volume with minimum loss. In the following sections, the discussion will focus primarily on internal counters. Window counters will be discussed in Section 13.3.

Internal gas proportional counting was introduced independently at a number of laboratories in 1952–1954. Important development work was carried out in Groningen (Vries and Barendsen, 1953). They used CO_2 as the counting gas and it became the preferred choice. Their system became the model for a large number of laboratories.

The main development period of the gas proportional counting technique was from 1953 to the mid 1960s, when liquid scintillation counters, mass produced for fast growing tracing studies and delivered ready for use, gave gas counting increasing competition in the radiocarbon dating field. After 1970, most new dating laboratories preferred liquid scintillation counters. This slowed down further development of the gas counter technique. The application of the AMS technique (Section 16.1) to radiocarbon dating in the early 1980s lessened still further interest in investing resources for improving gas counting. Although some improvements have been introduced in later years, progress in gas counting has been slow.

The intensive and highly successful development of ultralow-level germanium spectrometers, discussed in earlier chapters, has now brought us a better understanding of the background components of all types of low-level beta/gamma systems and opened up possibilities to considerably improve their performance. A renewed development effort can therefore bring us better gas proportional counting systems in the near future.

A new analysis of the background of gas counters is presented below. An equation that gives the total sensitive mass and makes it possible to quantify the dominating muon background component for counters of any size (at varying overburden and thickness of inner absorbing layer, i.e., the pure lead or mercury layer between the guard and sample counter) is derived. Furthermore, it is shown how background variations due to atmospheric pressure fluctuations and seasonal changes in the mean production height of muons can be used to estimate the background contribution of muons, neutrons and contamination. It is shown how the neutron background can be determined with acceptable accuracy by measuring the background with different types of gases.

Finally, the background of gas proportional counters is compared to that of germanium diodes on the basis of sensitive mass, i.e., in cpm per kg. This is of interest as we know that the background of the germanium detectors comes practically only from cosmic rays, muons and neutrons and they therefore set a standard for other types of beta/gamma detectors. This comparison gives interesting results, but needs further work.

13.2 Design of internal gas proportional counters

13.2.1 Conventional counters

The earliest proportional counters, before their use in radiocarbon dating, were primarily applied to the study of spectra of low energy beta emitters and X-rays. It was important in this work to have the same electric field gradient all along the anode wire in order to secure an essentially constant gas multiplication factor, even close to the ends of the anode. Many early radiocarbon counters followed this principle, which either made them more complex or increased the dead volume at their ends.

Modern gas proportional counters are of a relatively simple design. Dead volume should be minimized by having end walls of quartz, or of metal shaped in such a way as to avoid regions of high electric field. Both types work well and give a total detection efficiency of 85–90%, even up to 95%. Their construction is quite simple. A counter made by Robinson (1976), shown in Figure 13.1, is an example of a solid, simple construction. The tube and end pieces are made of electrolytic copper. As the counter was made for a negative high voltage applied to the cathode, the gas inlet is through a quartz tube. The coupling capacitor for picking up the pulses is at ground potential, minimizing the risk of spurious pulses. The anode is of stainless steel or tungsten, with a diameter of 25 or 50 μm. The parts are assembled with epoxy cement. CO_2 is very sensitive to electronegative impurities. The surface of the cement facing the gas must be kept as small as possible in order to minimize outgassing. Hedberg, at the dating laboratory in Stockholm (personal communication), minimizes the epoxy surface by using end pieces of stainless steel, slightly larger than the inner diameter of the copper tube. Before assembly they are cooled in liquid nitrogen and then pushed into the copper tube. The stainless steel gas inlet is then pressed into the end piece. This gives a very solid construction. Assembling a counter takes less than one day.

Gas proportional counters have been made of copper, stainless steel, brass and quartz. A comparison of the background of a large number of counters, discussed in section 13.6, has shown that all materials give a similar background. For many years quartz was considered the purest available material for the counters. It was widely used in spite of its unfavorable mechanical properties and the fact that its inner surface had to be coated with a thin metallic

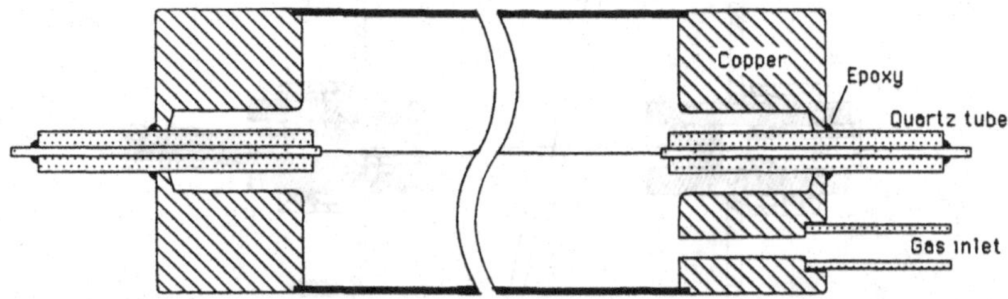

Fig. 13.1. A typical gas proportional counter for radiocarbon dating. From Robinson, 1976.

layer in order to make it conducting. Today we know, as discussed below, that copper, brass and steel are just as suitable.

13.2.2 Oeschger counters

In typical gas proportional counters, 70–80% of the background pulses come from electrons released by gamma radiation from the counter's wall (Section 13.8). The background can therefore be considerably reduced if there is only a thin wall between the sample counter and the annular guard counter (Houtermans and Oeschger, 1958). The design of these counters, usually called Oeschger counters, is shown in Figure 13.2. In the original counter of Houterman and Oeschger, the wall was made of metallized polystyrol film with a thickness of 6.9 mg/cm^2. The annular guard counter space has 21 equidistantly spaced anode wires. As it would be difficult to fill the inner counter separately, both counter elements almost always share the sample gas. The inner part has a volume of 1.5 L and the annular space 0.90 L. The sample gas fills both parts so that only 62% of the sample is actually being measured. In later versions, the volume of the guard element is usually 25–30% of the total volume. W.R. Shell (1970) has given a good description of the construction of these counters and their applications.

Measurements with an external ^{60}Co gamma source showed that the central counter had a sensitivity of only 20% compared to a counter of the same size with a thick wall. This reduction factor depends on the energy of the gamma radiation, falling with decreased energy. We can expect a similar reduction in the background count rate of Oeschger counters.

This arrangement is only practical for low-energy beta-active samples: tritium and ^{14}C. The background gamma radiation releases electrons predominantly from the outer metal wall, much less frequently from the thin inner wall. The electrons coming from the outer wall must first penetrate one of the annular guard counter elements, before entering the sample detector, giving simultaneous pulses in these elements. The pulse in the sample counter

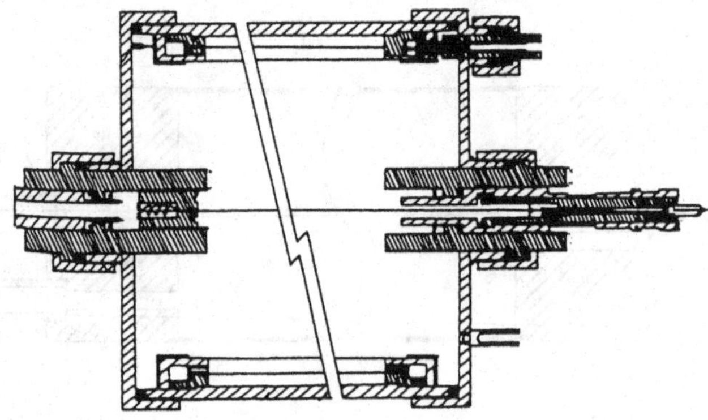

Fig. 13.2. An Oeschger counter with only a thin foil between the sample counter and the annular guard counter. (From Povinec, 1978).

will be eliminated from the counting channel by a veto signal.

The number of electrons ejected from the thin wall is small because of its low mass per area (g/cm^2). The wall must, however, be thick enough to stop most beta particles coming from the sample in the inner counter element and impinging on the thin wall. ^{14}C beta particles from disintegrating atoms in the central sample detector, which can penetrate into the annular guard counter, will activate both counter elements, and their pulses will therefore not be counted in the ^{14}C channel. This reduces the ^{14}C counting efficiency. An optimum wall thickness is about 1 mg/cm^2 for tritium and 10 mg/cm^2 for ^{14}C.

Figure 13.3 shows the background of an Oeschger counter as a function of the thickness of the inner wall. Assuming that a ^{14}C counter has a thickness of 10 mg/cm^2, this arrangement will decrease the background by a factor of about 3.5 when the working pressure is 1 atm. At higher gas pressure, the background increases, but the component coming from the wall will remain the same. At 3 atm the background reduction would be smaller, probably about 2.5, because of the increased background contribution from the gas. With optimum wall thickness, the loss in detection efficiency due to ^{14}C beta particles escaping from the sample element into the guard counter, penetrating the thin wall, is probably 5–10%, depending on pressure. This loss has apparently not been measured directly. This loss can be measured in an Oeschger counter filled with a relatively strong ^{14}C sample and the count rate recorded with and without the veto signal.

A comparison of the background of a large number of gas proportional counters, discussed in Section 13.6, shows that Oeschger counters generally have a low background, but hardly better than good conventional counters. Considering their more complex structure, their advantages seems to have been overestimated in the past.

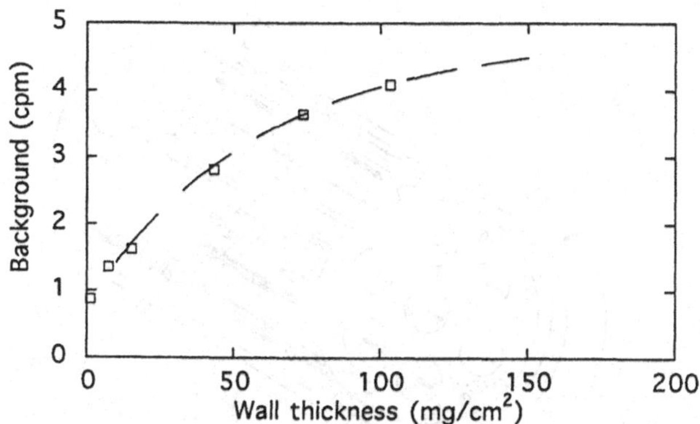

Fig. 13.3. The background of an Oeschger counter at varying wall thickness.

13.2.3 Multielement gas proportional counters

The sensitivity in gas proportional counting can be improved by increasing the size of the counter and by working at higher pressure. With a conventional counter construction, increasing the diameter and/or the pressure would require higher voltage across the counter. This may cause isolation difficulties and electron attachment may occur in the gas.

These problems can be avoided by dividing the sensitive volume with thin wires into several wall-less counter elements. A counter of this type, a *multielement counter*, is shown in Figure 13.4. It has seven detector elements of the same dimensions arranged in a hexagonal form and separated from each other by cathode wires. Its background can be reduced by eliminating simultaneous pulses from three elements, as most of these come from energetic electrons. The increased sensitivity of these counters must outweigh the disadvantage of their complex structure. The practical use of these counters seems to have been limited.

13.3 Gas counters for external solid samples

In the mid 1950s a need arose for Geiger counters with low backgrounds to measure weak beta-active fallout samples. Libby's technique of surrounding the sample counter with an array of guard counters, enclosed in a heavy shield, was adopted. The result was somewhat disappointing. The decrease in background was less than one had hoped for, no doubt due to contamination in both the sample counters and in the glass envelopes of the guard counters most often used in these systems. Furthermore, the systems were bulky. Good results were, however, obtained some years later when compact systems, consisting of flat sample- and guard counters lying close together, were designed. These counters were made of acryl plates and the counting gas flowed through them continuously.

In the mid 1970s these counters were greatly improved by using the new possibilities of

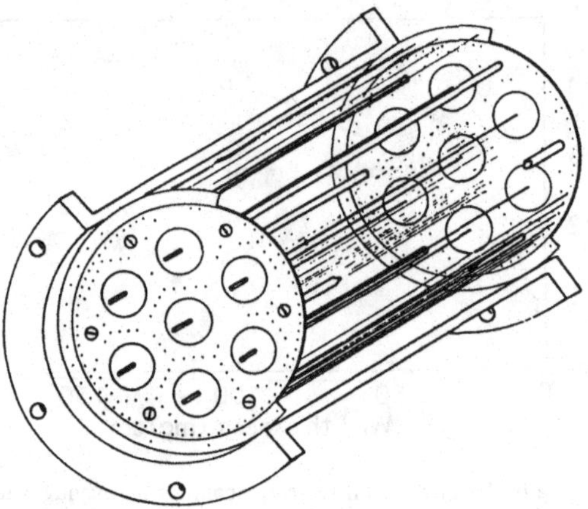

Fig. 13.4. A multielement, also called wall-less counter (Vojtyla and Povinec, 1992).

cheaper electronics. The author developed an improved version of the early counters, where an array of 5 to 10 sample detector elements was made from a single plate of plexiglass. These systems are described in Section 15.7.2.

The window diameter is usually either about 25 mm or 50 mm, although larger counter elements are produced. The background of these counters is 0.03–0.04 cpm per cm^2 of window area (Maushart, 1986). This background is no doubt dominated by secondary radiation. In order to test this, the author made a counter with a single detector element with a window diameter of 25 mm. Its background was measured inside the well of a large NaI crystal where it was shielded on all sides, except from below, by 2.5 cm of NaI. The NaI scintillation guard detector was shielded by 10 cm of lead. This shielding arrangement reduced the background from 0.20 cpm in the normal multi-detector system with a flat guard counter, to 0.032 cpm inside the NaI well. This demonstrates that the background of these multicounters can be reduced considerably by using an external guard counter system. We can expect, with careful selection of radiopure materials, a background of about 0.02–0.03 cpm with a 25 mm window diameter elements in surface laboratories.

13.4 Passive and active shield

Main shield. Libby thought that lead was unsuitable due to contamination for the main shield and used iron. Most laboratories followed his lead and used 20 cm of iron for the main shield. Frequently the shield was made somewhat thicker, especially its top. In addition to this, many laboratories used an inner shield as discussed below. The majority of laboratories therefore seem to have a shielding thickness well above its optimum value (Section 11.3.2). Figure 13.5 shows the shielding arrangement of typical low-level gas proportional counting systems.

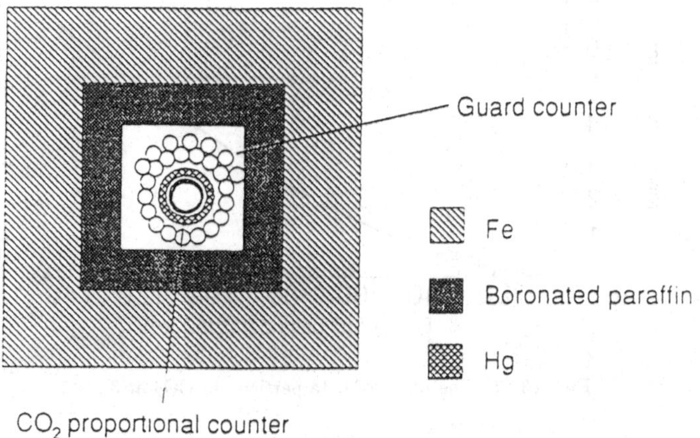

Fig. 13.5. A typical shield of a low-level gas proportional system.

Neutron absorbing layer. H. de Vries (1956) introduced the technique of inserting a neutron absorbing layer of paraffin loaded with boron for thermalizing and capturing the neutrons. The paraffin is generally between layers of iron, but sometimes it is inside the main shield. It is usually 10 cm thick, but only partially surrounds the sample counter. The background contribution of neutrons in gas proportional counters will be discussed in Section 13.9.

Inner shield. Most systems have an inner shielding layer, between the guard and sample counter, of 2–4 cm of either mercury or lead. The latter has been preferred in all recent systems. The combined effect of the guard system and inner shield significantly reduces the background contribution of secondary radiation (electrons, photons and neutrons) as discussed in Section 13.4.

Active shield. All low-level gas proportional counters have a guard counter system inside the main shield, which can be of any of the three types described in Section 11.8.1. No gas proportional system has yet been constructed with external guard counters, despite their obvious advantages. This is discussed in more detail in Section 13.13.

13.4.1 Counting characteristics and working voltage

The sizes of the pulses are proportional to the energy deposited by the energetic electrons in the gas of the counter, provided that space charge effects are small. As we are not interested in the energy spectrum, this effect is anyway of little consequence. The higher energy electrons, for example those of the higher part of the ^{14}C spectrum, may penetrate into the wall before they have lost all their energy. Figure 13.6 shows the range of electrons in CO_2 at a pressure of 3 atm. We must bear in mind that the path is tortuous.

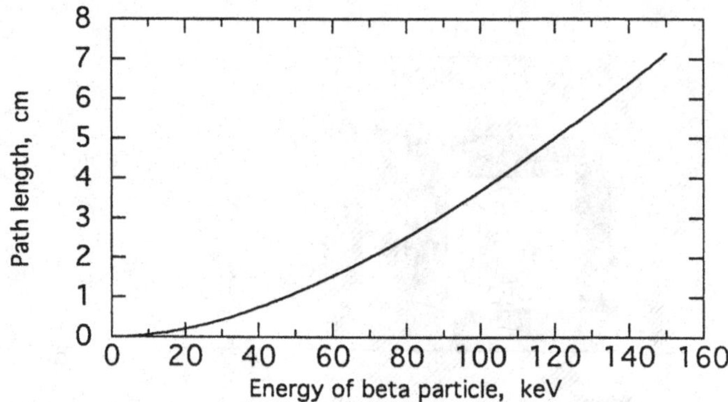

Fig. 13.6. The range of beta particles in CO_2 at 3 atm.

The threshold voltage V_{dis} of the discriminator circuit of the counting unit determines the smallest pulses that are counted. All pulses above this level are registered. The pulse size is determined by (a) the energy deposited in the counter, (b) the gas multiplication factor M, and (c) the amplification factor A. The energy deposition corresponding to an output pulse of V_{dis} will be denoted by E_{dis}.

In radiocarbon dating, the lower discrimination level V_{LL} is not set to accept beta particles of some predetermined energy. The approach is empirical. First the value of the lower level discrimination voltage V_{LL} is selected, i.e., the pulse size in volts at the input of the discrimination circuit. V_{LL} is fixed at a value well above the noise level of the detector-amplifier system. The most suitable value of the high voltage is then determined by plotting *characteristic counting curves*, i.e., the count rate of output pulses larger than V_{LL}, versus the high voltage, both for the ^{14}C beta particles (anticoincidences) and the muons (coincidences) with an absolutely pure CO_2 sample. Figure 13.7 shows typical counting characteristic curves. The working voltage is chosen on the flattest part of the ^{14}C plateau, which usually rises by about 1–2% per 100 volts increase in high voltage. It is very time-consuming to plot the ^{14}C characteristic curve because of the small count rates. Therefore, an external gamma source is used, usually with ^{226}Ra or ^{60}Co.

If there is a trace of electronegative contamination in the CO_2 sample, all pulses suffer the same relative decrease in size, both the muon pulses and the ^{14}C pulses. If such samples are counted at the same high voltage as pure samples, the detection efficiency, i.e., the number of counts per disintegration of an ^{14}C atom in the sample, will be less as more energy must be deposited by the beta particles in order to give a pulse size V_{LL}. This is normally corrected for by counting the contaminated sample at somewhat higher voltage δV, which increases the gas multiplication factor by a factor M, which restores the pulse height spectrum to that of a normal sample. A maximum correction voltage of 100 V is usually allowed. If more than this is needed, the sample is repurified.

The simplest procedure is, however, to work at a fixed high voltage for all samples and determine the correction from the position of the muon peak. A pulse height analyzer is actually not needed. The ^{14}C counting window can be divided into two sub-windows with almost equal count rates for pure samples. The ratio of counts in the two channels for a sample is then used to determine the correction needed to determine the count rate had the sample been pure. This method has been used in the high precision work at the dating laboratory in Heidelberg and Groningen.

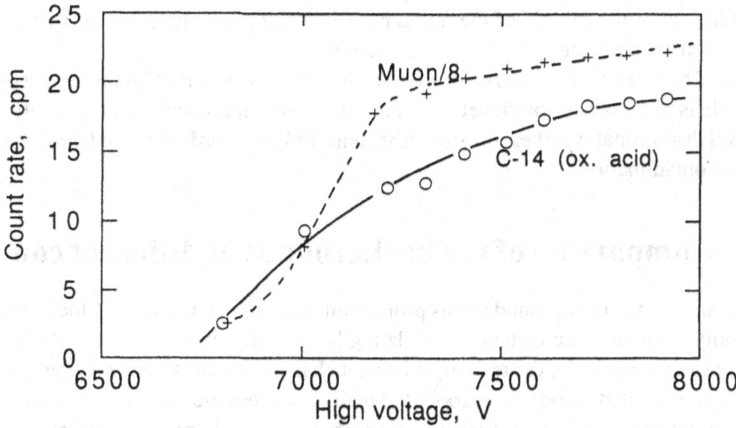

Fig. 13.7. The counting characteristic of a CO_2 gas proportional counter taken with oxalic acid standard and muon pulses (coincidences). From Nydal 1962.

Let us look at what determines the slope of the plateau of the ^{14}C curve. This is rarely discussed in the literature. Figure 4.3 shows the energy spectrum of ^{14}C. In its lowest part, below 10 keV, about 1% of the particles lie in each 1.0 keV energy channel. When the high voltage is increased, the pulse size increases exponentially and ^{14}C beta particles of lower energy are shifted up into the counting window, increasing the pulse rate. When the counting gas is CO_2 at 2.0 atm, the pulse size increases by a factor of 2 for each 220 volts when the anode diameter is 0.05 mm (Zastawny, 1966). We start at a high voltage where 90% of the beta pulses are counted (all above 10 keV). When the high voltage is increased by 100 V all pulses are increased by 37% in size. Pulses of 7.29 keV beta particles will be shifted up to the discrimination level and the count rate increases by 3.0%. If a further 100 V are added, the count rate will increase by an additional 2.0%. The slope of the plateau thus simply reflects that beta particles with ever lower energy reach the counting window. The slope of the curve decreases until spurious afterpulses begin to contribute to the count rate.

At what beta particle energy value (keV) does the discrimination level lie? This is hardly ever discussed in reports on these systems. It can be estimated in the following way. The muons give a broad peak. Its middle corresponds to muons that pass vertically through the axis of the counter. The energy deposited in these events can be calculated. A counter of the type shown in Figure 13.7 was filled with CO_2 to a pressure of 2 atm and its diameter was 6 cm. These muons therefore pass through 0.022 g/cm^2, and as the relativistic muons lose 1.8 MeV per g/cm^2, they deposit an energy of 39 keV in the gas. If the discrimination voltage is increased to a voltage Vμ, where the muon count rate is half of its plateau value, we have for this counter

$$E_{LL} = 39(V_{LL}/V_{\mu}) \tag{13.1}$$

Using a pulse height analyzer makes the selection of the operation point more simple and direct. The coincidence spectrum can, for example, be measured for about 10 minutes and the channel of the middle of the muon peak is determined. Its eventual shift to lower values is used to select the high voltage correction.

It should be noted that there is no need to set the discriminator level at the lowest possible value. This is not done in low-level ^{14}C counting using liquid scintillation counters. The lower level discriminator is then usually set at about 18 keV in order to avoid the disturbance of tritium contamination.

13.5 Comparison of the backgrounds of different counters

Before analysing the background of gas proportional counters it is useful to look at an interesting comparison of gas counters in ^{14}C dating laboratories made by W. G. Mook (1983). Since this study was made, the technique has not changed much, so this comparison is still valid. Mook collected information about the main characteristic parameters of gas proportional counting systems in radiocarbon dating laboratories. He received information from 49 laboratories with 174 counters. The smallest had a volume of 5 mL and the largest 5 L. CO_2 was the most common counting gas, but CH_4 or C_2H_2 were used in some cases. As the pressure varied from 1 to 5 atm, Mook normalized the background to 1 atm (B_N) in order to be able to compare its value to that of other counters of similar size. He assumed that the background increased 15% for each additional atm in pressure.

A number of the counters in Mook's report have been excluded in order to make the set of counters more homogeneous and to concentrate on ordinary counting systems; all Oeschger type counters and counters with a liquid scintillation guard detector have been excluded. The rest of the counters have been divided into three groups:

1. Ordinary counters in surface laboratories, which is the largest group.
2. Counters in underground laboratories (with more than 10 m water equivalent of earth layers or rock above, according to Mook's criterion).
3. Five counters in the cellar of the Physics Institute in Trondheim which have a background that deserves special attention.

The backgrounds of these three groups of counters are plotted versus their volumes in Figure 13.8. The scattering of background values for counters of similar volume in surface laboratories is large, considering the great effort presumably invested in each system. Why is this so? A systematic effort, based on the methods and experience of the development of modern germanium spectrometers, can no doubt improve this technique considerably.

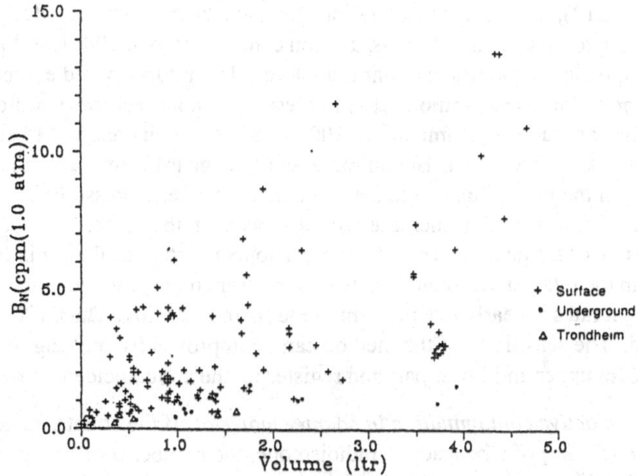

Fig. 13.8. A comparison of ordinary gas proportional counting systems. Data from Mook (1983).

13.6 Background components

After the foregoing general discussion of the background (Chapter 9) and of means to reduce it (Chapter 11) the discussion here can be brief.

$B_\gamma(Ex)$ *from external gamma radiation* is negligible, considering the thickness of the shield always used. Usually this is more than sufficient.

$B(Ct)$ *from contamination in the shield* and components inside it. According to Figure 13.8, this component seems often to be significant. Just as in the case of low-level germanium spectrometers, gas proportional counters can be improved by the careful selection of materials.

$B_\gamma(Rn)$, *from radon* diffusing into the shield. This is a very subtle background source as it can vary widely in time. Its effect on gas proportional counting systems has not been given sufficient attention in the past, and the background contribution of radon has only recently been measured quantitatively. Hedberg and Theodorsson (1996) measured the sensitivity of two gas proportional counters used for radiocarbon dating. The larger counter had a volume of about 1.0 liter, a background of 0.76 cpm and a count rate of 18 cpm for modern carbon. The free space between the counter and the central lead shield was estimated as 0.2 L. 100 Bq/m^3 of radon increased the background count rate by about 0.05 cpm. This radon concentration gives 0.020 Bq in the 0.2 L free air space around the counter. 1.7 photons are emitted per disintegration of each radon atom. The photons come essentially only from radon decay products. This corresponds to a detection efficiency of 2.3%. Considering the energy spectrum of the gamma lines and the detection efficiency of Geiger counters

(Knoll, page 212), the radon concentration may have been underestimated.

According to these measurements, a radon concentration of 100–150 Bq/m^3 would give a significant contribution to the count rate. Every laboratory should evidently have some check on the radon concentration. It is interesting to note that the guard counters can be used as a check of radon contamination. 100 Bq/m^3 gave an increase of 15 cpm over a mean guard count rate of 2000 cpm. But an increase in the guard count rate can also come from fluctuations in the muon flux due to a change in atmospheric pressure. This can, however, be corrected for, as a similar increase will also occur in the coincidence count rate. If the total count rate of the guard is recorded every 4 hours the statistical error is 10 cpm. Most of the pulses in the total guard count rate are coming from muons, so the ration of total to coincidence count rates is nearly independent of the cosmic ray flux. Thus, 100–200 Bq/m^3 can be detected. The sensitivity of the method can be improved by splitting the guard counter system into an upper and lower part and registering the anticoincidences between them.

B_β, *from beta active contamination in counter material.* If the counter material has a concentration a (Bq/kg) of a beta active radioisotope, the number n of beta particles escaping out of a wall with a surface area A cm^2 into the counting gas is according to Equation (9.2)

$$n = B_\beta = Aa/2\mu \tag{13.2}$$

A counter with a volume of 1 L will have an inner surface area of about 800 cm^2 and a good one will have a background of 1.0 cpm. B_β should then be less than 10% of this or less than 0.1 cpm. If the maximum energy of the beta particles is 1.0 MeV, the value of μ is 17 cm^2/g, and a should be less than 0.7 Bq/kg, or less than 0.3 Bq/kg if E_{max} =2 MeV. According to measurements made with ultralow-level germanium spectrometers, all materials commonly used for gas proportional counters have contamination an order of magnitude below this level. It therefore seems probable that a small or negligible background contribution generally comes from contamination in counter material.

$B_\gamma(Sr)$, *from secondary gamma radiation.* This is usually the main background component in systems in surface laboratories. It can be reduced to a certain limit with a thick inner shield. Modern germanium spectrometers, operating in laboratories with a small overburden, all have the same total background per mass of germanium, about 100 cpm/kg when the overburden is small. This background comes primarily from the cosmic muon secondary radiation. In a similar way we should be able to construct low-level gas proportional counting systems with the same predictable background. This is discussed in Section 13.13.

Neutron component. H. de Vries (1956, 1957) demonstrated that neutrons contributed significantly to the background of his CO_2 counters. By inserting a 10–15 cm thick layer of paraffin mixed with boron he reduced the background of a 2.0 litre counter from 5.5 cpm to 3.0 cpm in a laboratory with only a thin roof above. Most radiocarbon laboratories subsequently followed his lead.

Theodorsson (1991) has estimated that the neutron component is about 0.5 cpm per liter of CO_2 in an iron shield in a laboratory with only a small overburden and no paraffin. The inner shielding layer, as well as overburden, will reduce this component. Furthermore, it will no doubt be higher in a lead shield than in that of iron. This estimate rests on a weak basis. Monte Carlo simulation calculations supported by better measurements are needed. The determination of the neutron component is discussed in Section 13.9.

13.7 Background: wall and gas components

The gamma background pulses come not only from photons interacting directly with gas molecules in the sample counter, but also from interactions in the counter wall. Electrons which receive high kinetic energy (mainly through gamma Compton collisions) may be able to escape out of the wall, into the counting gas where they initiate pulses. This only occurs when the Compton event occurs close to the inner surface of the counter, because of the short range of the electrons. The analysis below is an extension of work carried out at the University of Bern (Oeschger, 1963, Oeschger and Loosli, 1977).

We look at a gas proportional counting system where the guard counter veto signal eliminates all direct muon pulses. The background B can then be described by the following equation:

$$B = B_\gamma(g) + B_\gamma(w) \tag{13.3}$$

where $B_\gamma(g)$ and $B\gamma(w)$ are components due to gamma ionization in the gas and in the wall respectively. In addition to these components there is a contribution from fast neutrons colliding with nuclei of the gas atoms (Section 13.9).

Gas background component. $B_\gamma(g)$ is proportional to the mass of the gas, M_g, and the primary ionization density, i_g (cpm/g), i.e., the number of ionization events produced in the gas per gram and per minute. Therefore

$$B_\gamma(g) = i_g M_g = i_g PV\rho \tag{13.4}$$

where

M_g is the mass of the gas,
P is the pressure,
V is the sensitive volume of the counter ,
ρ is the density of the gas (g/cm^3).

An experimental verification of the linear relation between the background and mass of the gas in a counter is shown in Figure 13.9.

It should be noted that i_g is a measure of the gamma flux seen by the counting gas, independent of the counter shape and the material it is made of, and almost independent of the gas atomic number up to $Z = 18$ (argon), as Compton interaction dominates. This parameter, if measured carefully, could allow valuable comparisons of similar systems. It only calls for one additional background measurement at, for example, half the normal working gas pressure. A pulse height analyzer, which records both the coincidence and anti-coincidence spectra, should be used in order to set the discrimination level corresponding to the same energy deposition at all pressures.

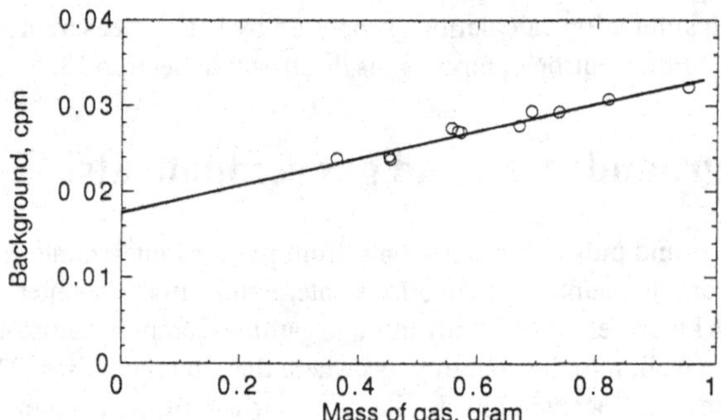

Fig. 13.9. Background of a gas proportional counter as a function of mass of gas in counter. From Kalt, 1962.

Wall component. $B_\gamma(w)$ is proportional to i_w, the primary ionization events per minute per gram in the wall material, and to the inner surface area A of the counter. It is independent of the pressure and type of gas. We can therefore write

$$B_\gamma(w) = ti_w A \tag{13.5}$$

and

$$t = B_\gamma(w)/(i_w A) \tag{13.6}$$

The parameter t has the dimension of thickness, here measured in g/cm^2. It is equivalent to a wall layer thickness where the number of ionization events, i_w, is equal to the number of electrons emitted per cm^2 of the wall. This layer will be called the *virtual wall layer*. Its thickness depends primarily on the energy distribution of the electrons, as both their range as well as the Compton coefficient depend only weakly on atomic number. The distribution of the transmitted energy to the electrons in the wall depends primarily on the gamma spectrum, less on the counter material. The gamma background can now be written

$$B = B_\gamma(g) + B_\gamma(w) = i_g M + ti_w A = i_g(M + t(i_w/i_g)A) \tag{13.7}$$

In cases where the Compton effect is dominant the ratio i_w/i_g will be close to 1.0 and we can then write:

$$B = i_g(M_g + tA) \tag{13.8}$$

It is natural to interpret $(M_g + tA)$ as the total sensitive mass m_{sens} of the counter:

$$m_{sens} = (M_g + tA) \tag{13.9}$$

Let us look at a typical counter in order to see the relative quantity of sensitive mass in the gas and in the wall. A 1.0 liter counter will have a cathode area of about 800 cm^2. Taking

0.020 g/cm^2 for t (see below), the sensitive mass in the wall will be 16 grams. If the counting gas is CO_2 at a pressure of 3 atm the mass of the gas will be 5.5 g, or only 25% of the total sensitive mass. The gas proportional counters are thus burdened by a large extra mass, sensitive to background gamma radiation.

Thickness of the virtual wall layer. Oeschger and Loosli (1977) used a value of 0.020 g/cm^2 for t, based on the assumption that the mean range of the primary ionization electrons in the wall was probably about 0.08 g/cm^2 and that t should be 1/4 of this. No direct experimental support was given for this estimate.

As t is an important parameter we will look carefully at experimental evidence for estimating its value. It can be estimated in three different ways.

1. The most direct way is to accurately measure the background of a gas proportional counter at varying gas pressure. Few good measurements of this type are found in the technical literature. Figure 13.9 shows the result of a good measurement (Kalt, 1986). The slope of the line is equal to i_g. $B_\gamma(w)$ is the background value at the intercept of the line with the y-axis. Assuming $i_g = i_w$, the value of t is 0.019 g/cm^2. Table 13.1 gives some further values. There is considerable scattering in the values of t, but it is useful to show the range and what type of information is needed to evaluate the data.

Table 13.1. Information on five low-level gas proportional counting systems used to find the thickness of the virtual wall layer.

	Uppsala	Groningen	Seattle	Seattle	Bern, undergr.
i_g, cpm/g	0.063	0.20	0.107	0.039	0.018
B(w)/A, cpm/cm^2	18×10^{-4}	16×10^{-4}	18×10^{-4}	8.7×10^{-4}	
t, mg/cm^2	28	8.5	17	20	19
Muons, cpm/cm^2	0.74	0.75	0.70	0.54	0.068
Inner shield	3 cm Hg	3 cm Pb	2 cm P	2 cm Pb	2 cm Pb
Wall material	Quartz	Quartz	Quartz	Quartz	Copper
Overburden, mwe	1	1.5	1.5	5	70
Neutron shield	Yes	Yes	Thin	Yes	No

2. The energy distribution of the electrons in a copper wall is close to that of the background spectrum of germaium diodes because of the small difference in atomic number (Cu: 29, Ge: 32). This spectrum (Figure 12.12) can be approximated crudely by a beta spectrum with a maximum energy of 0.7 to 1.0 MeV. The thickness of the virtual layer will, according to Equation (5.9), be $t = \mu/2$, where μ has a value between 17 and 28 cm^2/g and t will then be 0.018–0.030 g/cm^2. Instead of using these approximations, a more accurate value can be found by using the known background spectrum of germanium diodes and calculating the escape probability of the electrons by the Monte Carlo method.

3. The background measurements made with an Oeschger type counter can be used to estimate t. This counter (Section 13.2.2) has a thin wall between the sample and guard detector in order to decrease the wall component of the background. The background was measured with a wall of varying thickness (Figure 13.3). By approximating, as done above, the energy distribution of the electrons released in the thin wall by a beta spectrum, the background component from the thin wall will be proportional to $(1 - exp(m/M))$. This gives a good approximation for the background curve if $\mu = 0.060$ g/cm^2, giving a value of $t = 0.030$ g/cm^2.

The three entirely different methods for estimating the virtual wall layer thickness show that it has a value of 0.025 ± 0.005 g/cm^2.

13.8 Neutron background component

H. de Vries discoverd that cosmic ray neutrons gave a significant contribution to the background of his counters. When he inserted 10–15 cm thick layers of paraffin mixed with boron between the plates of iron in his shield the background of a 2.0 L counter, working at 3 atm of CO_2, was reduced from 5.3 to 3.0 cpm, or by 0.23 cpm per gram CO_2. When his counter was filled with propane at 2.0 atm, the paraffin layer reduced the background from 12.5 to 5.3 cpm or by 1.0 cpm/g. Propane thus gives 4.4 times higher neutron count rates than CO_2.

Recoil broadened fast neutron peaks in the background spectra of low-level germanium spectrometers were discussed in the last chapter (Figure 12.5). We now look at the same kind of pulses in counters filled with CO_2 and C_3H_8 (propane), respectively. It was shown in Section 11.7 how we can derive the energy spectrum of these pulses in materials other than germanium. Figure 13.10 shows the calculated spectrum of recoiling ^{1}H and ^{12}C atoms in gas proportional counters. The spectrum of pulses from ^{16}O recoiling atoms will be similar to that from ^{12}C atoms, but shifted to somewhat lower energy values. It is evident that the main part of the recoil pulses of both ^{12}C and ^{16}O fall into the ^{14}C energy window (usually about 5–150 keV), but only about 40% of the ^{1}H pulses do.

We now look at the relative probability of these neutron collisions with ^{1}H, ^{12}C and ^{16}O atoms. The smoothed neutron cross sections for these atoms are shown in Figure 13.11. We assume in the following that the mean cross section is that of a 0.5 MeV neutron, which is about 3.5 barn for both ^{12}C and ^{16}O and 5 barn for ^{1}H. When the background is measured with CO_2 and propane at the same pressure, the three carbon atoms in a propane molecule will give nearly the same contribution to the count rate as the three atoms in CO_2. The two gases have the same molecular mass (44) and their gamma background is therefore almost the same. The difference in background comes predominantly from the difference in the number of neutron pulses. In an open window (all pulses above 5 keV), the ratio of the neutron component in propane to that in CO_2 is $(3 \times 3.5 + 8 \times 5)/(3 \times 3.5)$ or 4.8. In a ^{14}C window, 5–150 keV, only 40% of the recoil pulses from ^{1}H are counted and the ratio becomes 2.9.

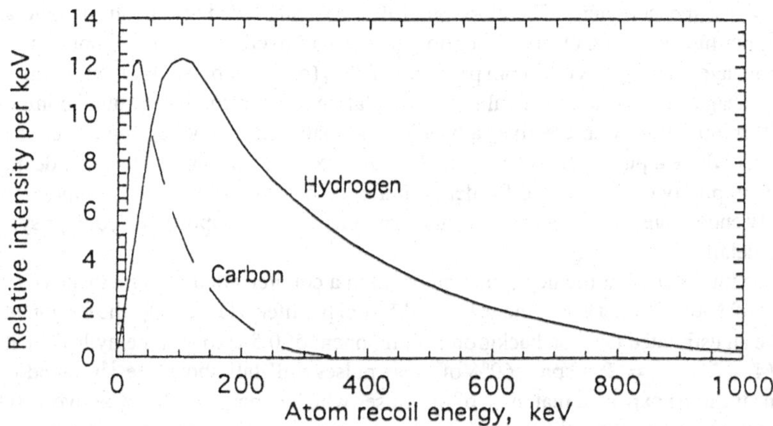

Fig. 13.10. The calculated recoil energy spectrum of cosmic-ray neutrons colliding with ^{1}H and ^{12}C nuclei.

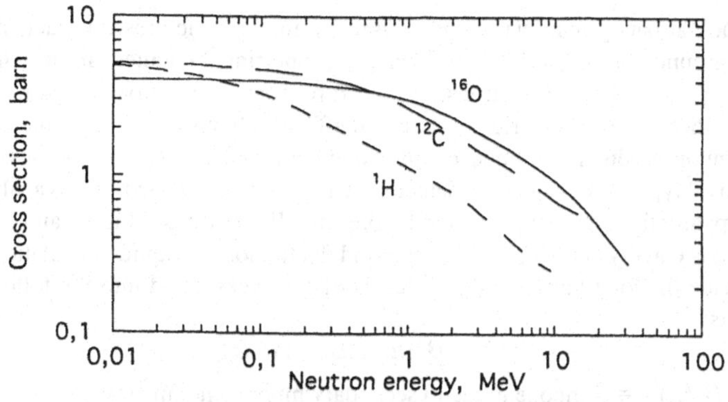

Fig. 13.11. The smoothed neutron atomic cross section for ^{1}H, ^{12}C and ^{16}O (left) and the energy spectrum of cosmis ray neutrons (evaporation spectrum).

Nydal et al. (1977) measured the background with both propane and CO_2. From his and de Vries' measurements Theodorsson (1991) has calculated that neutrons give about 0.5 cpm per liter STP of CO_2 in a laboratory with small overburden and without a neutron absorbing layer. This gives 0.27 cpm per gram of CO_2 (or for carbon or oxygen). This result rests on meager data. Using this value for calculating the neutron background component in liquid scintillation counters gives a background that seems to be too high. Better measurements, where a pulse height analyzer is used to fix the counting window, are desirable.

The simplicity of the method for determining the neutron background component will hopefully encourage scientists to study this component in gas proportional counter systems in more detail.

Let us finally look at the neutron component in a counter filled with methane (CH_4) to pressure of 3 atm. If we assume the value of 0.5 cpm per liter STP of CO_2, the carbon atoms will give an estimated neutron background component of 0.5 cpm and the hydrogen atoms $0.5 \times (4 \times 5/3.5) = 2.9$ cpm. 60% of these pulses will fall above the ^{14}C window, or 1.7 cpm. It can be expected that most of the pulses which deposit an energy above 150 keV in the methane come from the neutrons. This component can therefore be determined in all measurement runs, both when the background and unknown samples are measured, and it should have an atmospheric pressure coefficient of -10% per cm Hg. Most ^{14}C counters using methane as the counting gas have, however, a paraffin neutron absorption layer, suppressing greatly the neutron component.

13.9 Background atmospheric effects

We now look at background variations caused by atmospheric pressure fluctuations. The main background components of low-level gas proportional counters are usually those of secondary gamma rays and neutrons, produced by cosmic ray protons and muons in the shield. Variations in atmospheric pressure will affect both of these. Furthermore, the mean height of muon production influences the muon component. These variations, which are common to all types of beta/gamma detectors, will now be examined. As an analysis of this kind has apparently not been presented before, it will be discussed in detail. It should be noted that it is equally applicable to background fluctuations in liquid scintillation counters.

We divide the long term average value B of the background into the following three components:

$$B = B_\gamma(\mu) + B_n(p) + B_0 \tag{13.10}$$

where: $B_\gamma(\mu)$ = component due to secondary muonic gamma rays,
$B_n(p)$ = component due to neutrons produced by protons,
B_0 = constant background component.

We denote the fractions of the three components by f_μ, f_n and f_0, where

$$f_\mu + f_n + f_0 = 1 \tag{13.11}$$

When atmospheric pressure (P, cm Hg) increases by δP the cosmic-ray particles must traverse a larger mass of air and absorption will increase. The relative background atmospheric *pressure coefficient* for the muonic component is -2.0%/cm Hg and -9.6% for the neutron component. Furthermore, there will be slow seasonal variations in the muonic flux at sea level due to variations in the altitude of the mean muon production height (usually taken as the 100 mbar level). The height of the 100 mbar level over Belfast in the period 1978–1979 is shown in Figure 13.12. It lies higher in the summer than during the winter due to a higher mean temperature in a vertical column of the atmosphere. When the mean muon production height is higher, more of the muons decay before they reach sea level. The muon flux decreases by 0.53% for every km increase in the 100 mbar height.

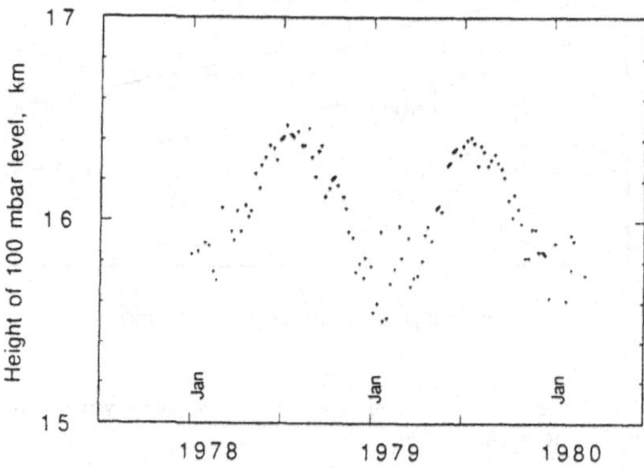

Fig. 13.12. The varitions in the 100 mbar level over Belfast. From Pearson, 1983.

The deviation δB from the yearly mean value of the background is therefore given by

$$\delta B/B = -(f_\mu 0.02\delta P + f_\mu 0.0053\delta H + f_n 0.096\delta P) \tag{13.12}$$

If the background is determined during a period when H is nearly constant, for example within a period of two months, the results, when plotted as a function of pressure, will lie close to a sloping line. When the same kind of measurements are taken during a different season the line will have the same slope, but will be shifted vertically. The slope is determined by $(f_\mu 0.02\delta P + f_n 0.096\delta P)$ and f_0.

If we measure the ratio $\delta B/B$ during the part of the year when H has its highest and lowest value (in summer and winter) at days when $\delta P = 0$ (or corrected for deviation from mean pressure) we can determine f_μ:

$$f_\mu = \delta B/B/(0.0053\delta H) \tag{13.13}$$

δH can be found from the nearest regional meteorological observation station. When f_μ has been found, f_n can be found from the slope of the line.

In order to illustrate the accuracy of such measurements let us look at hypothetical background measurements during summer and winter (when H is maximum and minimum respectively). This is calculated for a system discussed below, which has a mean background of 1.8 cpm that has an atmospheric pressure coefficient of -2.0% per cm Hg at constant H and where the difference in mean muon flux during summer and winter is 3.6%. We assume that the background is measured over a weekend (64 hours) at an atmospheric pressure between 74.2 and 77.8 cm Hg. Figure 13.13 shows the calculated results.

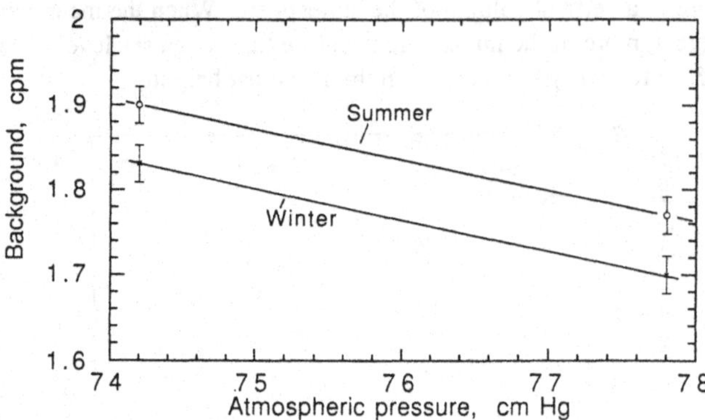

Fig. 13.13. Result of a hypothetical background measurement at varying atmospheric pressure at maximum and minimum height of the 100 mbar level.

Measurements of this kind can evidently give a good estimate of the relative size of the background components in Equation (13.10).

If background measurements are taken periodically over a whole year, the mean atmospheric pressure recorded, and the background plotted against the pressure, a result like the one shown in Figure 13.14 emerges. The lateral spread comes partly from the omission of the seasonal muonic variations and partly from statistical errors. The result can nevertheless be used to find the slope of the line discussed above, but it is better to confine the measurements to a part of a season where δH varies slowly.

If the result of each background run is corrected for the deviation of P from 76 cm Hg, and the result is plotted as a function of time, we see the seasonal muon variations due to changes in H. This is illustrated by Figure 13.15. The regularity of the curve reflects the long term stability of the counting system. If the value of H is known for each measuring period in a series of this kind, a least squares deviation analysis gives the best values of the relative fractions of the three background components. Using these values we can then find, with highest possible accuracy, the background value that should be used in the measurement of each unknown sample, rather than base the value on a few background measurements.

The most common practice for determining the background value to be used in the measurement of individual samples is to use the mean background values of two preceding and two subsequent measurements corrected for atmospheric variations. Instead of using the pressure for correcting background, the correction is more often based on the coincidence

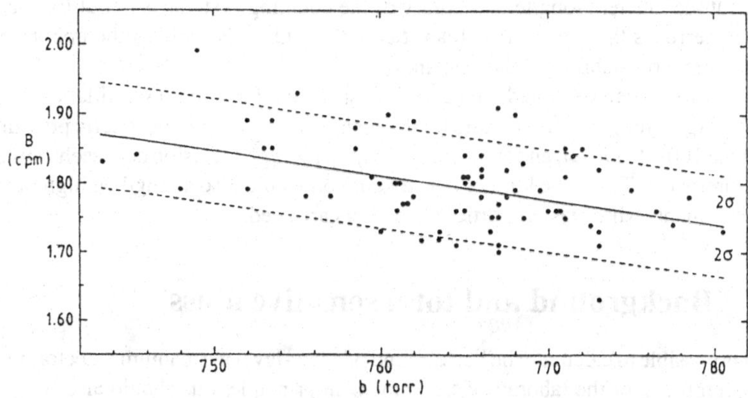

Fig. 13.14. Background count rate as a function of atmospheric pressure, taken over a long period (Høakonsson, 1982).

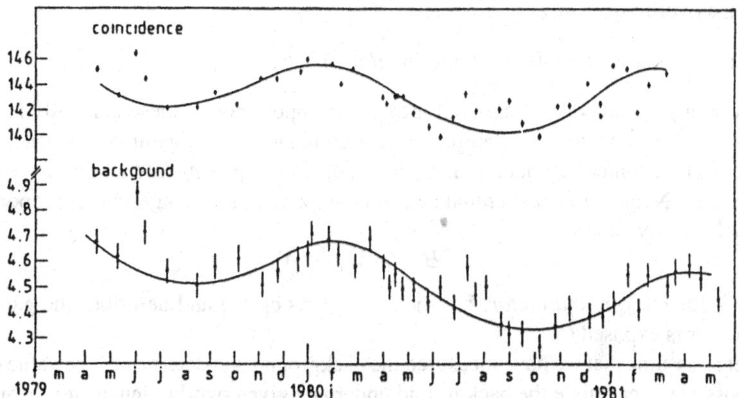

Fig. 13.15. Seasonal variations in background corrected for fluctuations in atmospheric pressure (Hertelendi 1987).

counting rate, or rather its deviation $\delta n_{co\iota}$. As the background is usually measured twice a month, the value of H does not change much during this period. This is a satisfactory method, but the inherent long term stability of the counting systems is not utilized and the background error is therefore larger than it could be. One also misses the opportunity to check the long term stability of the system.

Pearson (1983) demonstrated the nearly full stability of his liquid scintillation counting system (vintage 1960) over three years by taking into account the atmospheric pressure, the height of the 100 mbar level, and the solar activity, in his high precision dendrachronological ^{14}C measurements. The same kind of data evaluation would also strengthen high precision ^{14}C measurements when gas proportional counters are used.

13.10 Background and total sensitive mass

Today, it is possible to calculate the background of a low-level germanium detector if its size and the overburden of the laboratory are known. In principle, this should also be possible for gas proportional counter systems. An attempt to do this is presented below.

We assume that the background comes essentially only from secondary cosmic gamma rays, predominantly produced by muons. The background for a given system then depends on four parameters:

1. the size of the counter,
2. the mass of the gas,
3. the overburden (m_{ob}),
4. the thickness of the inner shield (in g/cm^2).

The type of gas and counter material are of less importance as these generally have a low atomic number in which the Compton effect dominates in the gamma interaction. Let us first look at a counter without an inner shield, operating in a surface laboratory with small overburden. Neglecting an eventual neutron component, its background is proportional to the total sensitive mass:

$$B = b(M_g + tA) \tag{13.14}$$

where b, the *background factor*, has the dimensions cpm/g and describes the gamma flux the counter is exposed to.

Let us assume that we have measured the background and determined the value of b. It is then possible to calculate the background under any given overburden, as it is proportional to the muon flux density:

$$B = B(m_{ob}) = bA_\mu(M_g + t.A) \tag{13.15}$$

An expression for A_μ is given by Equation 7.14.

We now add an inner lead shield with a thickness of s (cm). The fraction of background pulses that are not delayed, $(1 - f_d)$, is reduced by a factor of $e^{-x/s}$ according to Section 11.8.2, and B is then given by

$$B = B(m_{ob}) = bA_{\mu}(M_g + t \cdot A)(f_d + (1 - f_d)e^{-x/s}) \tag{13.16}$$

13.11 Estimate of the background factor b

In order to calculate the background of a given counter we must know the value of the important parameter b. This can be found from the background of good low-level counters. In order to avoid the influence of inner shield attenuation, only counters without this layer are considered. These are, however, rare. Data for four such systems are given in Table 13.2.

Table 13.2. Data on four low-level gas proportional counting systems used to find the value of the background constant b in Equation (13.14).

Laboratory	1. Norway	2. Hungary	3. England	4. Iceland
Volume, liters	1.5	0.83	0.030	0.065
B, cpm	3.2	3.75	0.28	0.59
B/m_{sens}, cpm/g	0.122	0.306	0.17	0.103
m_{ob}, mwe	2.5	1.5	0.5	2.8
b, cpm/g	0.16	0.36	0.18	0.15
Ref.	Nydal et al., 1977	Csongor et. al. 1986	Loosli et al., 1986	Theodorsson, 1991

There is a good agreement in the value of b between three of the laboratories. Laboratory no. 2, however, gives a value more than twice as high as the others. The systems in laboratories nos. 1–3 have cylindrical counters used for radiocarbon dating. The system in laboratory no. 4 consists of a small, flat main detector sandwiched between two somewhat larger guard counters of the same type, all working in the Geiger region (Theodorsson, 1991). Their thickness is 1.0 cm. The b value of laboratory no. 4 is presumably more reliable as the background was measured with the same system both in a surface laboratory and deep underground (100 mwe). The small residual background component due to contamination was determined from the underground measurements (0.22 cpm) and then subtracted from the background in the surface laboratory before calculating b. In the case of the other three systems one must assume that $B_{(Ct)}$ is small compared to $B_{(Sr)}$.

If we assume a value of 0.025 ± 0.005 g/cm^2 for t, the value found here for b is 160 cpm/kg. If we use a value of 0 .030 g/cm^2, as the study with an Oeschger counter with varying wall thickness indicated, b would have a value of 110 cpm/kg. It is interesting to compare these values to the background count rate per mass of low-level germanium diodes. The latter is about 100 cpm per kg of germanium according to Section 12.10. Because of the small difference in the atomic number of copper and germanium, we would expect nearly the same value.

The result of this comparison is promising, and it indicates that the background of gas proportional counters is predominantly from secondary gamma radiation produced by cosmic rays in the shield. Further work along these lines is needed, for example taking into account the neutron component.

13.12 Background of counters of different sizes

Finally, the influence of counter size will be considered. Its cathode area A can be expressed as a function of the ratio c between the length and diameter (cylindrical counter) and the effective counter volume V:

$$A = \pi/2(4V/c\pi)^{2/3}(1 + 2c) \tag{13.17}$$

The ratio c has a value from about 3.5 to 12, usually about 6. If this latter value is used, the cathode area is given by

$$A = 7.3V^{2/3} \tag{13.18}$$

where V is given in cm^3 and A in cm^2. This equation will give the cathode area with less than 15% error for the whole range of values that c normally takes. The background is now given by

$$B = bA_\mu(PV\rho + 0.146V^{2/3})e^{-x/2.0} \tag{13.19}$$

The background described by this equation is compared to measured background of the counters in Mook's study, that is, for counters without a central shield ($x = 0$) and with a 4-cm-thick lead shield ($x = 4$). It is assumed that there are three concrete floor plates above, or an overburden of 1.5 mwe ($A_\mu = 0.81$). A value of $b = 0.16$ cpm/g is used and $f_d = 0.2$. Only CO_2 counters with a volume less than 2.5 L are considered. The result is shown in Figure 13.16. The two calculated background curves seem to describe reasonably well the general trend of scattering of background values, but some of the counters evidently have an additional background component from radiocontamination and neutrons. A closer study of individual systems would no doubt give interesting information.

13.13 Improved systems with external guard counters

The development of modern low-level germanium spectrometers has brought us a wealth of information that can help in designing optimum systems. Transferring and adapting this experience in low-level gas proportional counting will hopefully be rather simple. There are only two main points to be considered:

1. Only radiopure materials should be used.

2. External guard counters should replace inner guard.

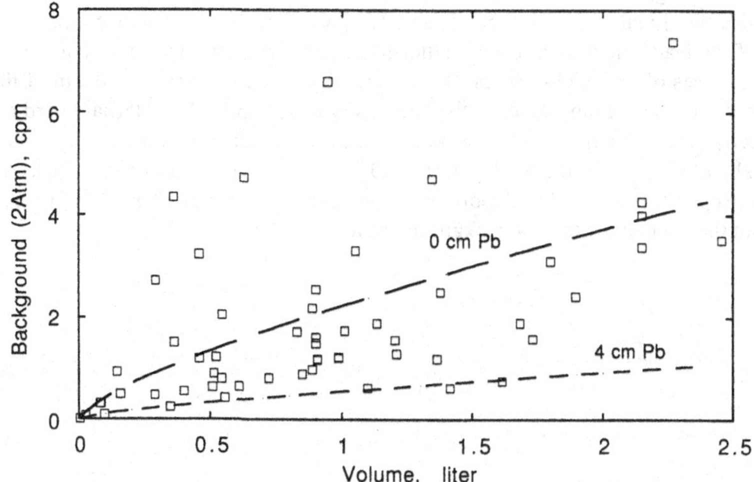

Fig. 13.16. Background of gas proportional counters referred to a working pressure of 2 atm and calculated background for systems without an inner shield and with a 3-cm-thick inner shield of lead. Data from Mook 1983.

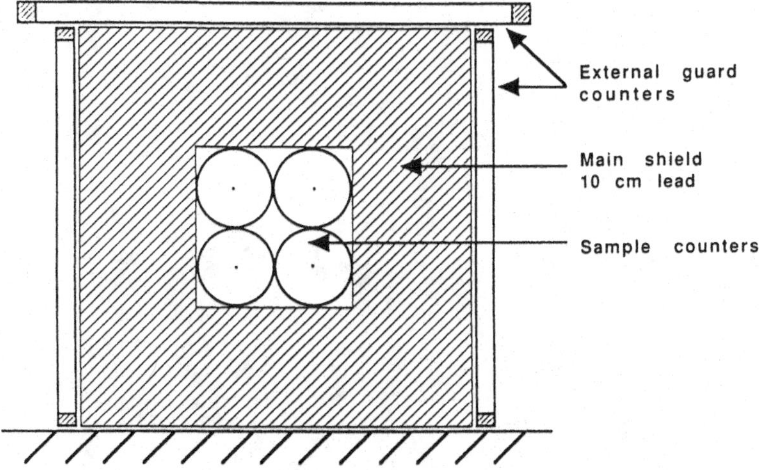

Fig. 13.17. A proposed system with 4 gas proportional counters and external guard counters.

Using external guard counters makes it easy to avoid radiocontamination, as the only material inside the shield will be the sample counters, which are made of pure copper. The inner layer of the lead shield must be of sufficient radiopurity. The flat guard counters should cover all sides of the shield except the bottom. Figure 13.17 shows a system of this type proposed for a dating laboratory at the University of Iceland. The external guard counters drastically reduce the mass of the system. With its four sample detectors the total weight of the shield will be about 1200 kg. For a 1.0 liter counter we can expect a background of about 0.6 cpm in a near-surface laboratory. With an overburden of 10 mwe where A_μ is 0.36 f_d about 0.1 we could expect a background of about 0.2 cpm.

Chapter 14

Liquid scintillation counting systems

14.1 Introduction

Liquid scintillation counters (LSC) for internal samples are used for assaying alpha and beta activity and they are the most widely used systems for measuring radioactivity. They are in some respects ideal detectors as the sample is mixed homogeneously with the detecting medium, eliminating problems due to absorption. Their detection efficiency is 100% for alpha particles and close to that for energetic beta emitters, but it falls at lower energies. The process from radiation interaction to a counted pulse is, however, more complicated than for other nuclear detectors and there are more interfering factors.

Liquid scintillation counters were tested as early as 1954 for the measurement of natural radiocarbon using organic liquids found in nature, e.g., ethanol. This early work gave promising results, but this technique did not become of practical value in radiocarbon dating until a method for synthesizing benzene from CO_2 with high yield was developed in 1960-1963.

The background of the early one phototube systems was quite high. The coincidence liquid scintillation counters, with two phototubes, introduced in late 1950s, reduced the background almost by an order of magnitude. Large scale production these systems started due to strong demand from scientists working with radioactive tracers in chemical and biomedical studies. These new systems could be used in radiocarbon dating without modifications and they quickly gained wide acceptance in dating work, in spite of their somewhat high background compared to gas proportional counters. In the 1970s they became the preferred method in new radiocarbon dating laboratories.

In the mid 1980s the use of liquid scintillation counters for the measurement of alpha activity increased rapidly after a method had been developed for separating alpha from beta pulses by pulse decay analysis and microprocessors had brought us cheap multichannel analyzers. This technique will be described in Section 14.7. In the following sections, we focus on the assay of beta activity.

When sample material is limited, and the volume of benzene therefore small, the background was uncomfortably high. A Finnish firm, Wallac, therefore designed a system dedicated to low-level counting, the Quantulus, which was marketed in 1984. Although primarily designed for dating work, it has been used for a large variety of environmental studies. A little later, Packard introduced a simpler low-level system, a modification of the firm's general models, where the background is reduced considerably in an ingenious way through pulse decay analysis.

Practically all development work on liquid scintillation counters has been carried out by commercial firms that naturally have limited interest in publishing the many details of the technique. The scientists and engineers, involved in such work, generally try to find, frequently in a highly empirical way, the shortest path to an acceptable solution. Only a small part of the results of their studies is published. Scientists, working in university laboratories or at research institutes, usually look more closely at the elementary processes, trying to understand quantitatively the basic phenomena and finally the results are published. For this reason, there is a conspicuous difference in the information we have on the background of germanium spectrometers and gas proportional counters on one hand and liquid scintillation counters on the other hand.

The following discussion on the background of liquid scintillation counters therefore inevitably rests on a basis that leaves much to be desired. Some of the statements and conclusions presented in the following sections should therefore be taken with reservations, but the author prefers each time to give the most plausible explanation he can find, rather than leave the questions open. This will hopefully stimulate further studies and new analysis of the data.

14.2 Scintillation detection and design of systems

14.2.1 Photocathode efficiency and dark pulses

The sample is usually dissolved in the scintillator, but in the case of ^{14}C dating, the liquid, benzene, is also the sample. It is usually in a 20 ml standardized flask (vial) of glass or plastic, close to the cathode face of the phototubes. It is important to collect as much as possible the emitted light on the cathode of the photomultiplier tube. Reflecting surfaces or mirrors are used for this purpose.

The early type of scintillation counters, shown in Figure 14.1 had a single photomultiplier tube which sensed the emitted light. When an energetic charged particle deposits energy E_a in the scintillator the number N_{ph} of photons created is proportional to E_a, $N_{ph} = SE_a$, where S is the scintillation efficiency, i.e., the number of photons produced per keV absorbed (Coursey and Mann, 1980). A fraction f_a of the photons is absorbed in the scintillator. Absorption in the sample is usually called quenching. A fraction G (light collection efficiency) of the remaining photons fall on the cathode of the phototube. The number of photons N_{ph} collected by the photocathode is then

$$N_{ph} = S\,E_a(1 - f_a)G \qquad (14.1)$$

The absolute value of G is not known, it can only be estimated. In order to enhance the size

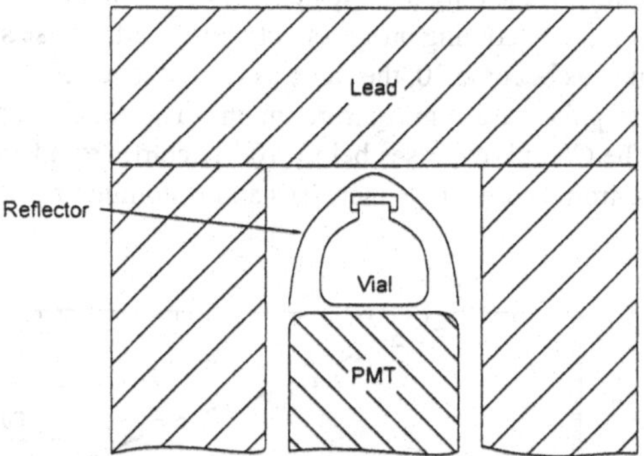

Fig. 14.1. A single photomultiplier liquid scintillation counting system.

of the signal pulses compared to phototube noise, it is important to minimize quenching and strive for highest possible value for G. The number h of photoelectrons emitted per keV deposited in the scintillator, the cathode coefficient, is then:

$$h = N_e/E_a = \varepsilon G N_{ph}/E_a \tag{14.2}$$

where ε is the photocathode efficiency, typically having a value of 0.25-0.28 electrons/ photon. The value of h describes a fundamental quality factor of a liquid scintillation counter system and it determines the attainable detection efficiency of low energy beta emitters. A typical value for a good system is 1.0 photoelectron per keV (Horrocks 1964, Einarsson and Theodorsson 1989).

The phototubes give not only pulses from light flashes in the scintillator. When the tube is kept in complete darkness, so-called dark or internal pulses appear. Most of them are of similar size as they are initiated by single electrons released from the photocathode, usually through thermal agitation. The spectrum therefore shows a broad peak with a tail of larger pulses, primarily due to statistical distribution in the number of secondary electrons released at the first dynode. Some of these internal pulses, however, come from scintillations produced by cosmic rays and radioactive contamination in the glass of the phototube. The anode output pulse is sometimes measured in units of average size of pulses given by a single photoelectron, V_{pe}.

Let us look at the background spectrum of internal pulses in these single tube systems. Figure 14.2 shows the integral spectrum of pulse size (in units of V_{pe}) for a phototube in a surface laboratory and below an overburden of 70 mwe, where the muon flux has been reduced by a factor of 20 (information from EMI). Most of the thermal electron pulses lie below 2 to 3 V_{pe}, with little difference, as to be expected, between surface and underground. From 2 to 10 V_{pe} there is still some difference in the pulse rate at surface and underground. In this range the number of Cerenkov pulses are independent of overburden, they must therefore come from beta particles in the phototube envelope. Above 10 V_{pe}, the underground curve falls more rapidly, indicating that the number of beta induced Cerenkov pulses, compared to muon pulses, are becoming progressively fewer. At a pulse size of 100 V_{pe} the ratio surface/underground has become 20, the same as the reduction in the muon flux, indicating that all the Cerenkov pulses are coming from muons. The conclusion is that in surface laboratories, most of the Cerenkov pulses below 10 V_{pe} come from beta particles, above that the fraction coming from cosmic rays increases and is dominating at 50 V_{pe} and above.

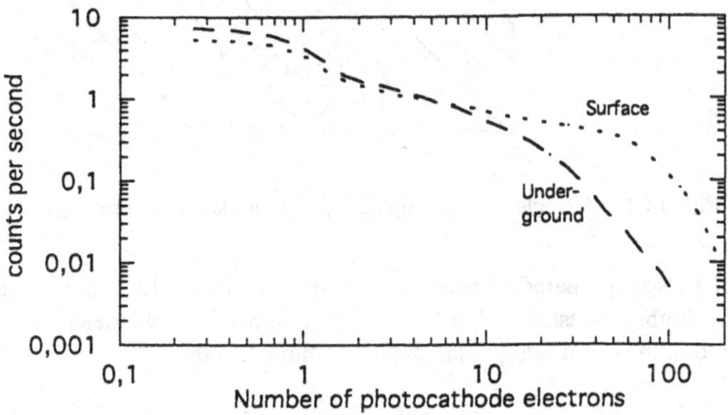

Fig. 14.2. Integral spectrum of size of internal pulses for a phototube in a surface laboratory and at an overburden of 70 mwe (information from EMI).

In order to suppress the internal pulses in the early single phototube systems, the discriminator was set high enough to eliminate most of them, probably at about 5 V_{pe}. As the beta particles of tritium and low energy beta particles of ^{14}C give scintillations that only release a few photoelectrons, this seriously impaired the counting efficiency, especially for tritium.

14.2.2 Two phototube systems

The number of thermally released electrons was typically many thousands per minute. A method for suppressing them was found late in the 1950s. A second phototube was added, also facing the vial, and pulses were only accepted as a valid signal when they came simultaneously (in coincidence) in both channels. By setting the coincidence condition as low as 10-20 ns, accidental coincidences are insignificant in spite of the large number of thermal pulses. Now, the discriminator level of each tube could be set below the single photoelectron peak, securing maximum detection efficiency and low background without getting a high background. Figure 14.3 depicts schematically the main units of these systems. The minimum average scinillation required to give a valid pulse now corresponds to 3 released photoelectrons. If only two are released in average, they may both appear in the same tube.

Figure 14.4 shows schematically the detector unit and shield of these systems. The sample and phototubes are generally shielded by 3-5 cm of lead. Practically all these systems have automatic sample changing. There is therefore a circular opening in the shield above the vial measuring position for automatically inserting the samples. With a 5 cm thick shield a large part of the external gamma radiation then probably comes through this hole and a thicker layer of lead will then not help much.

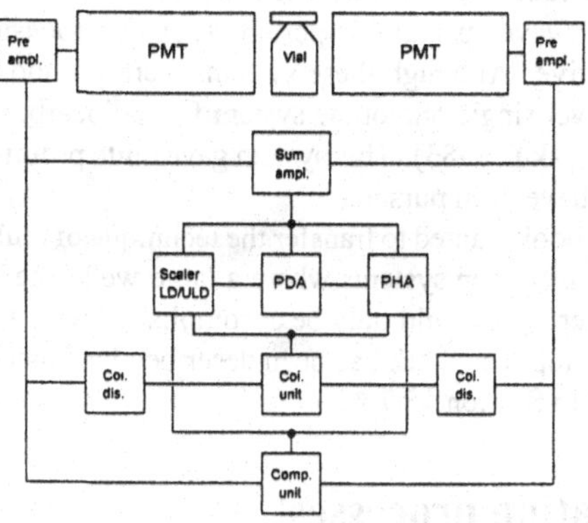

Fig. 14.3. Schematic diagram of a coincidence system.

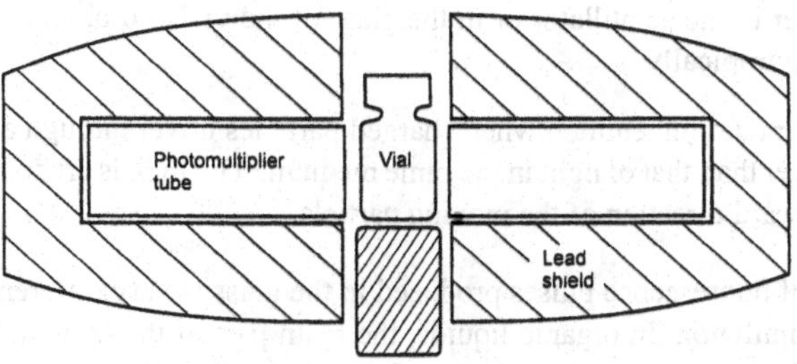

Fig. 14.4. Detector unit and shield of a coincidence system.

14.2.3 Single phototube systems

The coincidence systems were introduced to suppress the background in the low energy range due to thermal photoelectrons. Phototubes have been improved considerably over the four decades they have been used in nuclear research. Today, their photoefficiency is higher, the emission of thermal photoelectrons is much lower and their quality is more uniform.

In modern liquid scintillation counters with automatic sample changers, the additional cost of a second phototube does not increase the price of the systems significantly. When other requirements, however, become important, such as dimensions, the single phototube system may be attractive. Although these systems were abandoned because of their high background, a low-level single phototube system for radiocarbon dating was presented in 1983 (Pei-yun and Ting-kui, 1983). This system gave quite promising results, but its development seems not to have been pursued.

The author of this book wanted to transfer the technique of multicounter Geiger systems to low-level liquid scintillation systems where a large well type NaI detector was used as an active guard counter. This could only be accomplished with a single phototube system. The result of this development work has been described by Einarsson (1992). The system, Kvartett, is described in Section 15.7.4.

14.3 Scintillation processes

The process from the ionization events to the pulses coming from the anode of the phototube and their eventual electronic separation into true signal and background pulses is a good deal more complicated in liquid scintillation counters than in germanium spectrometers and gas proportional counters.

We first look at the scintillation process. Scintillation flashes are not only produced in the liquid scintillator but also in the glass of the vial and the envelop of the phototube. There are two types of interactions that result in scintillations:

1. Fluorescence, produced when charged particles ionize and excite atoms along their path, either in the scintillator or in the glass of the vial and phototube. This light is emitted isotropically.

2. Cerenkov radiation, emitted when charged particles travel through a medium with a speed larger than that of light in the same medium. The light is predominantly emitted in the forward direction of the moving particle.

The shape of fluorescence pulses produced in the glass is very different from that produced in the scintillator. In organic liquids, the main part of the light signal decays very fast, having a decay time constant of about 3 ns. A small fraction of the excited organic molecules are lifted to energy levels with longer decay times, hundreds of nanoseconds. The intensity of this component is strongly influenced by secondary solutes. It can be enhanced or suppressed. In the present context we are interested in suppressing it. In glass, the pulses are weak and their decay is controlled by energy levels with different decay times, with an effective value of the order of 100 ns. These slowly decaying pulses can therefore be separated from fast pulses.

Valid beta scintillations in the liquid sample have a very short decay time, whereas the flourescence background scintillations from the glass have a long one or they have a main component of fast scintillations (Cerenkov) followed by a slowly decaying component (glass fluorescence). This can be used to reject electronically a large fraction of both of the latter type scintillation pulses, as discussed in more detail in Section 14.6.

14.4 Background components

When we in the following speak of the background of liquid scintillation counters, we refer to its value in a ^{14}C counting window with about 70% counting efficiency with 5 ml of benzene. It is generally difficult to find in the technical literature strictly comparable background values as they are given at a counting efficiency anywhere from 60% to 85%, which can give a factor of 2–3 difference in background.

We start the discussion by presenting an overview of the past and present state of the technique. Figure 14.5 shows the distribution of background values for 14 commercial liquid scintillation counting systems used in radiocarbon dating laboratories. The background values of earlier systems (1965-1985) are taken from articles in *Radiocarbon*. Typical background values for two models of modern systems, from Wallace and Packard, are also shown in the figure. In the following discussion we assume that the ^{14}C background count rate with 5 ml of benzene of a good liquid scintillation counter of the old type is about 4 cpm.

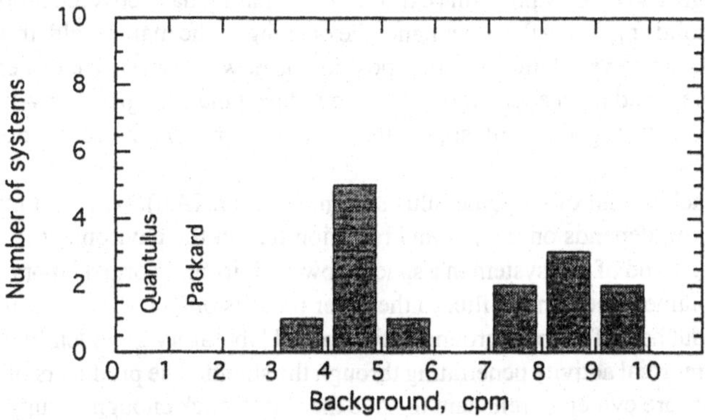

Fig. 14.5. Distribution of background values for 14 commercial liquid scintillation counters used in radiocarbon dating and typical value of two modern systems.

We divide the background B into four components in a similar way as in Section 9.2:

$$B = B_\gamma(Ex) + B(Ct) + B(\mu) + B(Neut) \tag{14.3}$$

where the terms on the right side of the equation represent the background components of:

1. external gamma radiation from radioactivity in surrounding materials,
2. internal contamination,
3. muons and their secondary radiation and
4. neutrons.

These components will be discussed in the following and their relative contribution estimated in systems where no special precautions for suppressing the background have been taken except shielding the detector unit by lead. Methods to reduce the background will be described in the next section.

$B_\gamma(Ex)$, *from external gamma radiation.* The thickness of the shield of commercial liquid scintillation counters is seldom given in their specification. The lead thickness is usually 3-5 cm (Otlet and Polach, 1990), which is not enough to reduce this component to an insignificant value. We must remember that these systems were primarily designed for work where low background is of secondary importance.

Direct evidence for insufficient shielding is seen in background values of a liquid scintillation counter system operated in Harwell over many years (Otlet and Polach, 1990). Normal low potassium vials with 14 ml of scintillator were used. In 1968 the system had a background of 14 cpm. When it was moved to a new location, apparently with lower radioactivity in the building materials, the background fell by 2.0 cpm. Some years later an extra layer corresponding to about 2 cm of lead was added above the main shield. This reduced the background by 0.5 cpm. This extra layer can hardly have covered more that about 25% of the 4π solid angle. On the other hand, the opening of the main shield, through which the sample vials are lowered into counting position, is now shielded. We can estimate that if a 4π extra shield had been added it would have reduced the background by 1.5–2.0 cpm. These improvements, together with some other, had now brought the background down to 6.0 cpm.

Even the background of the Quantulus system (Section 14.6), with its minimum lead thickness of 7 cm, depends on the gamma radiation flux in the environment. Figure 14.6 shows the background of this system in a special low-activity and normal laboratory at varying benzene volume. The gamma flux in the latter room is only 1/20 of that in the normal laboratory. About half of the background in the normal laboratory is evidently from gamma radiation of primordial activity penetrating through the shield. The producers of liquid scintillation counters are evidently reluctant to have the shield thick enough to suppress $B(Ex)$ to an insignificant value, as they do not want to make these systems too heavy.

General liquid scintillation counter systems have a thinner lead shield than the Quantulus, 3–5 cm of lead. For our standardized sample of 5 ml of benzene, with a background count rate of 4.5 cpm, it can be assumed that about 2.0 cpm comes from external gamma radiation. It is desirable to measure the background of a liquid scintillation counter with a lead shield of varying thickness in a room with a known radiation level, for example given in

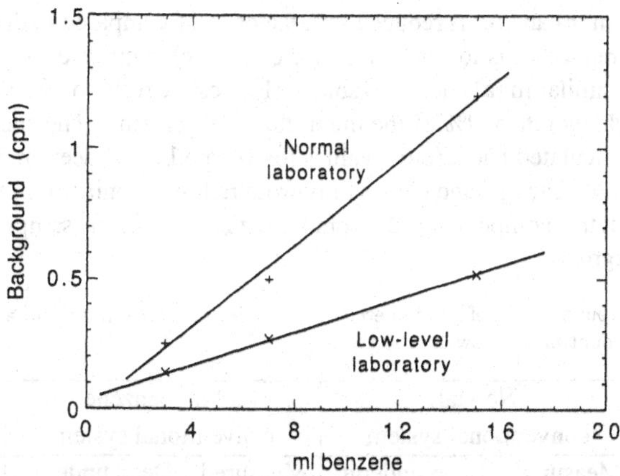

Fig. 14.6. The background of Quantulus for varying volumes of benzene in a normal and a low-activity laboratory.

nSv/h. Furthermore, the penetration of the external gamma radiation into the shield should be measured. Today, this is quite simple, using a small CsI scintillation unit as discussed in Sections 11.3.2 and 11.4.1. The phototubes must be removed when this measurement is made. Measurements of this kind will hopefully be done in the near future.

$B_\gamma(Ct)$, *background from contamination* in shield and components inside it. One source dominates, i.e., that of potassium, thorium and uranium in glass (vial and phototube envelope) and in ceramic supports inside the phototube. The background contribution depends on the proximity of the contamination to the sample and photocathode. The background pulses are produced by three different processes:

1. beta particles escaping out of the wall of the vial into the scintillator,
2. beta particles producing Cerenkov radiation, either in the vial or the glass envelope,
3. gamma radiation absorbed in the scintillator.

At present we have insufficient knowledge about the size of this component. A study carried out by Schotterer and Oeschger (1980) gives useful information. They measured the background of a commercial liquid scintillation counters at three locations:

(a) in the basement of the physics building of the University of Bern, with an overburden of 5 mwe,

(b) in a room in the underground facilities below the building (overburden 70 mwe) and

(c) in the underground ^{14}C laboratory with floor, walls and ceiling made of low-activity concrete, reducing the gamma flux by a factor of 6 inside the room.

The muon flux in (b) and (c) is reduced by a factor of 11 compared to the basement laboratory and the gamma flux is lower by a factor of 6 in (c) compared to (b). The gamma flux is presumably similar in (a) and (b). Table 14.1 gives the result of their measurements. In the underground laboratory 9% of the muon flux still remains. The second and fourth columns give the calculated background values for a hypothetical deep underground location where the residual background comes only from radiocontamination in the glass. The results indicates that this component is 2.5 cpm for the 22 ml benzene sample, or about 13% of the surface background.

Table 14.1. Background (cpm) of a commercial liquid scintillation counter and a modified system in different rooms. Counting window 15-150 keV.

Laboratory	No vial conventional system		22 ml benzene conventional system		49 ml benzene Special system
	Measured	Deep undergr. Calculated	Measured	Deep undergr. Calculated	Measured
Basement	13.3		19.9		7.8
Underground	3.2	2.3	10.9	10.1	
^{14}C laboratory	1.5	0.6	3.3	2.5	1.6

In a second experiment the two-tube system was removed from the lead shield and put inside a plastic tube which was shielded by 10 cm of lead. A special 50 ml vial of radiopure delrin filled the space between the two tubes and utilized their whole end surface area. It is interesting to note that the background is lower in this special system compared to the conventional one, in spite of its nearly two times larger benzene volume. This may be explained by a difference in vial material, as it can be assumed (although not explicitly stated in the paper) that a glass vial was used in the commercial system. If this is correct, about half of the residual background count rate per ml of benzene comes from radiocontamination in the glass vial and half from the photomultiplier tube.

Direct beta pulses from radioactivity in vial material. These pulses can either come from energetic beta particle induced Cerenkov pulses in the glass or from beta particles escaping from the wall of the vial into the scintillator. Plastic vials are free of radiocontamination. Their use in low-level counting poses, however, serious problems as counting times are long, which can lead to serious sample loss through evaporation. Furthermore, in precision work the vials must be washed and reused, which is difficult with conventional plastic vials. Teflon vials have been used a good deal in the past in work of this kind, but most scientists seem to have given them up and accepted the somewhat higher background of glass vials. Figure 14.7 shows the spectra of background and ^{14}C standard (benzene) measured in glass and teflon vials in a Wallac Rackbeta 1217. The spectra of the standard show that the light collection efficiency is nearly the same for both vials. The glass vial gives a large background contribution at low energy coming from beta contamination. The energy scale is logarithmic. The end point of the ^{14}C spectrum gives the energy channel of 150 keV. It would be very interesting to know the energy spectrum of the beta contamination in the glass vial, but the figure does not allow any such conclusion. This a typical example of the serious disadvantage of the logarithmic energy scale so common in liquid scintillation counting systems.

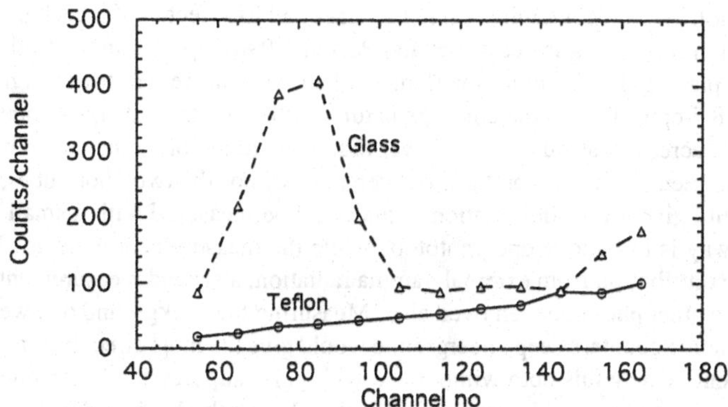

Fig. 14.7. Spectra of background and ^{14}C standard measured in glass and teflon vials in a Wallac Rackbeta 1217. The glass vials give a large background contribution at low energy. The energy scale is logarithmic. From Rauert et al., 1988.

The concentration of radiocontaminants in glass vials is rarely given despite the fact that it is relatively simple to measure. Low potassium borosilicate glass is most frequently used, but its thorium and uranium content is similar to that of ordinary glass. With 5 ml of benzene in the vial the effective inner surface area facing the liquid, from which beta particles can enter the scintillator, is about 15 cm^2. We assume that the number of escaping particles is equal to the number of disintegrations in a layer with a thickness of 0.02 g/cm^2 (Section 9.2.5). The beta activity in the wall will then give about 0.02 cpm per Bq/kg. These beta particles simultaneously excite glass flourescence giving a slow pulse in the scintillator, which can be used to reject these background pulses.

Bowman (1989) measured the difference in background of 5.5 ml of benzene in teflon and low-potassium glass in two different systems and got 0.33 cpm and 1.12 cpm respectively. The former result was for a LBK (Wallac) "Kangaroo" system that rejects a good deal of the Cerenkov pulses. It would increase the usefulness of studies of this kind if the contamination in the glass is measured. Today, this is easy to do with a germanium spectrometer.

Cerenkov scintillations. Energetic beta particles from primordial radioactivity in the vial and the envelope of the phototube, produce Cerenkov scintillations that may be detected by both tubes. These particles produce simultaneously slowly decaying glass fluorescence that can be used to suppress these pulses.

Gamma absorption in scintillator. Photons from the radiocontamination in glass may be absorbed in the liquid sample, giving a fast pulse. When these photons come from nuclei in the uranium or thorium series, a beta particle is emitted simultaneously, which excites glass fluorescence with slow decay. This may be used to reject the pulse. When, however, the photon comes from ^{40}K, no beta particle is emitted as the nucleus disintegrates through electron capture.

Information on radiocontamination in the glass and its effect on the background of liquid scintillation counters is scarce. Otlet and Polach (1990) reported in the study described above that replacing the phototubes with new alkali tubes decreased the background from 10.0 cpm to 8.5 cpm. Due to the complex nature of the background signal, one cannot be sure that this decrease was due only to lower radiocontamination in the new tubes.

A measurement of the photon flux between the faces of the two phototubes, where the vial sits, would give useful information. This could be measured with a small CsI diode. The easiest way is to remove one phototube while the measurement is made. In order to separate the contribution from external gamma radiation, a second measurement should be made with the other phototube removed also. Measuring the background of a well shielded liquid scintillation counters deep underground would give the most direct information about this component. Doing this both without and with pulse suppression based on analysis of slow decay component, would show how effective this method is in reducing scintillation pulses produced in the glass.

$B_\gamma(S_T)$, *component due to secondary gamma radiation* interacting with scintillator molecules, is probably not a dominating component, as in germanium diodes and gas proportional counters, but it is a significant one. It can only be reduced by the application of an anticosmic counter or by operating the systems deep underground. This radiation is mainly produced by muons in the shield, as discussed in Sections 9.2.6 and 12.5. It can be assumed that the secondary gamma ray flux, to which liquid scintillators and germanium diodes respectively are exposed to inside their shields, is nearly the same. We can therefore assume that the photons give nearly the same number of pulses per unit mass for both detectors, about 100 cpm/kg, according to Section 12.10. This would give about 0.4 cpm for a sample of 5 ml of benzene. These pulses have a fast decay and can only be suppressed by anticosmic counters. The secondary radiation can also give Cerenkov scintillation and glass flourescence through energetic Compton electrons released in the glass of the vial and in the phototube envelope. Pulse decay analysis can reduce the number of these pulses.

Muon background component (direct). The muons can give background pulses either by direct ionization in the scintillator or through Cerenkov scintillation in the glass. We look first at direct pulses in a sample of 5 ml of benzene. In a laboratory with small overburden the muons traversing the liquid scintillator give about 4 cpm. Most of the pulses produced are much larger than those normally of interest and fall above the sample counting channel. These muons will also go through the glass of the vial and produce a slow fluorescence pulse, which helps further in reducing this muon component in systems with pulse decay analysis.

Muons going through the glass of the vial above the liquid or through the envelope of the phototube produces both Cerenkov and glass flourescence. These scintillations may be detected by both phototubes and thus be recorded. Their background contribution can, however, as discussed above, be reduced by rejecting pulses accompanied by the slowly decaying glass flourescence. Muons hitting the ends of the phototubes may give Cerenkov scintillations that will be detected. The flat end face is naturally most sensitive, but it seems likely that the cylindrical part of the tubes close to their front end is also sensitive. If we make the simplifying assumption that the glass in a width of 1.0 cm of the end of the two phototubes and vial contain the glass mass that is sensitive to Cerenkov pulses produced by muons, this gives an area of about 15 cm^2 through which about 12 muons pass per minute. In view of the background of the old liquid scintillation counters without pulse decay analysis, this is evidently an overestimate. Counting background pulses that are in coincidence with a small flat guard counter above the special liquid scintillation counter unit mentioned above would give valuable information about the muon/Cerenkov background contribution.

Neutron component. Neutrons contribute to the background of gas proportional counters. They must therefore also give background pulses in liquid scintillation counters. The recoil pulse height spectrum to be expected has been discussed in Section 13.9 and it is shown in Figure 13.10. The range of the protons with a maximum energy about 2 MeV is only a fraction of a mm in the scintillator so they lose all their energy in the vial. Most of these recoiling protons give a pulse height that is above the region of general interest and will therefore be eliminated by the upper level discriminator. Recoiling carbon atoms, on the other hand, would have a maximum energy of about 450 keV and most of their pulses would be registered in a typical ^{14}C counting window.

In order to evaluate the neutron background contribution in 5 ml of benzene, we can use the estimate of the neutron component in a CO_2 gas proportional counter in a surface laboratory which gives 0.27 cpm per gram (Section 13.9). We can then expect a neutron background component of 1.0 cpm in 5 ml of benzene, or about 20% of the total background. This value is probably too high as discussed in next section. No systematic study seems to have been made of a neutron background component of liquid scintillation counters. It could be determined with a special detector unit inside a 10 cm thick lead shield by measuring the background both with and without an inner boronated paraffin layer.

14.5 Atmospheric background fluctuations

When well shielded liquid scintillation counter systems are brought deep underground, where the cosmic ray flux is reduced to an insignificant value their background is reduced to 20–30% of its surface value (Calf and Airey, 1982, Einarsson and Theodorsson, 1991). For commercial systems, where component from external radiation is significant, this reduction factor is smaller. This shows that the cosmic rays contribute up to about 3/4 of the total background, the rest coming from external and internal primordial radioactivity in well shielded systems. The cosmic ray component is modulated by atmospheric parameters as discussed in Section 13.10. One cm increase in the atmospheric pressure reduces the muon component by 2.0% and the neutron component by 10%. Furthermore, an increase in the height of the 100 mb level reduces the muon component by 5.3%/km. These background fluctuations are discussed in detail in Section 13.10. The same analysis applies directly to the background of liquid scintillation counters.

Corrections for these fluctuations are practically never applied directly to individual measurements in liquid scintillation counters. Instead, a batch of 20-30 samples are measured in a rotating mode for 100 minutes each, or to a predetermined number of counts in case of weak samples. Typically each batch usually has 4 background samples, 5 modern standards and 15 unknown samples. Each cycle takes about one day and the total counting time for the batch is about 3 weeks. The background samples take about 30% of the total counting time. The average count rate of all four background samples is subtracted from the mean count rate of each unknown sample. This simple measuring routine is natural in a system with automatic sample changing.

This counting mode gives a good compensation for atmospheric background fluctuations. It has, however, two disadvantages. In order to get sufficient accuracy, more time is spent on background samples than in the method described below and control is not kept on the long term stability of the system.

More accurate background values can be obtained by using the inherent long time stability of these systems as demonstrated by G.W. Pearson (1983) in his high precision work in establishing the dendrachronological correction curve in radiocarbon dating. In this way a higher background precision is obtained with shorter counting time. All background measurements over a whole year are then corrected for deviation from standard atmospheric pressure (76 cm Hg). This gives amplitude of the seasonal variation of the muon flux. The correlation between corrected background and height of the 100 mbar level can now be determined. When this has been found the background value of each sample measurement can be calculated by taking into account the mean atmospheric pressure over the counting time and the height of the 100 mbar level. This gives more accurate background values with less total measuring time than batch counting. The corrected background value thus determined should only show slow long time variations with a period of 11 years, following the sunspot frequency. There can however occur short time intensity variations during periods of unusual solar activity, as discussed in Chapter 7. Pearson (1983) demonstrated nearly full stability of his liquid scintillation counting system (produced 1960) over a period of three years.

Pearson found a seasonal background variation of 0.20 cpm due to a 0.64 km summer/winter difference in the height of the 100 mbar atmospheric pressure level. This difference in height should give $0.64 \times 5.3 = 3.4\%$ seasonal variation in a background that has been corrected for variations in atmospheric pressure. This shows that the mean background contribution of the muons is $0.20/0.034 = 5.9$ cpm of a total of 7.9 cpm. According to this, the muon background component of his system is 74% of the total background, in good agreement with the estimate given above.

Pearson found further that the background decreased by 2.0% for each cm Hg increase in atmospheric pressure. From this we can calculate that the neutron background component should be 0.8 cpm and the constant residual background (from primordial radioactivity) 1.2 cpm. According to the estimate in last section we should expect a neutron background component of 3.0 cpm in Pearsons 14 ml samples. This indicates that the estimate of neutron background in gas proportional counters may be too high. These results, however, must also be taken with reservations due to the uncertainty in the 0.20 cpm for seasonal variation.

14.6 Background reduction

There are six methods to reduce the background of liquid scintillation counters:

1. Massive shield, usually of lead.
2. Counting window setting.
3. Pulse amplitude comparison.
4. Pulse decay analysis.
5. Masking vials.
6. Anticoincidence discrimination.

The fist method, surrounding the detection system has already been discussed. A counting window, consisting of two discriminators, the lower and the upper, has always been used. A more recent variation of it will be described. Pulse amplitude comparison has been known for some 25 years, but is probably not used in all systems. Pulse decay analysis, introduced by Packard in the mid 1980s and developed further since that time, brought the low-level liquid scintillation counting technique systems with an important background reduction by relatively simple means. Anticoincidence shielding was tried in the 1970s in various laboratory systems but with limited success. The introduction of the Quantulus system by Wallac (1984), with its annular liquid scintillation guard counter, presented a breakthrough in low-level liquid scintillation counting.

Counting window. The counting window, i.e., the pulse size interval determined by the low level and high level discriminators, influences both the background and the counting efficiency. These two parameters are empirically related. For a given accuracy one should select a counting window that gives the highest number of measured samples during a given time interval. The Figure of Counting Capacity, FCC = S/(1+B/S), should be maximized, as discussed in Section 1.6. Usually the lower level discriminator is factory set where practically all the smallest pulses, i.e., single photoelectron pulses, trigger the coincidence discriminators. The discriminator after the summing amplifier can always be set by the user. It may be set for example at a level where pulses corresponding to a total of 10 photoelectrons in the two tubes. A useful background decrease may be obtained if the discriminator of both channels, before the coincidence circuit, are set at a value of corresponding to pulses where, for example, at least three photoelectrons are released in each phototube, as many of the small pulses from a Cerenkov scintillation in one phototube only gives few photoelectrons in the opposite tube.

Pulse amplitude comparison (or disparity ratio). This method of background reduction also uses the fact that Cerenkov pulses produced in one of the phototubes, usually give much fewer pulses in the opposite tube. Figure 14.3 depicts schematically a modern liquid scintillation counting system with pulse amplitude comparison and pulse decay analysis. A special circuit is added in modern systems that calculates the ratio of the larger pulse to that of the smaller and if this ratio exceeds a given value the pulse is excluded from the counting channel. This technique is least effective in tritium counting as the pulses of each phototube are inititated by a few photoelectrons, with large Poisson scattering. In this case, plastic vials, which are free of radiocontamination, are a better choice. Kaihola (1993) found that 8 ml of water and 12 ml of a scintillator gave a background count rate of 2.7 cpm in a glass vial and only 0.42 cpm in a teflon vial. As high precision is rarely needed in tritium measurements, plastic vials can be safely used.

Pulse decay analysis. The counting channel is set to accept only fast decaying pulses, pulses where the anode signal has fallen practically to zero after 10 ns. As discussed above, slowly decaying glass flourescence is excited at the same time as Cerenkov scintillation, produced by muons, electrons or beta particles in the glass of the vial or the phototube envelopes. The tail of the slowly decaying fluorescence is composed of a series of individual pulses of similar size, all initiated by single photoelectrons. The presence of this tail can be identified and its size measured either by counting the single photoelectron pulses (digital analysis), starting for example 10 ns after the appearance of the initial pulse (the Packard method), or by integrating the tail and compare it with the fast pulse (analog analysis). If the tail (or a burst of afterpulses) is some specified fraction of the main pulse, that may depend on its size, the pulse is rejected. Both the digital and analog analysis should lead to nearly the same result. This method is apparently quite effective for the reduction of Cerenkov pulses from glass of vials and phototube envelopes. Its use requires, however, a good knowledge of the underlying principle as described below.

When benzene is counted a small part of the molecules excited by the sample beta particles are lifted to an energy level with a long decay time (triplet states), of the order of tens of nanoseconds, as will be discussed in more detail in Section 14.7.3. Some of the true beta pulses may then be classified as background pulses by the decay analysis circuit, reducing the detection efficiency. Secondary solutes can suppress the excitation of triplet states. Cook and Anderson (1992), who studied this in detail, recommended a combination of butyl-PBD and bis-MSB as secondary solutes in benzene. In the Packard low-level liquid scintillation counting systems, the reduction of muon Cerenkov pulses can be enhanced by an annular cylinder surrounding a slim 7 ml vial, made of a plastic scintillator guard with a relatively long decay time. The two phototubes receive also the flourescence light from this guard, but its pulses are easily recognized by the long decay time.

The effect of this simple extra scintillator is apparently to enhance the size of the slow pulse from muons and energetic electrons that traverse this cylinder before passing the sample. This shows that without this extra scintillator, the tail of some of the Cerenkov pulses is too weak to give a rejecting signal. This is especially important when the discriminator must lie at lowest possible value, as for example in the case of tritium measurements.

Masking vials. A significant number of the Cerenkov pulses produced in the phototubes can be rejected by pulse amplitude comparison. The number of photoelectrons produced in one phototube from Cerenkov radiation can be reduced significantly if the part of the vial above the surface level of the sample is masked by an opaque layer, for example black tape. This transmission of light between the phototubes is called *cross-talk*.

Anticoincidence background suppression. The advantage of the early liquid scintillation counting systems lay in their favorable price and reliability as they were mass produced for a large market and the systems were continuously improved, using the latest electronic technique. Their weakness, compared to gas proportional counting, lay in their high background, which evidently came from the cosmic rays. It therefore represented an important step in the progress of low-level liquid scintillation counting when a Finnish firm, Wallac, at the initiative and through the involvement of Henry Polack, started the development of a system dedicated to low-level counting, using an active guard for suppressing the cosmic ray background component. This system, the Quantulus, was presented in 1984 (Kojola, 1985).

Figure 14.8 depicts a cross section of the detection unit, active guard and the shield of Quantulus. The minimum thickness of the asymmetric shield is 8 cm. The guard detector is a long annular liquid scintillation counter where the two phototube sample detection system is inside an axial acentric tube in the guard counter. This arrangement gives a background reduction of a factor of 10. The guard counter does not suppress Cerenkov pulses produced by beta particles of primordial radioactivity in the glass. The system, however, presumably incorporates background pulse rejection through pulse decay analysis and pulse height comparison, although this is not given in its specifications.

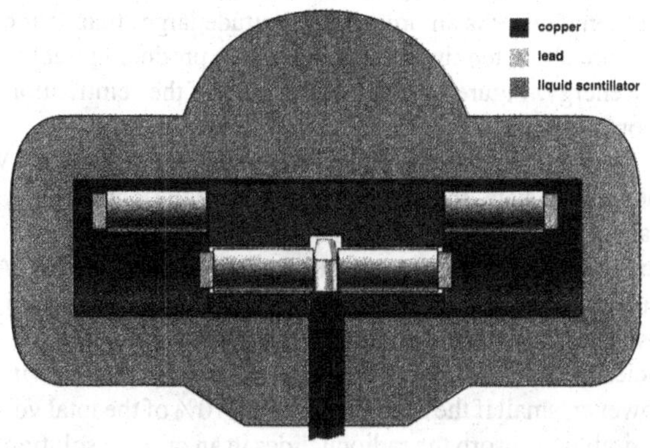

Fig. 14.8. Cross section of the detection unit, active guard and shield of the Quantulus.

14.7 Liquid scintillation alpha counting

The assay of weak alpha activity with liquid scintillation counters competes in many cases successfully with Si(Li) diodes, which have very high energy resolution, low background and relatively low price. The main advantages of liquid scintillation counting are:

1. Practically 100% detection efficiency.
2. Useful energy resolution.
3. Very low background.
4. Large samples.
5. Automatic sample changing.

The basis of modern alpha liquid scintillation counting, that is alpha spectroscopy and discrimination of alpha/beta pulses, was laid late in the 1950s. The exploitation of these techniques, however, required complicated electronics and they first became generally available in commercial systems in the 1980s when the technology of microprocessors and large scale integrated circuits had been well developed.

14.7.1 Scintillation of alpha samples

Because of their very short range in the scintillator, alpha particles lose all their energy in the liquid sample. The resulting pulse size is proportional to the deposited energy. The scintillating yield (photons per keV of absorbed energy) of charged particles falls with increasing ionization density. Because of the large mass of the alpha particles compared to that of electrons, their ionization density is an order of magnitude larger than that of electrons. Alpha particles are therefore about ten times less effective in producing light as beta particles and electrons of same energy. Figure 14.9 shows the ratio of the scintillation yield of these particles as a function of energy.

Unlike beta, alpha particles are emitted in monoenergetic groups. When all decaying nuclei end in the ground state, the alpha particles have the same energy. The spectrum of monoenergetic alpha emitters therefore show corresponding peaks.

Pure aromatic counting samples give the highest scintillation yields. After chemical separation the nuclides to be measured are, however, frequently in an aqueous solution. In this case a water miscible scintillation cocktail must be used rather than one that offers highest scintillation efficiency. The addition of the aqueous solution will inevitably cause quenching, which is, however, small if the addition is only 5-10% of the total volume of the sample. It is therefore desirable to absorb the radionuclides in an organic solution, that usually gives small quenching. This is often possible through liquid-liquid extraction. The concentration of the radionuclide to be measured is frequently 10 to 100 times higher in the extracting organic liquid, which is subsequently dissolved in the scintillation cocktail. Organophilic extractants such as alkyl/aryl phosphoric acids give negligible quenching.

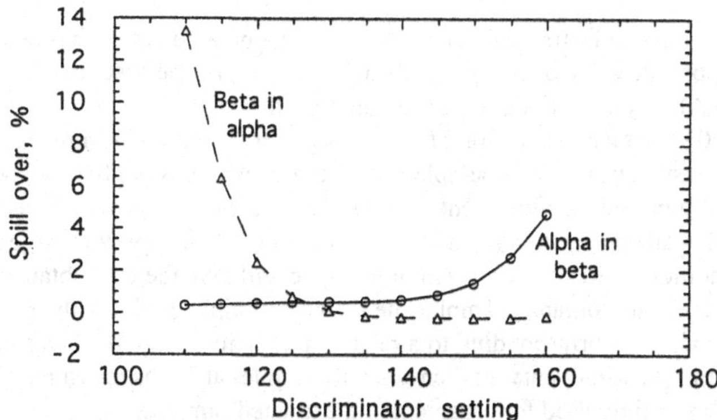

Fig. 14.9. The ratio of scintillation yield of beta to alpha particles as a function of energy.

Pulse decay separation also depends on the scintillation cocktail and does not work on all sample compositions and is influenced by secondary solutes.

14.7.2 Energy resolution

In the assay of alpha radioactivity the separation of pulses of alpha particles of different energies is a valuable asset. The resolution R of the detectors used is therefore an important parameter. It is defined as

$$R = FWHM/H_0 \tag{14.4}$$

where FWHM is full width of the peak at half maximum and H_0 is the pulse size at the peak centroid. The width of the peak produced by monoenergetic alpha particles is determined by the number of electrons, N_{pe}, released from the cathode of the phototube. Their number have a Poisson distribution. This leads to a resolution given by

$$R = 2.35\sqrt{F/N_{pe}} \tag{14.5}$$

where F is the Fano factor and N_{pe} is the number of released photoelectrons (Knoll, 1988, page 116). According to Equation (14.1) N_{pe} is given by

$$N_{pe} = \varepsilon\, S\, E_a(1 - f_a)G$$

In order to achieve best possible resolution, the user should select a scintillating cocktail with highest possible scintillation efficiency and minimize the quenching effect of the dissolved sample. He must also take into consideration that the composition of the sample may affect the possibility to separate alpha and beta particles, as discussed in the next section. The producer of the system has no doubt done his best to maximize light collection efficiency, but, as discussed below, it is affected by the type of vial.

One of the most important features of modern liquid scintillation counting systems is their automatic sample changing. This requirement limits the freedom of the designers to maximize G, the light collection efficiency. A manually operated system has been designed where highest possible value of G is probably achieved, giving the best possible resolution. This is the PERALS system produced by an Amercan firm, Ordela.

Figure 14.10 shows a cross section of its detector unit. The liquid sample is in a culture tube immersed in silicone oil in the sample compartment, which is sandwiched between the face of the phototube and a hollow light reflector. Light collection is maximized both with this close sample/cathode geometry and by eliminating liquid/air gaps with large refraction-index discontinuities. Evidently, a system of this type will give the best obtainable resolution. With optimal, pure organic, samples the FWHM of alpha peaks with an energy of 5 MeV is about 230 keV, corresponding to a relative resolution of 4.6%. The small sample size (about 1.5 ml) is a disadvantage when aqueous samples are to be measured. These systems are therefore mainly used for organic liquid extracted samples.

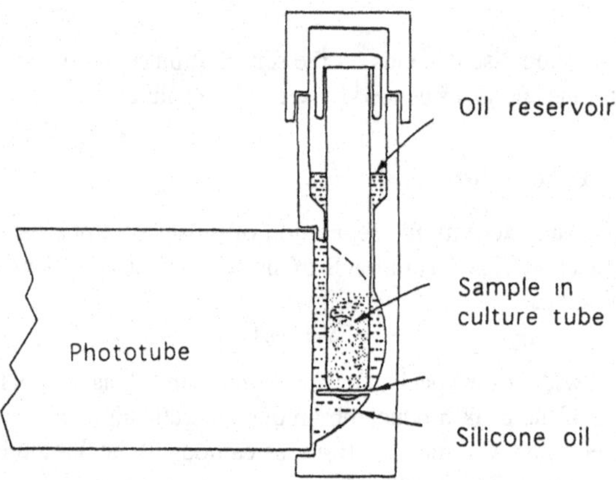

Fig. 14.10. Cross section of the detector unit of PERALS. From McDowell and McDowell, 1994.

Conventional liquid scintillation counting systems have lower light collection efficiency, severely influenced by the type of vials used. Yu-Fu et al. (1990) report a FWHM value of 280 keV for alpha energy of 5.76 MeV for a sample in an extractive cocktail, using teflon vial and the Quantulus system. HDEHP extractant was used in a toluene/naphtalene scintillator. It is interesting to see how the peak width depends on the type of glass:

Type of vial	Teflon	Polyethylen	Etched glass	Glass
FWHM, keV	280	290	340	390

The FWHM depends on the amount of HDEHP in the scintillator as seen in the following results:

HDEHP, %	5	10	15	20
FWHM, keV	280	300	325	360

14.7.3 Pulse decay analysis

In spite of their much higher energy, the alpha pulses are superimposed on a background of beta pulses from the sample and from environmental radiation discussed in Section 14.3 because of the low scintillation efficiency. This background can be reduced by two orders of magnitude by pulse decay analysis, also called pulse shape analysis, or pulse shape discrimination, depending on the producer of the system.

When charged particles interact with the liquid, the molecules can be excited into different energy states having different decay times. The cocktails used in liquid scintillation counting have a fast component with a decay time of about 3 ns and a slow component, from triplet states (S3), with a decay time of about 200 ns. In stilbene, a much studied organic crystal, the fast and slow components have decay times of 6 and 370 ns. Most of the light is in the fast component. G.T. Wright discovered in 1956 that the distribution between the energy states depends on the nature of radiation. Electrons excite mainly the fast component, but the densely ionizing alpha particles and protons produce a considerably larger fraction of the slowly decaying triplet states. The decay of flourescence induced in stilbene by electrons, protons and alpha particles is shown in Figure 14.11.

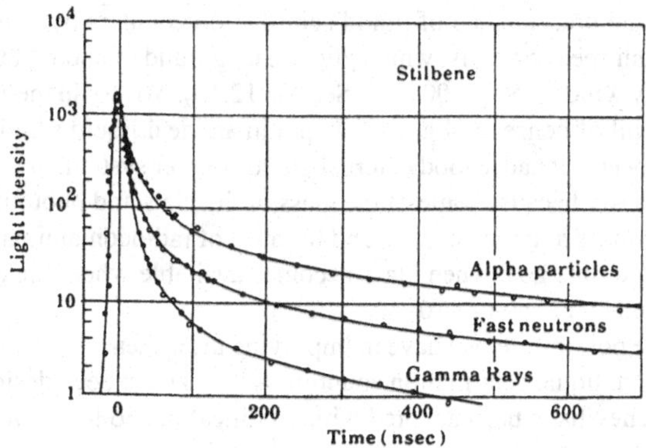

Fig. 14.11. The decay of flourescene induced in stilbene by electrons, protons and alpha particles.

In order to separate the alpha and beta/electron pulses, the difference in time from the instant a pulse rises above a given discriminator level until it falls below a second discriminator is measured. A hypothetical distribution of these time intervals for an alpha/beta sample is depicted in Figure 14.12. The separation point is usually selected where the alpha detection efficiency is still close to 100%, but the contribution of beta/electron pulses has been reduced to an insignificant value.

The separation curve depends on the type of scintillation cocktail, and a new curve must be measured for each type. The addition of naphtalene to conventional toluene and xylene based cocktails enhances the separation. Furthermore, quenching of individual samples may also affect the separation.

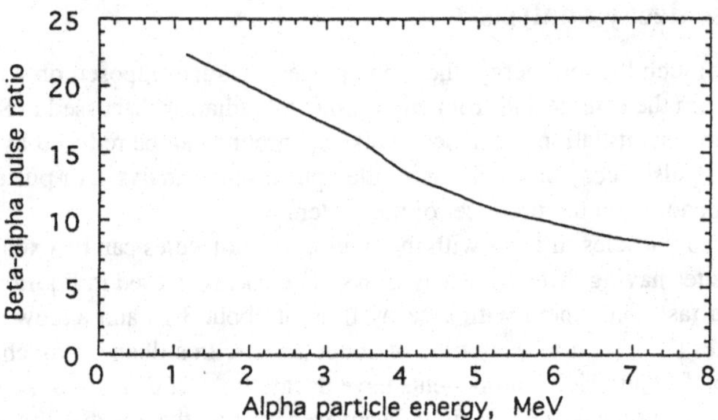

Fig. 14.12. Separation curves for alpha and beta pulses using pulse shape analysis.

14.8 Future prospects

Let us first look at the present state of liquid scintillation counters by comparing them to low-lewel germanium spectrometers, which give a background of about 100 cpm/kg of germanium in an energy window 50–1500 keV (Section 12.10). We could then possibly expect a background for 5 ml of benzene (4 g) of 0.4 cpm in an ideal liquid scintillation detector, and a factor of 10 lower if it had a good external guard counter system, or 0.04 cpm. We are far from this goal, partly because some of the glass of the vials and phototubes constitute a part of the effective mass and increases it, and because of radiocontamination in the glass. We might get closer to this goal when glass becomes available where the radiocontamination has been reduced by a factor of 10–20.

Let us look at the possibilities we have in improving the systems with the present type of glass. General modern liquid scintillation counting systems have been designed for a broad spectrum of uses. They have been adapted with modifications to low-level counting. It is only natural that the result is not satisfactory for some of present low-level counting work. Because of the long counting times we need a larger number of detectors. When each sample is counted for one day or longer, we can sacrifice automatic sample changing. We should look at a system with a 10 cm thick lead shield and anticosmic background reduction and of simple design.

The Kvartett system, described in Section 15.7.4, has demonstrated the possibilities of a multicounter system with single phototube detector units. The multidetector concept should be developed further. A multicounter system based on the single phototube type offers an attractive solution, similar to the gas proportional system described in Section 13.15. Figure 14.13 depicts this system. It has 9 sample channels. The background count rate would probably be similar to that of Quantulus. It would have a single guard counter sandwiched between two 5 cm thick layers of lead. Its total mass would be about 400 kg. We could expect still lower background by operating the system, for example, under an overburden of

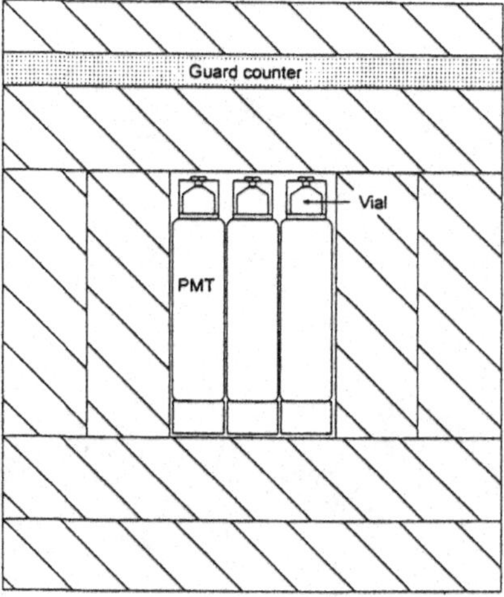

Fig. 14.13. A proposed multi sample liquid scintillation counting system with 9 sample channels.

10 mwe and with the new generation of radiopure photomultiplier tubes that we can expect to come on the market in a few years.

Recommended reading

- Introduction to alpha/beta discrimination on liquid scintillation counters. Charles Dodson. Beckman Instruments, Inc., 1991.

Chapter 15

Other low-level counting systems

15.1 Introduction

The three main detector systems used for the assay of weak beta and gamma radioactive samples have been described in the last three chapters: systems with germanium diodes, gas proportional counters and liquid scintillation detectors. In this chapter we discuss:

1. Si(Li) diodes.
2. ZnS scintillation detectors.
3. NaI scintillation counters.
4. CsI scintillation counters.
5. Phoswich detectors.
6. Multidetector systems.
7. Germanium diodes with a Compton guard.

Si(Li) diodes are equally important in measurements of weak alpha activity as the three detectors, described in Chapters 12–14, are for beta and gamma activity. However, as there are few interfering factors, Si(Li) diodes do not warrant a special chapter.

15.2 Si(Li) detectors

Unlike germanium, intrinsic silicon, i.e., where the impurities have been reduced to a level where they contribute insignificantly to the resistivity, cannot as yet be produced. The residual impurities, usually of the p-type, must be compensated for by acceptor atoms of the same concentration. Lithium drifting has been the main process for this compensation. Highest purity silicon is usually of p-type, where acceptor atoms dominate the conductivity. Donor atoms must therefore be added to compensate for this impurity conductivity. Lithium, which tends to form interstitial atoms in the silicon crystal, is usually applied.

Lithium from a heavily doped surface layer is drifted by an electric field at elevated temperature (about 40 °C) into the Si crystals. In recent years, the technique of ion implantation has been used increasingly. In this production method, a beam of accelerated monoenergetic lithium atoms with a well defined range are buried in the silicone crystal, leaving a closely controlled lithium concentration that accurately compensates for that of the impurity atoms at each depth. The deposition depth of the lithium ions is varied by changing the acceleration voltage. This process is more expensive than diffusion, but the implanted diodes have various advantages over the diffused type, the most important being that their surface can more easily be cleaned. Diodes of both these types are designated Si(Li).

At the n-p junction there is a repulsion between the majority carriers in each layer and a depleted region is formed that widens as reverse bias is applied. This region is the sensitive layer, and its thickness is from 100 to 1000 μm. The cost of the diodes is nearly proportional to the depletion depth. Table 15.1 gives the range of alpha and beta particles in silicon. Because of the large bandgap of silicon, the leakage current of the Si(Li) diodes is very small and they can be operated at room temperature. Furthermore, the migration rate of the Li atoms is negligible.

The Si(Li) diodes are mainly used for the assay of alpha activity, but they are also useful for spectrometric measurements of beta activity and low energy photons. In the latter case the low atomic number is an advantage as fewer beta particles will be backscattered.

Table 15.1. Range of alpha and beta particles in silicon.

Range μm	Electron keV	Alpha MeV
100	15	15
300	31	55
500	45	85
700	52	105
1000	73	130

The alpha particle resolution of the Si(Li) diodes is very high, and is highest for small diodes in which the noise is lowest. The FWHM of the peaks is about 12 keV for the smallest diodes and 35 keV for the largest. In order to utilize fully this good resolution, the energy loss of the alpha particles before they hit the diode must be minimized. The diode and sample planchet are therefore always in an evacuated chamber in order to eliminate energy loss in the gap between the sample and diode and the sample layer is made as thin as possible. All the alpha emitting nuclides can form metallic ions that can be deposited in a very thin layer on a metallic plate, either by electrodeposition on stainless steel or through self-deposition on silver.

Silicon is one of the purest available materials and its contribution to the background of the diodes is negligible. Background alpha particles can therefore only come from the

the plate on which the sample is deposited, from contamination on the diode window from earlier samples or from radon and its decay products. New diodes have a background count rate of about 2 cpd/cm^2 for the energy range 3–8 MeV or about 0.04 cpd/cm^2 in a typical alpha window of 100 keV. This low count rate can only be maintained by careful use of the systems. High activity samples should, for example, never be counted with diodes intended for the lowest activity except for short periods.

Atoms can be ejected by recoil from the sample planchet and adsorbed on the diode window. If these decay product atoms are radioactive, they will increase the background of the diode. A low residual gas pressure is usually maintained in the counting chamber, and the recoiling atoms are stopped before they reach the detector. As they have a positive charge when formed in decay, they can be pulled back to the planchet if the sample holder is connected to a low negative voltage of about 10 volts. This can greatly reduce the contamination problem. Atoms can also sublime from the sample, especially those of polonium. A low residual pressure also mitigates this problem.

15.3 ZnS alpha scintillation detectors

The fundamental principles of ZnS alpha detectors are described in Section 8.4.5. The detector units come in two different forms: (1) scintillation vessels (or cells) for the measurement of radon and (2) flat ZnS screens close to a photomultiplier tube, for the measurement of solid alpha samples.

Radon scintillation cells, Lucas cells. Scintillation cells for the measurement of radon have been used for four decades. They are made in a variety of sizes and shapes. These cells are descendants of the scintillation cell introduced by Lucas (1957), shown in Figure 15.1. A detailed report on the long experience of their use has recently been published (Lucas, 1995). The cells are usually made of metal with a quartz window facing the photomultiplier tube. The inner side, except the bottom facing the phototube, is coated with 20–40 mg/cm^2 of ZnS. Their volume is 50–200 cm^3 and they have a detection efficiency of about 0.7 counts per alpha disintegration.

During the last 15 years, large cells have been developed in order to increase sensitivity. Their operating parameters have been studied by Cohen et al. (1982) and Stoop et al. (1993). Cohen et al. reported that the light collected by the photocathode decreases by a factor of 2 for each 7 cm increase in length. This limits the length of the chambers to about 50 cm. For the final size Cohen et al. chose a length of 38 cm with a thick scintillator coating (100 mg/cm^2), which also served as a reflector. A cell with a MgO reflector on the outer side of the acryl chamber, or under a thin ZnS layer on the inner side, gave poorer results than a thick ZnS layer. The total volume of their final chamber was 3.0 liters, its detection efficiency 0.5 counts per alpha disintegration and the background 0.5 cpm.

Usually there is an air gap between the photomultiplier face and the window (bottom) of the scintillation cell. This is more convenient when the cell is frequently removed for filling. Optical coupling, using silicon oil, increases the light collection efficiency by about 40%. For small cells, where all alpha pulses are well above the discrimination level, this is

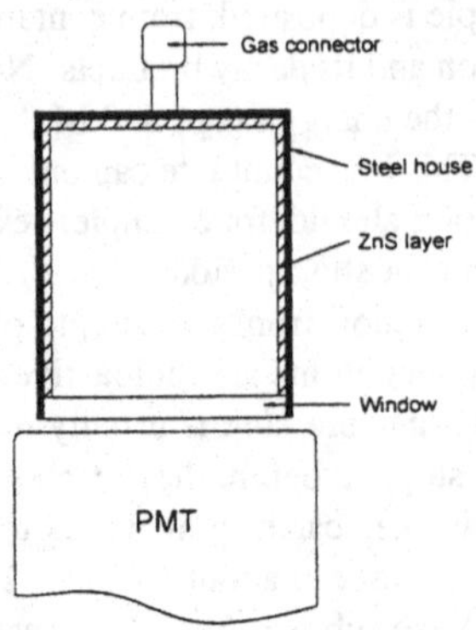

Fig. 15.1. A modern version of a Lucas cell.

of little consequence. For large cells it is, however, better to have the cell optically coupled to the phototube.

The very low background of ZnS alpha scintillation cells is achieved by eliminating the small pulses coming from other charged particles, mainly from cosmic Cerenkov radiation in the phototube and muons passing through the ZnS layer, so that only the larger pulses of alpha particles are counted. The reported background count rate per unit surface area of the ZnS layer (cpm/cm^2) varies by an order of magnitude. The carefully constructed acryl cells made by Liu et al. (1993), which were discussed in Section 8.4.5, probably give the lowest background, 1.0×10^{-4} cpm/cm^2. This is a factor of 2 lower than the alpha background of Si(Li) diodes described in the last section, and also lower by the same factor than the emission of alpha particles from stainless steel plate II, described in Section 8.2. This difference is probably not significant.

Most ZnS scintillation chambers give 5–10 times higher background. Liu et al. (1993) tested ZnS from various sources and found a difference of an order of magnitude in its radiopurity. A thorough study of lowest obtainable background seems not yet to have been carried out. When the cells are made of acryl, which is generally completely radiopure, the background pulses can only come from alpha contamination in the ZnS. There is no reason to have the ZnS layer thicker than 10 mg/cm^2. However, the crystal size of the ZnS may limit the minimum obtainable thickness, and care must be taken to secure close to full ZnS coverage in order to obtain 100% detection efficiency.

It is interesting to try to estimate the radiocontamination in the ZnS when the background is 1.0×10^{-4} cpm/cm^2. We assume that every alpha particle in a layer of 10 mg/cm^2 is detected, and that 4 alpha particles are in the radioactive series. The concentration of the alpha contamination is then about 40 mBq/kg. This low concentration is difficult to measure by its gamma activity.

As when working with Si(Li) diodes, great care must be exercised in maintaining the low background of good cells. Each counted radon sample will deposit its decay products through plate out on the wall, and the long lived activity of the long lived ^{210}Po will gradually build up there. With a constant count rate of 1.0 cpm for one year the background will increase by 0.012 cpm, or 20 counts per day, due to ^{210}Po. If a sample giving 1000 cpm is counted for one hour it will increase the background by 0.001 cpm. Special short period studies with strong samples can be made, but care should be exercised.

Cohen et al. (1983) report that it is necessary to repeatedly pump and flush the cells with air after measuring strong samples, as radon adsorbs onto the inner surface of the cells.

ZnS scintillation counters for solid samples. These detectors consist of a plate of plastic material, covered with a thin layer of ZnS in optical contact with the face of the phototube. The sample is on a plate of similar diameter, 2–3 mm below the ZnS layer. The study of Liu et al., described in Section 8.4.5, shows that a maximum pulse height with the transmission geometry is obtained using the thinnest layer of ZnS tested, 3.6 mg/cm^2, although the maximum range of ^{241}Am alpha particles is about 9 mg/cm^2 in the phosphor.

As for the Lucas cells, we can expect from radiocontamination the same background count rate per cm^2 covered by the ZnS, about 1×10^{-4} cpm per cm^2. Such low background levels seem, however, not to have been achieved. If there is an air space of 3 mm between the sample and the ZnS screen, a typical radon indoor air concentration of 50 Bq/m^3 would give a background contribution of 5×10^{-4} cpm per cm^2. For the lowest background, the detector unit should therefore be kept in radon free air.

The advantage of these detectors is their simplicity, low background and large detecting surface, which can be of nearly the same diameter as the phototube used. Today, plastic foils covered with ZnS are commercially available, so these detector units are easy to assemble.

15.4 NaI scintillation detectors

Sodium iodide crystals were the main gamma radiation detectors for nearly two decades, but they were gradually replaced by germanium diodes in the 1970s. The main advantages of NaI detectors are: (a) their high efficiency for gamma radiation due to the high atomic number of iodine ($Z = 53$), (b) the crystals can be made in large sizes and different shapes, and (c) they are moderately priced. They have, however, two serious disadvantages: (a) the width of their gamma peaks is about 40 times larger than those of germanium diodes and (b) the detector units have significant radiocontamination in the glass of the phototubes. The NaI scintillation counters therefore have poor selectivity and rather high background.

The radiocontamination from the glass envelope and ceramic supports of the photomultiplier tubes generally contributes 30–50% of the background of a well shielded NaI detector unit in a surface laboratory (Section 11.4.5). It has recently become possible to produce glass where the concentration of primordial radioactivity has been reduced by a factor of 10–30 compared to earlier types. In the near future, we can therefore expect phototubes with greatly reduced radiocontamination and NaI scintillation units with much lower backgrounds, possibly an order of magnitude lower than that of present units of the same size. It is, however, not easy to predict accurately how much the background reduction will be as it is more difficult to reduce the contamination in the ceramic parts of the phototubes. With the new phototubes, well shielded NaI systems with partial or complete external guard counters may in specific cases become competitive with germanium spectrometers when single radioisotopes are measured. The contribution of the primordial contamination may then fall from about 30% of the background of a NaI unit shielded by 10 cm of lead to about 3%. After this improvement the background can probably be decreased by a factor of 5 with an external guard counter system.

15.5 CsI scintillation detectors

The scintillating properties of CsI(Tl) have been known almost as long as those of NaI(Tl). Early measurements showed, however, that CsI crystals gave pulses that were only about half the size of pulses from NaI for the same energy absorption. This is not caused by the emission of fewer photons, but is a result of mismatching between the wavelength spectrum of the CsI scintillations and the spectral sensitivity curve of the phototubes, as discussed below. Because of this mismatching, the energy resolution of the CsI crystal units is lower. Gamma scintillation detectors with NaI crystals, rather than CsI, were therefore developed at a rapid rate, although the latter had various advantages.

The decay time of CsI depends on the type of particle causing the ionization and pulses from electrons, protons and alpha particles can be distinguished by pulse shape analysis. Therefore, CsI crystals have been used in various high energy studies.

During the last decade the manufacturing techniques of large area, low capacitance, Si photodiodes have been greatly improved. These diodes have a quantum efficiency of about 70% in the spectrum interval of the emitted light from CsI. The noise, coming partly from fast fluctuations in leakage current and partly from the amplifier, is below the size of gamma pulses, except those of the smallest energy. These new detector units have interesting advantages compared to their close relatives, the NaI scintillation counters with a photomultiplier tube:

1. The CsI units are free of radiocontamination.
2. The CsI units are compact as the diodes occupy little space.
3. Their gamma efficiency is high because both atoms have high atomic number (53 and 55) and the mass density of CsI is 4.51 g/cm^3, compared to 3.52 for NaI.
4. They are rugged, both because of favorable mechanical properties of CsI and the compactness of the units.
5. They tolerate well rapid changes in temperature.
6. CsI hardly needs encapsulation as it is only weakly hygroscopic.
7. The diode voltage is only 50–100 volts and it hardly needs stabilization. The diode current is a fraction of a μA.

The main disadvantages of the CsI scintillation units are:

1. The noise in the diode and amplifier sets a limit to its use at low energies. This noise increases with the leakage current, i.e., the area of the diode.

2. The maximum sensitive area of the diodes is small compared to the area of the photocathodes photomultiplier tubes used for NaI crystals of similar size.

3. The units are a little microphonic.

The scintillation efficiency (photons per MeV absorbed in the crystal) is slightly higher in CsI than NaI (Table 15.2). Its emission spectrum is poorly matched to the spectral sensitivity of photomultiplier tubes, but well matched to that of the silicon photodiodes. Figure 15.2 depicts the emission spectra of CsI and NaI and the spectral sensitivity of silicon PIN diodes and a photomultiplier tube with a bialkali photocathode. The figure demonstrates the mismatch between the light from the CsI crystal and the spectral sensitivity of the phototube and the good match with the diode.

Table 15.2. Comparison of gamma-detecting crystals.

	NaI	CsI	Ge
Atomic number	11 and 53	55 and 53	32
Density, gm/cm^3	3.67	4.53	5.32
Photons/keV	43	52	
Max. emission, nm	415	550	
Approx. price, \$/cm^3	2	1.5-2	

These new units have, despite their limitations, interesting possibilities in low-level counting, primarily because of their radiopurity. Figure 15.3 depicts a CsI/diode scintillation unit and its built-in preamplifier, produced by an American firm, eV Products. The diameter of the crystal is 4 cm, its length 4 cm, and its mass 0.23 kg. It is coupled to a 28 $\times$ 28 mm^2 diode. A pulse height spectrum of ^{137}Cs measured with this unit is also shown.

Studies have been carried out at the Max-Planck-Institute for Physics and Astrophysics in Munich, Germany, with CsI/diode units in order to explore their possibilities in low-level counting (Kilgus et al., 1990). Crystals of various sizes were studied, the largest having a mass of about 1.0 kg. The resolution of the small crystals, having a volume of a few cm^3 and a small diode for light detection, was mainly limited by the statistics of the emitted photoelectrons. The large crystal (1 kg) with a 1.8$\times$1.8 cm^2 diode had a resolution of 150 keV (FWHM), which was almost independent of the energy of the gamma photons. In this unit the resolution is evidently limited by the electric diode noise.

Weak alpha peaks in the spectrum of the 1 kg crystal showed that it was contaminated by thorium-chain radioactivity corresponding to about 2 $\times$ 10^{-18} g Th/g Cs, but no ^{238}U

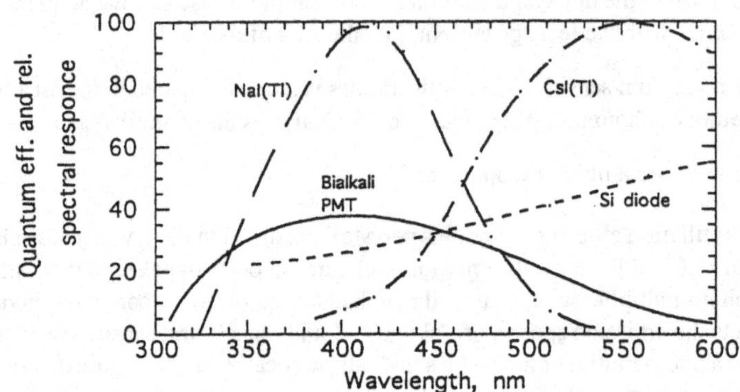

Fig. 15.2. Emission spectra of CsI and NaI and the quantum efficiency of silicon PIN photodiodes and a photomultiplier tube with a bialkali photocathode.

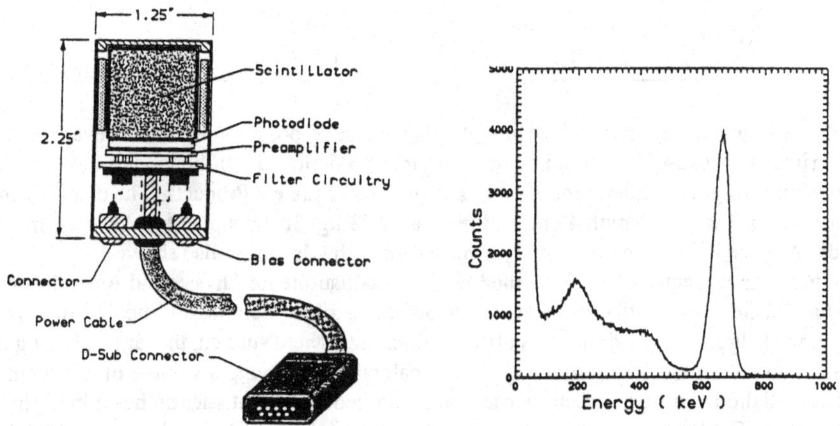

Fig. 15.3. Cross section of a CsI/diode unit (left) and its ^{137}Cs spectrum (right).

series activity was detected at an intensity of two orders of magnitude below the thorium contamination level.

In the second phase of this work, improved crystal units were studied (Kotthaus, 1992). Most of the measurements were made on cylindrical crystals (diameter: 7.5 cm, height: 5 cm, mass: 1.0 kg) and a crystal in the form of a truncated square pyramid with a mass of 3.3 kg. The scintillating light was detected with 18×18 mm^2 PIN silicone photodiodes, optically coupled to the flat crystal front face. Crystal surfaces were polished and wrapped with several layers of white teflon tape for diffuse light reflection. For mechanical protection and electric shielding, the detectors were housed in thin-walled Al or Cu cans. The FWHM width of the 3.3 keV crystal units was about 150 keV, almost independent of energy.

Experiments showed that the energy resolution can be improved somewhat by adding a second diode. After a careful preparation of the crystal surface and the diffuse light reflector, the optimum peak width of the 898 keV ^{88}Y line was 95 keV (10.5%). It should be noted that the sum area of the two diodes, facing the large crystal, is only 6.5 cm^2 whereas a NaI crystal of similar size would have a phototube with a photocathode area of about 44 cm^2.

Background measurements were carried out in a surface laboratory and in a salt mine under an overburden of 420 mwe, where the cosmic-ray muon flux has been reduced by a factor of about 370 compared to its surface value. The CsI crystal with its two diodes was surrounded by 20 cm of lead, which was sealed with plastic foil to keep radon in the air from penetrating into the detector.

The background spectrum in the mine is shown in Figure 15.4. Above 2.0 MeV it is probably dominated by secondary radiation induced by cosmic muons in the lead shield. There is a weak, but statistically significant, peak at 2.61 MeV showing the presence of small amounts of thorium activity. Considering that the preamplifier is sitting close to the crystal, this comes as no surprise. The background spectrum of this large crystal shows no alpha peaks. The internal thorium contamination is less than 75 μBq/kg.

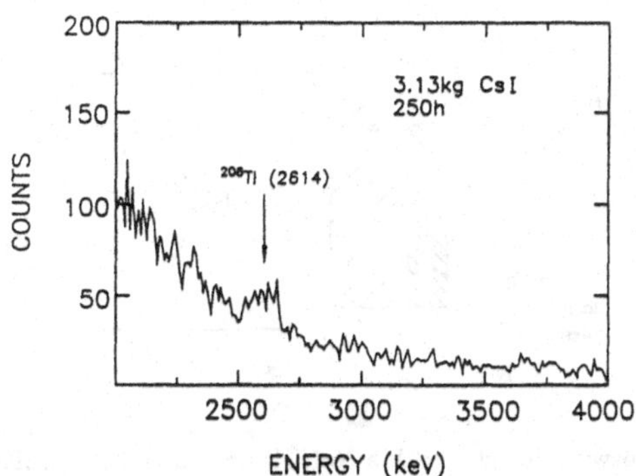

Fig. 15.4. The background spectrum of a large (3 kg) CsI/diode unit in a mine with an overburden of 420 mwe.

At the 13th International Radiocarbon Conference in Dubrovnik in 1987 a group of experts decided to try to measure the relative gamma flux in their laboratories, inside the shield of their gas proportional counters as well as outside it. In this way they hoped to gain a better understanding of the large differences that had been found in the background of gas proportional counters of similar sizes. It was assumed that these differences came mainly from small contamination of primordial activity inside the shield and that it could be detected by a 2"×2" NaI scintillation unit. The results were negative due to the radioimpurities in the NaI unit, which masked the weak activity that was to be measured (Theodorsson et al., 1992). The new radiopure CsI/diode units can now solve this problem. Because of their compactness it will usually be easy to locate them inside the shield without removing the sample detector.

The Cs/diode units may also be useful in the measurement of low-level gamma active samples, such as ^{137}Cs and ^{222}Rn. It seems probable that the price of the CsI/diode units will gradually decline and larger units will become available. It would be easy to design multisample systems of similar type as discussed in Section 15.7.5.

15.6 Phoswich detectors

The phoswich detector is a special type of a scintillation counter with a double scintillator layer of materials with different decay times (Figure 15.5). We focus here on the most common type where one side of a NaI crystal is optically coupled to the photomultiplier tube and a thin plate of acryl is coupled to the other side. These units are used for the assay of beta activity. The scintillation acryl plate is made thick enough to stop practically all the

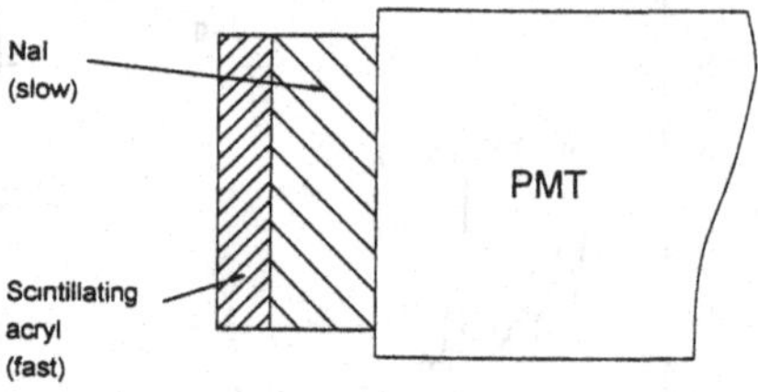

Fig. 15.5. A phoswich detector with a NaI(Tl) crystal and a scintillating acryl plate.

beta particles coming from the sample, giving only fast pulses. The NaI crystal serves as an anticosmic detector. Other energetic particles will penetrate both scintillators, giving a pulse with both a fast and a slow component. These pulses can easily be separated from those with only the fast component.

When soft X-rays are measured the thin plate is usually of CsI, which has a decay time of about 1 ms, roughly 4 times longer than that of NaI. This difference in decay time is enough for the needed pulse shape discrimination.

15.7 Multicounter systems

15.7.1 Early development of multicounter systems

Until early in the 1970s, the price of the electronic part of low-level counting systems was usually higher than that of the detectors. This changed with the introduction of large scale integrated circuits (late 1960s) and their general application in electronic systems. Single chips that had tens to hundreds of transistors and diodes now became available. With large scale integration with MOS and CMOS transistors, the number of components per chip rose gradually from hundreds to hundreds of thousands. These integrated circuits were typically sold at a price of $10–50. $ After the invention of the microprocessor in 1971 it became possible to produce ever more powerful computers and systems at prices that declined as time went on. The revolution in electronic and computer technologies has radically changed the design prerequisites of nuclear instruments.

In low-level counting, the sensitivity can be increased by giving each sample a longer counting time. In order not to lose counting capacity, the number of detectors must be increased. When they are of simple design, it now became advantageous to increase significantly the number of detectors by designing multicounter systems with an array of identical detectors.

These systems have various advantages. Usually they are compact, their size often increases only moderately with the number of detectors as they share a common shield. The electronic system is greatly simplified, compared to a group of earlier single detector systems. The detectors share a high voltage supply, multichannel analyzer and a PC-computer which does all the data processing. Furthermore, it is an important advantage in beta/gamma counting systems when one of a number of identical detectors always measures a background sample.

The development of systems of this type started in 1973 with a collaborative project of Risö National Laboratory in Denmark and the University of Iceland, when a low-level detector unit of 5 Geiger counter elements was made. This development work has been described by Theodorsson (1975, 1988). Multicounter systems have been made with detectors of different types:

1. gas counters for external samples,
2. internal gas proportional counters,
3. liquid scintillation detectors, and

4. Si(Li) diodes.

With better photomultiplier tubes, we may in the near future also see systems with an array of NaI and CsI scintillation units. Germanium diodes are too expensive to make a multicounter system an attractive choice, although significant amounts of lead could be saved.

15.7.2 Gas proportional multicounters for external samples

The Ris multicounters, mentioned above, are of simple and varied design. Figure 15.6 shows a cross section of one of these. They have 4–10 flat circular detector elements made from a single 10 mm thick acryl plate. The diameter of the elements is from 2–5 cm and they have a thin, aluminized mylar window. A flat rectangular guard counter is above the sample detector plate. This gives a very compact detector unit. The detectors can be operated either in the Geiger or proportional region. The former is a little simpler electronically. The counters are continuously flushed with argon containing 1% propane or some other suitable gas mixture. A typical operating voltage is 1250 volts. The plateau length is 200–300 volts and it has a slope of 2–3% per 100 volts. For a detector element with a window diameter of 2.5 cm, the total background in a surface laboratory and using a 10 cm thick lead shield is 4.0 cpm and the anti-coincidence background 0.16–0.20 cpm. These counters are used for

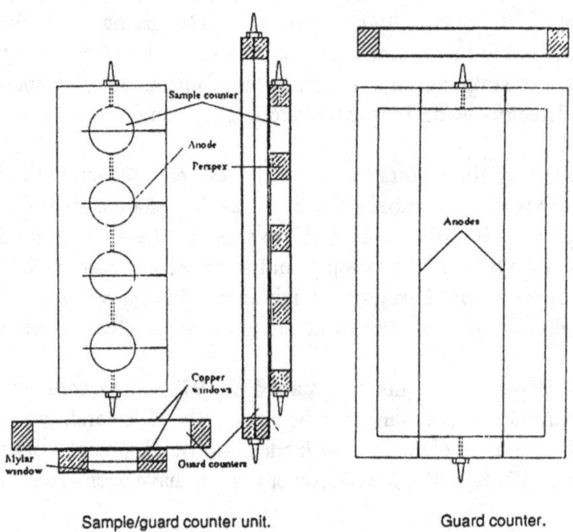

Fig. 15.6. A cross section through a multicounter.

the measurement of various types of solid alpha and beta active samples. They greatly increase the counting capacity, and standards and background samples can be measured more frequently. According to the multicounter principle we should whenever possible work with an abundance of detector elements so that we can better count each sample to the desired accuracy. Furthermore, it is desirable to continuously count the background with one detector element.

Today, the electronic part of these systems consists of a rather simple dedicated electronic unit and a PC-computer, which can just as easily serve ten detectors as one, replacing most of the electronics of earlier related systems. The computer is controlled by a sophisticated program that keeps continuous control over the proper function of the system by checking proper counting statistics, anticoincidences as well as coincidences, in short subperiods. Counter systems of this type are produced commercially by a number of firms. Their performance has been described by Maushart (1986).

15.7.3 Gas proportional multicounters for internal samples

A very well designed multicounter system with gas proportional counters for CO_2 was built in 1980 in Heidelberg (Schoch, 1980; Schoch and Mnnich, 1981) for high precision ^{14}C dating (Figure 15.7). The CO_2 counters are made of radiopure materials and are of a very

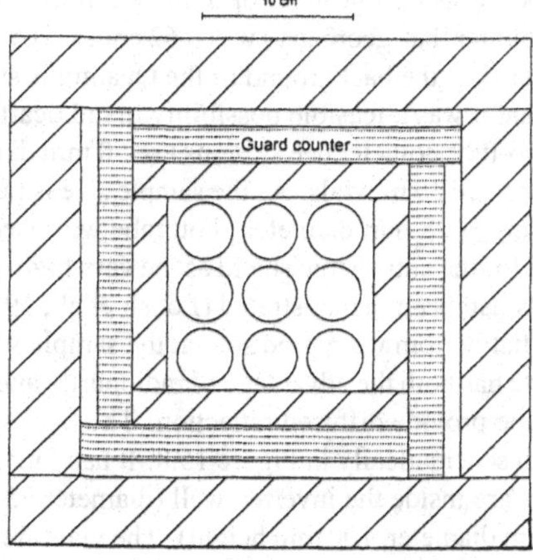

Fig. 15.7. A multicounter system with 9 gas proportional detectors used for radiocarbondating.

simple design. The tubes are of electrolytic copper with quartz discs at the ends. The system has 9 identical 4 liter sample detectors. They are all filled to the same pressure (2.2 atm) and operate on a common high voltage supply. They share a common guard counting system, consisting of 5 flat gas proportional counters. The main lead shield is 10 cm thick and there in an inner shielding layer of 5 cm of lead between the guard and sample counters.

A minicomputer (DEC PDP 11/63) was used for on-line data acquisition. Similar systems have since been set up at a few other ^{14}C dating laboratories.

15.7.4 Liquid scintillation multicounters

The positive experience of Geiger multicounters at the University of Iceland led to a study of the possibility of designing a similar system with liquid scintillation detectors. The goal was a compact system for low-level work with 4 sample detectors and manual sample changing, primarily intended for ^{14}C dating (Einarsson and Theodorsson, 1989). In its final form the system is also suitable for measuring very weak samples with beta/gamma emitters. When a substantial part of the total cost of a counting system lies in the detector unit, there is little advantage in making a multicounter system. It was therefore a prerequisite that the detector units could be made simpler and cheaper than in conventional liquid scintillation counting systems. It was evident that only a single-phototube system could fulfill the requirements of compactness and low price. At first sight this did not look a promising solution for the counting of weak samples, as the single phototube systems had been abandoned decades ago because of their high background.

At the Banff liquid scintillation counting conference in 1983, a Chinese liquid scintillation counting system for radiocarbon dating with a single phototube was described (Pei-yun and Ting-bui, 1983). Its lowest background for 5 ml of benzene with the vial inside a liquid scintillation guard counter (hat-geometry) was 0.62 cpm at 70% ^{14}C counting efficiency, which is not much higher than the background of the Quantulus system. This looked quite promising and proved that it was a feasible possibility in radiocarbon dating.

The conventional two-tube coincidence systems use 50 mm diameter tubes. This size is necessary for the standardized 20 ml vials. As the sample size is rarely more than 5 ml benzene in radiocarbon dating, a 28 mm diameter phototube was used in the system. A large NaI crystal was selected for the anticosmic shield as the very high background reduction efficiency of these crystals had been demonstrated (Loosli et al., 1986). Special quartz vials of conical shape with a flat bottom were used so that the sample would sit close to the face of the phototubes. Quartz has both the advantage of radiopurity and high light transmission. Three layers of teflon tape provide diffused reflection. The final system (Einarsson 1996), called Kvartett, is shown schematically in Figure 15.8. It has four identical liquid scintillation detectors. The vials are inside the inverted well (diameter 75 mm, depth 50 mm) of a large NaI crystal (150 mm diameter, 100 mm height). The vials are arranged in a tray which is placed precisely and closely over the faces of the phototubes. This assembly is lifted into the well of the NaI crystal. The entire detector and guard system is enclosed in a 5 cm thick lead housing, weighing 500 kg. This system and its performance was described at the 14th International Radiocarbon Conference 1991 (Einarsson 1993).

Figure 15.9 shows the anti-coincidence background spectrum for one of the liquid scintillation detectors of Kvartett. The very low background and the multisample facility make this system very sensitive for the measurement of various beta-active samples, such as ^{14}C, ^{90}Sr, ^{99}Tc and ^{137}Cs. Table 15.3 compares the ^{14}C sensitivity of Kvartett with that of two other low-level systems.

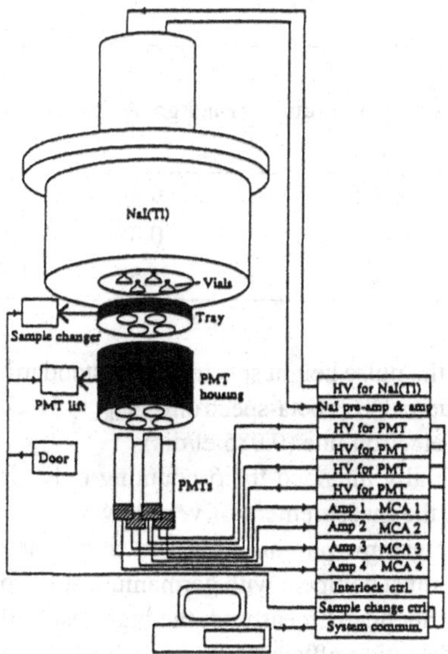

Fig. 15.8. Kvartett, a 4 detector low-level liquid scintillation counting system with a NaI guard counter

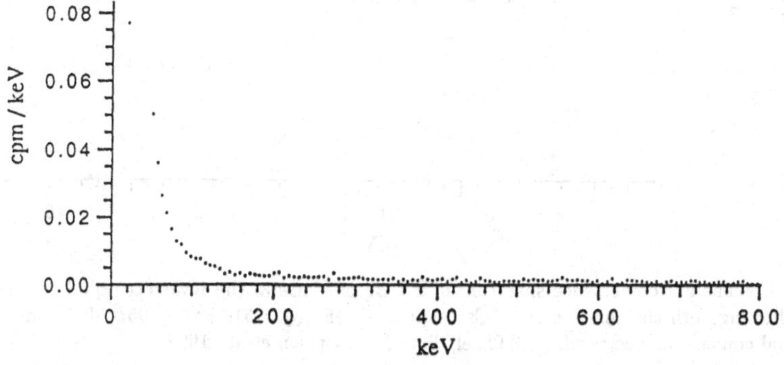

Fig. 15.9. The anticoincidence background of one of the detectors of Kvartett. Theodorsson, 1992.

Table 15.3. Background measurements with 3 ml benzene in vial (Einarsson and Theodorsson, 1995).

System	Backgr. tot., cpm	Backgr. AC, cpm	^{14}C counting eff.
Kvartett	1.5	0.44	70.0%
Quantulus		0.39	70.2%
Packard		1.53	70.0%

Figure 15.10 depicts the pulse height spectrum of a standard with about 2.5 Bq of ^{137}Cs, showing both the continuous ^{137}Cs beta-spectrum (E_{max} = 514 keV, 0.95 β/dis) and the 624 keV internal conversion electron line (0.085 el/dis).

The Kvartett was initially intended for β-counting only. It soon became evident that it could also be used for the measurement of very low levels of beta/gamma-emitting nuclides, but only when the gamma photon is emitted simultaneously with the beta particle. In these measurements it must compete with germanium spectrometers. When, after chemical separation, only one nuclide is measured, the high resolution of the germanium diode is not needed, but high detection efficiency and low background are important. The large NaI crystal of the Kvartett system has a much higher gamma detection efficiency than the germanium diodes.

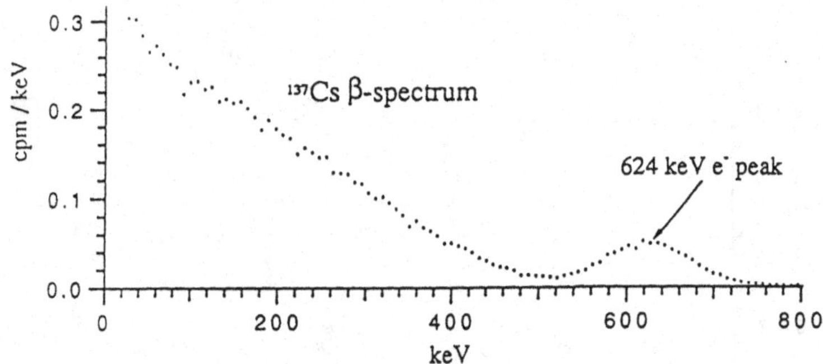

Fig. 15.10. The anticoincidence spectrum of 2.5 Bq of ^{137}Cs in the Kvartett liquid scintillation counter, showing both the continuous ^{137}Cs β-spectrum (E_{max} = 514 keV, 0.95β/dis) and the 624 keV internal conversion electron line (0.085el/dis). Theodorsson et al., 1992.

When simultaneous beta/gamma emitting nuclides are measured in Kvartett, the NaI crystal is used as the main detector. Used conventionally, the large crystal gives a hopelessly high background. The background can, however, be reduced by accepting only NaI pulses as a valid signal when a pulse comes simultaneoulsly from a liquid scintillation counting detector. The analyzer thus gives four NaI coincidence pulse height spectra, one for each sample. This coincidence technique decreases the background of the NaI crystal by a factor of 400–600 in the energy range of 500–1500 keV.

Figure 15.11 depicts the total background spectrum of the NaI unit (no veto signal). A weak annihilation line at 511 keV and a stronger ^{40}K line at 1462 keV are seen. Figure 15.12 depicts the greatly reduced background spectrum that is in coincidence with one of the liquid scintillation detectors. The sensitivity of this technique is demonstrated here by showing

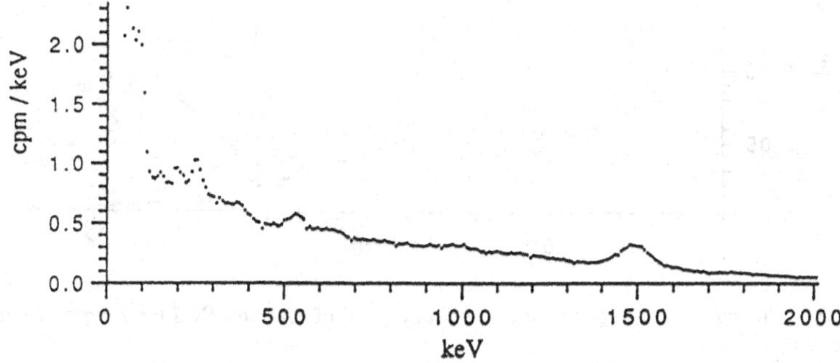

Fig. 15.11. The total background spectrum of the NaI guard crystal of Kvartett. Theodorsson et al., 1992.

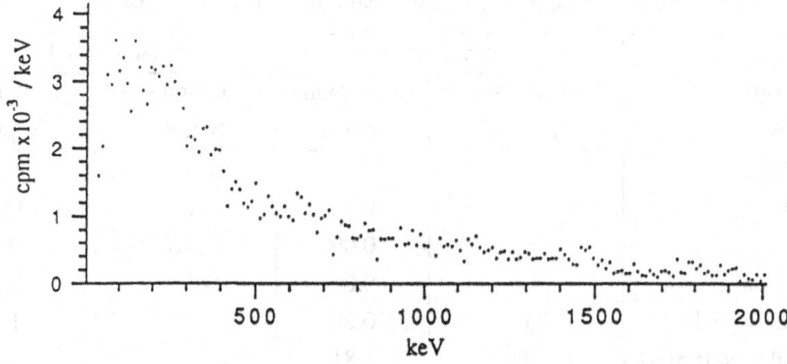

Fig. 15.12. The background spectrum of the NaI guard crystal of Kvartett in coincidence with pulses of one of the liquid scintillation detectors. Theodorsson et al., 1992.

the results of low-level test measurements of ^{134}Cs. The decay schemes of the ^{134}Cs and ^{137}Cs isotopes was shown in Chapter 4, Figure 4.4. In this study, 0.7 ml of aqueous standard solution was added to 2.3 ml of optifluor. Figure 15.13 shows the coincidence spectrum with

7 Bq ^{134}Cs in the sample detector vial. The spectrum shows the two main gamma lines of ^{134}Cs, 604 keV (0.97γ/dis) and 796 keV (0.87γ/dis) as well as their sum line.

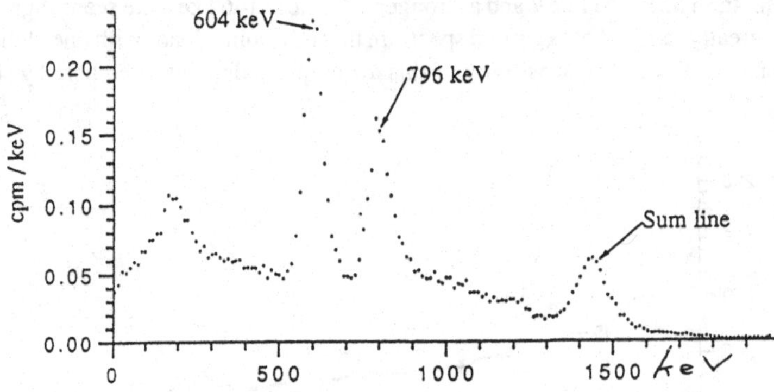

Fig. 15.13. The coincidence spectrum of the NaI crystal of Kvartett/NaI for 7 Bq ^{134}Cs in one of the LS detector vials.

Table 15.4 compares the counting efficiency (counts/disintegration) and background for Kvartett with the Ge-crystal (27% relative efficiency) at the State Radiation Control Institute in Iceland for ^{134}Cs (coincidence mode) and ^{137}Cs (anticoincidence mode).

Table 15.4. Comparison of Kvartett and a Ge detector for ^{134}Cs and ^{137}Cs.

	Kvartett		Ge Crystal	
Energy (keV)	Detection sensitivity (c/dis),%	Background (cpm)	Detection sensitivity (c/dis), %	Background (cpm)
^{134}**Cs**				
604,γ	11.0	0.13	2.2	0.6
790,γ	9.4	0.09	1.7	0.5
Sum line	4.4	0.06		
Total in peaks	24.9	0.28	3.9	1.1
Total spectrum	59.9	1.81		
^{137}**Cs**				
140–530, β	54	0.7		
530–765, IC-peak	8.5	0.25		
664, γ			2.0	0.5

These results demonstrate that Kvartett is a useful addition to the family of systems for the measurement of low-level beta and beta/gamma active radionuclides. It would, for example, be very useful in the measurement of β^+ emitting radioisotopes. Much further work is needed to exploit fully the potentialities of systems of this type, both in improving their performance and testing them in different applications. The gamma coincidence background can probably be decreased by almost an order of magnitude by applying an external guard counter and increasing the thickness of the shield to 10 cm. The system can be optimized for beta/gamma counting by using a larger NaI crystal.

15.7.5 Multidetector systems with CsI scintillation counters

Because of the radiopurity and compactness of the CsI scintillation counters they offer interesting possibilities for constructing simple multidetector systems. It seems probable that their price will decline in the future. The small size of the crystal units and limited low energy range is, however, a disadvantage. Their limited size can partly be offset by having a CsI unit on both sides of the sample. Figure 15.14 shows a proposed multicounter system with CsI scintillation detectors for four samples, where each sample is sandwiched between two CsI crystal units. It has a single, large flat guard counter. The total weight of the system will be about 500 kg.

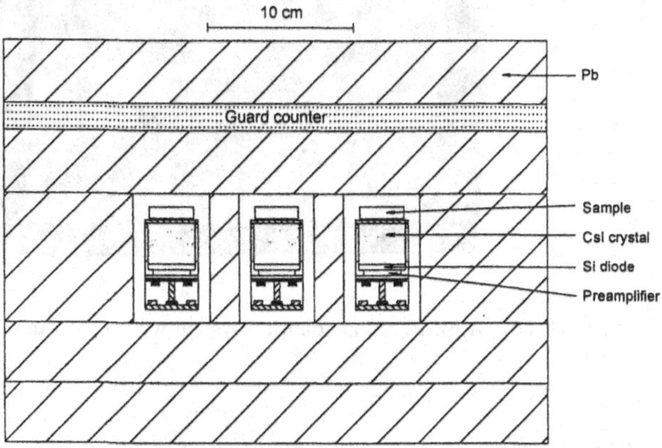

Fig. 15.14. A proposed multicounter system with CsI scintillation detectors.

15.7.6 Multidetector systems with Si(Li) diodes

Alpha activity must frequently be measured at very low concentrations, (a) from man-made radioisotopes, i.e., coming from tests with nuclear bombs and from the nuclear industry, and (b) from various primordial alpha emitters. Si(Li) diodes are used in most cases for these measurements. Their very low background and high resolution are a prerequisite for the needed sensitivity. The samples must frequently be counted for days, even a couple of weeks. A large number of detector units are therefore needed. Multicounter systems with an array of Si(Li) diodes are then used. These are produced by a couple of firms. Figure 15.15 shows a system of this type from the American firm EE&G and Figure 15.16 schematically depicts its electronics. The height of the sample holders can be adjusted and they are frequently connected to a bias of -12 volts in order to collect positive decay atoms on the sample planchet that have escaped from the sample through recoil. Each chamber usually has an independent vacuum door, but some systems have a common door.

The output of the main amplifiers is fed to a 4 or 8 channel multiplexer and then to a common analog to digital converter, which usually has 4096 channels, 512 channels for each detector. Finally a PC-computer with sophisticated software processes the data. Each detector has its separate vacuum chamber, vacuum gauge, variable bias supply, leakage current monitor, preamplifier and main amplifier.

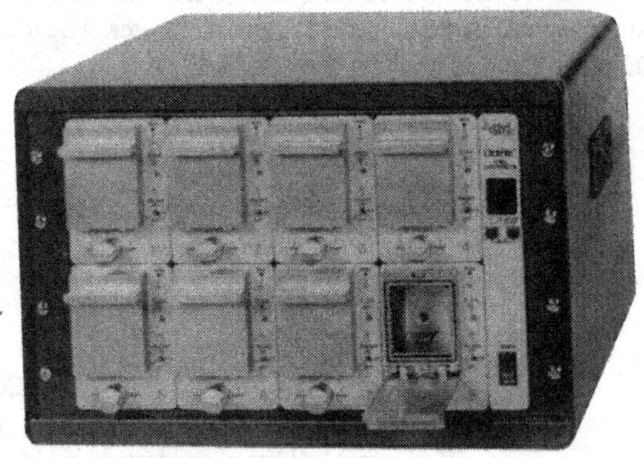

Fig. 15.15. A Si(Li) alpha counting system with eight sample channels.

15.8 Compton suppression spectrometers

One of the main advantages of low-level germanium spectrometers is that large samples can be measured with little or no preparation by chemical or physical methods. Environmental samples generally contain natural radioisotopes of various other radionuclides than those that are to be measured. Potassium is, for example, in all untreated organic samples. A significant, often dominant, part of the background is then due to Compton scattering of 1461 keV ^{40}K photons. This Compton continuum may obscure weak photopeaks, and thus increase counting errors and decrease sensitivity.

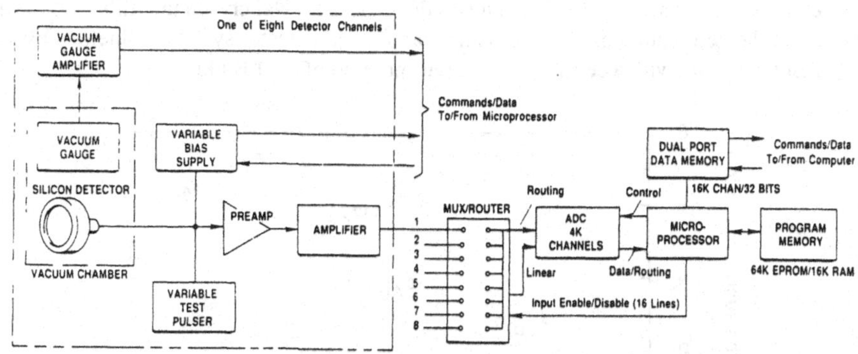

Fig. 15.16. A schematic diagram showing the electronic units of a multicounter system with 8 Si(Li) diodes.

The Compton continuum can be reduced by surrounding the germanium diode with a large gamma detector, which serves as an anti-coincidence guard. Compton scattered radiation from photons which have deposited a part of their energy may be captured by the large surrounding guard detector. When a pulse comes simultaneously from both detectors, the germanium pulse is blocked by a veto signal from the guard.

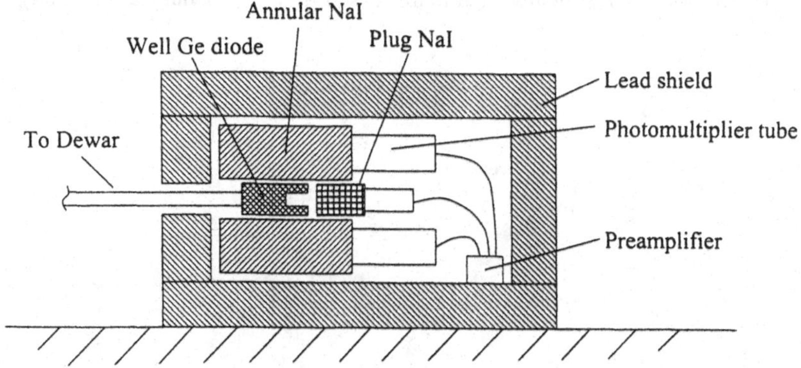

Fig. 15.17. A low-level germanium spectrometer with a Compton guard.

During the past three decades, several types of Compton guard detectors have been used. Large NaI crystals are most effective. We therefore focus on them. Figure 15.17 depicts schematically a system of this type, described by El-Daoushy et al. (1995). The sample detector is a well type germanium diode with a diameter of 5.0 cm and an active volume of 125 cm^3. A large annular NaI crystal, outer diameter of 30 cm, length 30 cm and inner diameter 7.8 cm, is used as a Compton guard detector and a NaI crystal unit fills the hole at the end of the germanium diode. The sample and guard detector system is shielded by 10 cm of Boliden lead, with a certified ^{210}Pb concentration of $\leq$34 Bq/kg.

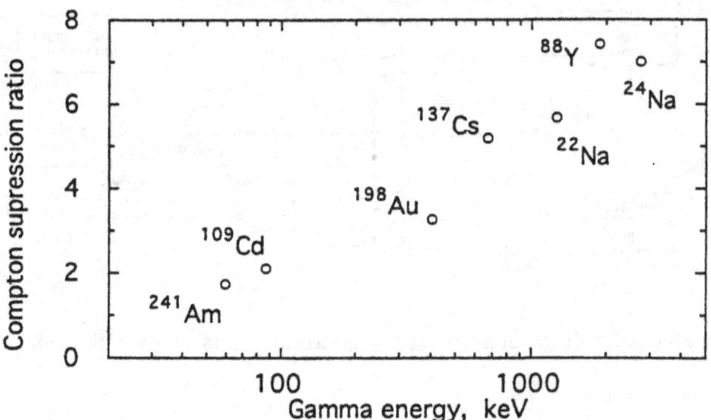

Fig. 15.18. Compton suppression factor of some radionuclides. From Chung et al., 1985.

The suppression factor is usually quoted as the ratio of the height of the Compton continuum without suppression to the height with suppression. The suppression factor increases with the energy of the gamma line. Figure 15.18 shows the Compton suppression factor of some radionuclides for a system similar to that described above (Chung et al., 1985).

Chapter 16

Non-radiometric methods and neutron activation analysis

16.1 Introduction

During the last 15 years a number of new methods have been developed for the measurement of weak radioactive samples. Three of these will be discussed in this chapter. Two of them depend on counting the radioactive atoms rather than their emitted particles. These methods are called non-radiometric, and are based on mass spectrometry. The third method uses neutron activation. Long-lived radionuclides are irradiated in a high-flux reactor where some of the atoms are transformed to short-lived ones, giving induced short-lived activity which may be an order of magnitude higher than that of the mother atoms.

The detection limit of the radiometric methods can be set, somewhat arbitrarily, at 1.0 mBq. When non-radiometric methods are used the total mass of the radioactive atoms sets the limit of detection. The number, N, of atoms in a sample with an activity of a Bq is

$$N = \tau a = \frac{t_{1/2}}{\ln 2} \cdot a \tag{16.1}$$

where τ is the mean life-time of the radionuclide. The mass, m, of the sample is

$$m = N\, M\, 1.67\, 10^{-24} \tag{16.2}$$

where M is the mass number of the radioisotope. Table 16.1 shows the mass of 1.0 mBq for some important radioisotopes.

One microgram (10^{-6} g or 6×10^{15} atoms at $M = 100$) of a large number of elements can be measured by a wide variety of modern methods, while 10^{-16} g can only be determined by the most sensitive. 1.0 mBq of uranium and thorium can be measured by a number of non-radiometric methods, but the same activity of ^{14}C and ^{137}Cs can only be measured with the most sensitive of these.

In addition to high sensitivity, the non-radiometric methods offer further advantages:

1. Less sample preparation.

2. Small sample size.

3. Rapid measurement and high throughput.

Radiometric methods are usually most sensitive for short-lived nuclides as the number of atoms disintegrating per second is high for a given total mass of the radioactive atoms. Therefore, non-radiometric methods are rarely used for radionuclides like ^{90}Sr, ^{137}Cs and ^{241}Am.

Normally, the radionuclide has to be determined in a mixture with its stable isotopes or together with its other radioactive isotopes. The methods must therefore be capable of isotopic analysis. ^{99}Tc is an exception as it has no natural isotope, and ^{99}Tc is found practically only in fission products.

Table 16.1. Mass of atoms of 1.0 mBq of some long-lived radioisotopes

Radionuclide	Half-life, years	Mass, g
^{232}Th	1.4×10^{10}	2.5×10^{-7}
^{238T}Th	4.5×10^{9}	8.1×10^{-8}
^{129}I	1.7×10^{7}	1.6×10^{-10}
^{99}Tc	2.1×10^{5}	1.6×10^{-12}
^{239}Pu	2.4×10^{4}	4.3×10^{-13}
^{14}C	5.7×10^{3}	6.1×10^{-15}
^{137}Cs	30	3.1×10^{-16}

16.2 Acceleration mass spectrometry (AMS)

Acceleration mass spectrometry (AMS), using the tandem technique, has become a very powerful tool for assaying of extremely low abundances, 10^{-15} to 10^{-10} grams, of radioisotopes with half-lives ranging from 10^2 to 10^8 years. The applications of this technique cover a wide range of areas including archaeology, geology, geochronology, hydrology, oceanography, sedimentary processes, cosmochemistry and environmental sciences.

AMS involves acceleration of ions of the atoms to be measured to an energy of 2–10 MeV in order to make it possible to strip off a few of their outer electrons when the atoms penetrate a thin foil or a layer of gas at low pressure. Subsequently the ion beam is analyzed mass spectrometrically in magnetic and electric fields. The advantages of the method are not only its extremely high sensitivity, but also the very small samples size, of the order of one milligram. The method has been used extensively for the measurement of ^{10}Be, ^{14}C,

^{26}Al, ^{36}Cl and ^{129}I. The measurement of other radioisotopes, e.g., ^{39}Ar, ^{41}Ca, ^{59}Ni and ^{90}Sr, is being studied.

Some tens of laboratories are engaged in AMS work and the worldwide annual AMS measuring capacity is presently (1996) about 25 thousand samples per year. The development of this method started late in the 1970s. All the early systems used accelerators that had been developed and used in nuclear research. With decreasing activity in the study of nuclear energy levels, these systems became available for other research projects. One of these was mass spectrometric measurement of minute amounts of various radioisotopes, where individual atoms are counted after a complicated atomic mass separation. The old accelerators were modified for these studies. Late in the 1980s, a few systems, the second generation, were built especially for AMS studies and now the third, considerably improved, generation of AMS systems is being produced and sold commercially.

The accuracy and throughput of these systems has continuously been increasing. The most important radioisotope measured is ^{14}C, mainly in radiocarbon dating. The third generation AMS systems can measure about 4000 samples per year with a mean activity of 50% modern carbon to a precision of 0.5%. A number of the AMS laboratories have shown that they can reach an overall accuracy of 0.2–0.3% in ^{14}C dating, which is equivalent to the highest accuracy in radiometric counting.

The AMS systems can be used for a variety of radioisotopes, but modification must be made in order to adopt them to each atomic species. We focus, in the following, on the most recent AMS systems for the measurement of ^{14}C. In 1988 the Woods Hole Oceanographic Institution (WHOI) received a $5 million grant for establishing an AMS facility mainly intended for the measurement of ^{14}C in oceanographic studies. In work of this kind, hundreds, even thousands, of sea water samples must be collected and processed. It evidently saves much work when the carbon dioxide needs only to be extracted from half a liter of water, compared to hundreds of liters when radiometric methods are used to measure the ^{14}C. This initiated the development of the third generation of AMS systems, carried out by US-AMS Corporation. Measurements with the new system started at Woods Hole in 1991 and the original goal of measuring 4000 samples per year at an accuracy of 0.5% was reached in 1993.

High Voltage Engineering Europe in Holland is continuing the development and production of these spectrometers. Their first system became operative 1994 in Groningen. Figure 16.1 shows this system. It can be divided into six sections:

1. Ion source.
2. Separator/recombinator.
3. Tandem accelerator with electron stripper.
4. High energy mass spectrometer.
5. Detector system.
6. Computers for data acquisition and system control.

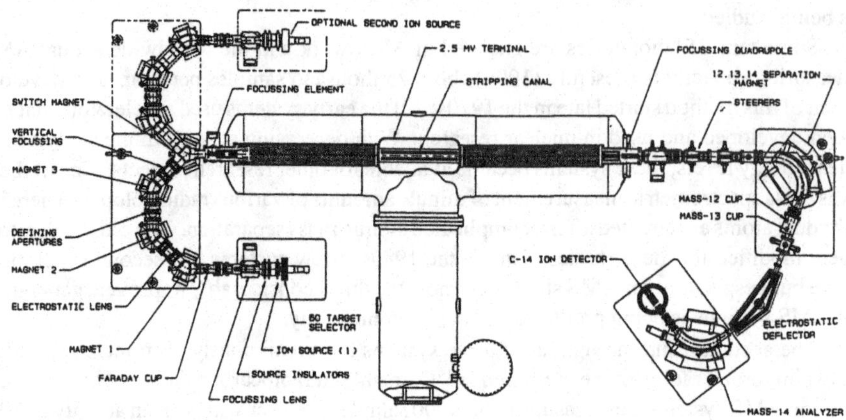

Fig. 16.1. The AMS system of High Voltage Engineering Europe in Holland.

Ion source. The carbon samples are in the form of graphite pellets. They are pressed into a hole, 2 mm in diameter, in the ends of 59 cylindrical aluminium holders on the periphery of a rotary carousel. The Woods Hole system has two separate ion sources. A beam of positive cesium ions are accelerated to impact on the graphite sample, resulting in ejection of C^- ions that are accelerated to an energy of about 40 keV. The accelerated C^- ions are focused to a narrow beam.

Separator/recombinator. The C^- beam is first deflected 90° by two magnets whereby the ^{12}C, ^{13}C and ^{14}C ions are separated into three beams about 2 cm apart. A chopper wheel reduces the ^{12}C beam intensity by a factor of about 100 in order to bring it to an intensity similar to that of ^{13}C. Two further magnets bend the beams by 90°, recombining the three masses a͏nd the beam is injected into the accelerator.

Accelerator. The high voltage terminal is maintained at 2.5 million volts from a Cockroft-Walton supply fed by a 35 kHz solid state oscillator. When the C^- ions have been accelerated they are sent through a 1 cm diameter 1 m long stripping canal, a tube with argon at low pressure. The ions lose here most of their outer electrons, the majority emerging as C^{3+} ions. These positive ions are then accelerated by the same high voltage.

High energy mass spectrometer and detector system. The mass analysis is carried out in the high energy 110° magnetic spectrometer. The ^{12}C and ^{13}C beams are measured individually by Faraday cups at their image points where a typical current is 0.5 μA. The ^{14}C ions pass through further a 33° electrostatic deflector and finally a 90° magnet. The C^{3+} ions are detected after dE/dx analysis in an ionization chamber with a thin window. This system eliminates all other ions, the machine background is practically zero. The background comes essentially only from contamination in sample preparation. It is typically 0.1–0.2% of modern carbon.

Computer control. Two PC-computers control the system. One is for ion source and sample control and one for control of the rest of the system. The system can run unattended through the whole 2–3 day measuring period.

For radiocarbon dating, the AMS technique has all the technical properties we could wish: small samples (normal size 1.0 mg), low background (0.1–0.2% modern carbon), high precision (error down to 0.2% for samples younger that one half-life) and high throughput. Its only disadvantage is the high price of the system.

16.3 Inductively coupled plasma-mass spectrometry

Inductively coupled plasma-mass spectrometry (ICP-MS) can be used for the analysis of most elements at a mass concentration of 10^{-9} to 10^{-6} g/g. It is a variant form of the mass spectrometric methods. It has been used for the measurement of stable isotopes for some 15 years, but until recently it has also been applied to the assay of minute concentrations of radioactive isotopes that have half-lives above about 1000 years. When the half life is more than 10 thousand years ICP-MS is usually more sensitive than radiation counting. Mass spectrometers of this type are produced by a number of firms. Koppenaal (1992) has given a review of the development of inductively coupled plasma mass spectrometry and Ross et al. (1993) have described its possibilities in measuring long-lived radionuclides.

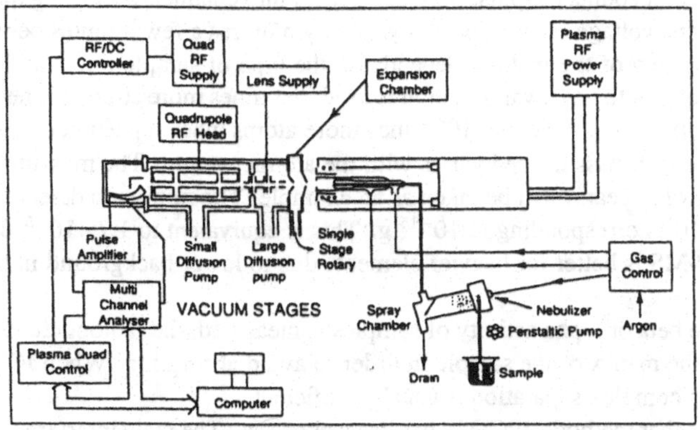

Fig. 16.2. Inductively coupled mass spectrometer.

Figure 16.2 shows schematically an inductively coupled mass spectrometer. Its basic units are:

1. Sample injector.
2. Plasma unit (ion source).
3. Ion lens chamber.
4. Mass separator.
5. Ion detector.
6. Data handling system.

The sample is introduced into the ionization (plasma) unit as a gas or in the form of small particles, e.g., by pneumatic nebulization, ultrasonic nebulization or electrothermal vaporization. The method depends on the sample to be measured. The ionization unit must be very effective in forming ions of all types of materials. It is a chamber where an intense plasma is sustained by inductive coupling to a high frequency power supply. The sample is injected into the plasma, which has a temperature corresponding to about 6000 K, where it is atomized and ionized. The plasma unit injects a stream of atoms through an orifice with a diameter of about 1 mm into the ion lens unit, which is at high vacuum. Most of the emerging atoms are neutral, but some are positive or negative ions. The negatively charged extraction lens attracts positive ions and separates them from the negative and neutral atoms. The accelerated positive ions are focused into the quadrupole lens, which consists of four parallel metal rods in a rectangular geometry. With a proper voltage applied to the rods, only ions with an M/Z value in a narrow range pass through the lens. A channel electron multiplier counts the positive ions that pass through the lens. The computer data acquisition and handling unit counts the pulses and sorts them into channels according their M/Z ratio in replicate lens voltage scans. The analysis only requires a few minutes per sample.

The sensitivity of the method depends on the type of sample injector. It is $10^7 - 10^8$ atoms when electrothermal vaporization is used, ten times more atoms are needed when ultrasonic nebulization is used and 100 times more atoms when a pneumatic nebulizer is applied. The sample mass needed varies with the atomic weight. The measurement of ^{239}Pu (half-life 2.4×10^4 years) will be taken as an example. The minimum detectable number of atoms is 2×10^7, corresponding to 10^{-15} g. This is equivalent to 1.7×10^{-5} Bq. The sensitivity of ICP-MS is better for heavier elements due to lower background in the high M/Z region.

When the beta or alpha activity of samples is measured, the radionuclides must be separated from the matrix of the sample in order to avoid absorption. When ICP-MS is used, partial or less complex separation is usually sufficient.

As do all other methods, IPC-MS has its weaknesses. The main interference comes from isobars, when there is an element in the sample with nearly the same M/Z ratio. This interference is rare at the heavier masses. The second interference occurs when the tail of a strong peak overlaps the adjacent M/Z peak, for example when a strong peak from ^{238}U overlaps the ^{237}Np peak. This interference depends on the resolution of the instrument, i.e., the width of the M/Z peaks. Figure 16.3 shows the scan spectrum of a solution containing 100 ng/ml of natural uranium and 10 ng/ml of ^{237}Np.

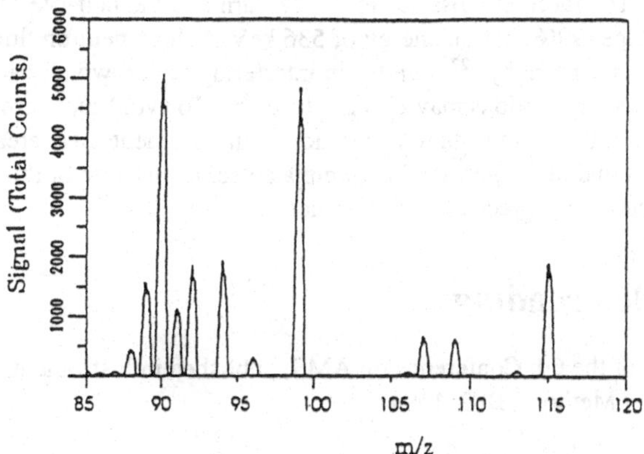

Fig. 16.3. Technetium line scan for an ICP-MS.

The stability of the instruments depends on how long they hold their calibration curve. In order to keep the accuracy within a few per cent, the spectrometer should be calibrated a few times every day. Instrument stability is affected by deposition of sample on the hole (cone orfice) of the plasma unit.

16.4 Neutron activation analysis

Finally, we discuss briefly neutron activation analysis of long lived radionuclides. The detection limit of some long lived radioisotopes can be improved by orders of magnitude by neutron activation analysis where the transformed nucleus has a short half-life. The count rate of the induced activity can therefore be much larger than that of the original activity. Sometimes the methods for the detection of the emitted particles may be more sensitive, for example when the activation product nucleus is gamma emitting with high yields and the parent nucleus is a beta emitter, e.g., ^{99}Tc and ^{135}Cs.

Taking typical conditions, we assume that the sample is irradiated for 15 hours in a flux of 10^{13} neutrons/second. We then get the following increase in specific activity (Rosenberg, 1993):

Radionuclide	^{99}Tc	^{129}I	^{135}Cs	^{237}Np
Half-life, years	2.1×10^5	1.7×10^7	2×10^6	2.1×10^6
Sens. increase	2600	10^6	350	3×10^5

^{129}I will be taken as an example of the advantage of neutron activation analysis compared to direct counting. ^{129}I emits beta particles, gamma radiation with an energy of only

39.6 keV, and X-rays. The most sensitive direct method is to count the beta particles in a liquid scintillation counter. The thermal neutron activation analysis is based on the reaction $^{129}I(n, \gamma)^{130}I$. The thermal cross section is 27 barn and the half-life of ^{130}I is 12.4 h. The main gamma line (99%) has an energy of 536 keV. At high neutron fluxes, absorption of three-fold neutron capture by ^{127}I can be an interfering factor, which varies as the third power of the flux. Other reactions may also produce ^{130}I. To avoid such errors, iodine must be separated from Cs, U and Te prior to irradiation. Other elements may create activity that increase the background of the germanium gamma detector. Because of this interference, a radiochemical separation is practically always needed.

Recommended reading

- Proceedings of the 6th Conference on AMS, published in Instruments of Nuclear Instruments and Methods, B82, 1994.

Chapter 17

Important applications of low-level counting

17.1 Radiocarbon dating

The technique of radiocarbon dating has gone through a continuous development during the nearly half century of its existence. It started with Libby's solid-carbon Geiger counters. Within a few years they were replaced by gas proportional counters that required smaller samples and were less sensitive to airborne contamination. The possibility of using liquid scintillation counters in ^{14}C dating was demonstrated in the mid 1950s, but the technique first became practical some five years later when a high yield method was developed for the synthesis of benzene from CO_2. Liquid scintillation counting is now more widely used than gas proportional counting in radiocarbon dating.

In 1978–1980 a new, entirely different, method was developed for the measurement of natural ^{14}C and various other long lived radioisotopes. This is accelerator mass spectrometry (AMS), described in Section 16.2. We now therefore have three methods for the measurement of ^{14}C in radiocarbon dating that compete on the base of accuracy, price, throughput and sample size.

The youngest method, the AMS technique, is probably most highly developed as a result of the very large resources that have been invested in it during the last 15 years. The third generation of AMS machines, which are now being produced, are based on the best that the mechanical and electronic techniques and automation can offer. During this period, the radiometric methods have been standing in the shadow of this promising technique. It has been difficult to acquire money for improving them as many experts have been saying that the radiometric techniques would soon be replaced by AMS machines. Although the development of gas proportional and liquid scintillation counting has been slow in the last 15 years, many improvements have been introduced. The most important of these are multidetector systems with either gas proportional or liquid scintillation detectors. PC-computers, allowing detailed data sampling and analysis, greatly improve and simplify the electronic systems.

High accuracy is more important in radiocarbon dating than in any other sector of radiometry. Late in the 1950s it became clear that the initial fundamental assumption of radiocarbon dating, i.e., that the $^{14}C/^{12}C$ ratio in atmospheric CO_2 had been constant in the past, was not entirely correct. H. Suess discovered that the massive burning of fossil fuel in this century had diluted the ^{14}C concentration and H. de Vries found that the $^{14}C/^{12}C$ ratio, corrected for decay, measured in tree rings from the past centuries showed small, slow fluctuations.

It became clear that the age of the dated samples could not be calculated from their measured count rate assuming a constant $^{14}C/^{12}C$ ratio in atmospheric CO_2 and using the known half-life of ^{14}C. Instead, a calibration curve, based on very accurate measurements of the $^{14}C/^{12}C$ ratio of dendrachronologically dated tree rings, has to be used. Hundreds of tree ring samples had therefore to be measured with high precision. A concentrated effort was initiated at a number of laboratories to do this work. Large samples (5–10 grams of carbon) were measured to an accuracy of 0.20–0.25%. A major part of this calibration work was carried out at the laboratories of M. Stuiver (Seattle, USA) and of G.W. Pearson (Belfast, N-Ireland). Stuiver used CO_2 gas proportional counters and Pearson a liquid scintillation counter. Other laboratories, such as those in Groningen and Heidelberg, have also contributed to these studies. The impressive result of this massive work is described in Radiocarbon, Volume 35 (no. 1), 1993. The results from different laboratories are practically identical even though two different counting techniques are used and samples were of different types of trees that had grown in different climates and continents.

When samples are dated the most common accuracy is 0.7–0.8%, corresponding to an error of about 70 years. Most users would no doubt like to have more precise age determinations. The accuracy is mainly limited by the counting capacity of the equipment, which is again limited by its price. Counting each sample for a longer time than normally allotted, thus increasing the accuracy, takes practically no extra work. The price per counting channel in the next generation of radiometric systems, that are technically within reach as described in Sections 13.15 and 14.8, will be much lower than in present systems. It therefore is probable that the general accuracy of moderately old samples will in the future be 0.25–0.35%. The comparison in Section 17.5 of the three methods will therefore be based on an accuracy of 0.30%. Although accuracy can be increased or counting time made shorter with large samples, the sample quantity in archeology and geology is usually limited and the counting samples generally contain 1.5 to 2.0 grams of carbon. The comparison of the methods will therefore be based on this sample size.

In the following sections each method, including sample preparation, will be described. Their relative merits and disadvantages will then be discussed. In this comparison we must remember that large resources have been invested in the AMS technique and much has been done to make sample preparation and measurement automatic with numerous checks on proper operation. The sample preparation and counter filling of radiometric methods is manual with limited automatic checks. Furthermore, various procedures are used that are not supported sufficiently by experimental work. We can therefore expect important further improvements in the radiometric methods in coming years if modest amounts of money are invested in the work.

For the gas proportional and liquid scintillation counting systems we assume that the improvements that are now within reach and have been discussed in Sections 13.15 and 14.8, will be used. Although accuracy can be increased or counting time made shorter with large samples, the sample quantity in archeology and geology is usually limited and the counting samples generally contain 1.5 to 2.0 grams of carbon. The comparison of the methods will therefore be based on this sample size.

17.2 ^{14}C dating with gas proportional counters

Sample preparation. A greatly improved method for sample preparation was introduced at the dating laboratory in Heidelberg (Germany) some ten years ago (Kromer, 1989). This method has been improved and simplified at the dating laboratory in Stockholm (Sweden) by M. Hedberg (personal communication). The samples are burned in a high pressure bomb and purified by adsorption on charcoal at $-80°C$. The burning and gas cleaning system can stand on a table of 2 m^2. A technician can prepare about 8 samples per working day with a single system.

Counting. It is assumed that the samples are counted in a multidetector system with 9 or 16 sample detectors with common external guard counters, as described in Section 14.15. Hedberg (personal communication) has built counters of very simple design that are both easy to make and assemble. The electronics can be quite simple. Each counting channel only needs a small dedicated unit with an amplifier and 3 or 4 discriminators and pulse counting chips, which are read periodically by a PC-computer. Compared to most present day systems, the improved one will be simpler to operate. All detectors share a common high voltage supply and the measurement is started immediately after the counters have been filled. A discriminator will split the muonic coincidence pulse height spectrum into two parts with nearly equal counting rates for pure samples. Small impurities in the CO_2 gas will shift pulses from the upper channel to the lower, changing the ratio of the number of pulses in the two channels, compared to a pure sample. The ratio will then be used to find a proper correction for the effect of an eventual sample of the impurity on the total number of pulses in the ^{14}C channel.

It is assumed that the net volume of the sample counters is 1.0 L, that they are filled to 3 atm with CO_2, and that their counting efficiency is 90%. Modern carbon gives 13.5 dis/min per gram of carbon. The background will be about 0.5 cpm. A 5700 years old sample (one half-life old) then gives 11.2 cpm. Counting this sample for one week will give a standard error of 0.33%, corresponding to ± 26 years. Measuring the same sample to 0.5% would take 2.5 days.

Some procedures in the gas proportional dating technique can presumably be improved. Storing samples before counting can be taken as an example. Most samples are at present stored for three weeks to allow eventual radon to decay to an insignificant level. A large number of high pressure steel cylinders are required for this storage. Is this storage really necessary, except for special types of samples with high probability of radon contamination? Is it not possible to count samples contaminated slightly by radon and correct for the increased count rate in the ^{14}C channel caused by radon? Radon and its decay products to the long-lived ^{210}Pb give both large alpha pulses, well above the ^{14}C counting window, and beta pulses, the largest part of which are counted in the ^{14}C window. In order to be able to determine a radon correction factor, all pulses where (for example) more than 1.0 MeV energy has been deposited should be recorded in a separate alpha channel. The ratio of alpha pulses and beta pulses from the radionuclides in the chain from ^{222}Rn to the long-lived ^{210}Pb is constant. A sample contaminated with radon is measured and the ratio of additional pulses in the ^{14}C channel and the alpha pulses accurately determined. This ratio can thereafter be used to find the radon correction from the number of pulses in the alpha channel in samples to be dated, provided that this correction is within a given limit.

17.3 ^{14}C dating with liquid scintillation counters

Sample preparation. Recent improvements in both sample preparation and in counting systems can greatly improve the overall performance of liquid scintillation counting in radiocarbon dating. Skripkin (personal communication), at the dating laboratory in Kiev (Ukraine), has developed a greatly improved system for the synthesis of benzene. It is both more compact than earlier systems and it saves much time in sample preparation.

Counting. The Kvartett multisample system, discussed in Section 15.7.4 has demonstrated that a liquid scintillation detector unit with a single phototube can give very low background, comparable to that of the Quantulus. This experience can be used to design a very attractive multicounter system of simpler design with a low background. It should be noted how simple each detector channel is. It consists of a 12 cm long phototube, 28 mm in diameter, on which the conical quartz vial sits in a tray, about 1 mm above the face of the phototube. Optical coupling would probably increase the light collection efficiency by a factor of 1.5–2 and lower the background by 20 – 40%. In the following we assume, nevertheless, that there is an air gap between the vial and the tube, as this makes sample changing easier.

An electronic system of simple design, similar to that used with the gas proportional counting system described above, will be used. The phototubes will share a common high voltage supply. Correction for eventual light quenching of samples will be made by splitting the ^{14}C spectrum into two halves, and the ratio of the number of pulses in the two channels used to determine the quenching correction factor.

Figure 17.1 depicts a proposed multicounter system. The detectors will be inside a common shield of 10 cm of lead with a single, large, flat guard counter sandwiched between the two 5 cm thick lead layers above the detectors, as described in Section 13.15. In view of the extreme simplicity and compactness of the detector units, it would be natural to have 9 such units in a system. Considering the simplicity of changing samples and the long counting times (one week or more), it would not take much extra work to operate the system in a remote laboratory with an overburden of 20–50 mwe. Easily accessible rooms of this type are found, for example, in underground stations. It is a simple matter to check the counting daily through remote reading the computer data.

Each sample will normally contain 3 ml of benzene (2.21 g carbon) that gives a count rate of 22.5 cpm for benzene of modern carbon at 75% counting efficiency. The background can be assumed to be 0.5 cpm in a surface laboratory. A 5700 year old sample will then give a standard deviation of 0.30% with a counting time of one week, corresponding to a standard deviation of ±24 years in age. This is nearly the same accuracy as obtained with the gas proportional system described in the last section.

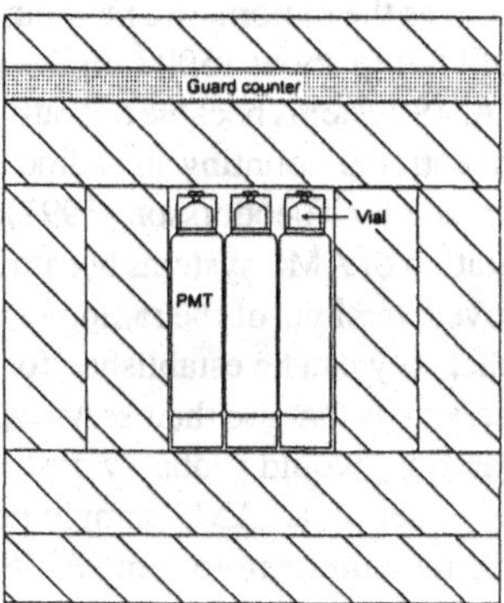

Fig. 17.1. A proposed multicounter system with nine single phototube liquid scintillation detectors.

17.4 ^{14}C dating with the AMS technique

The accelerator mass spectrometric (AMS) technique, described in Section 16.2, has now been used for more than 15 years in radiocarbon dating. The Woods Hole system was designed for the measurement of 4000 samples younger than one half-life to a precision of 0.4%. This goal has been met. Higher accuracy, even down to 0.20–0.25%, has been achieved with a number of AMS systems by prolonging measuring times. The precision of the AMS technique therefore parallels that of the best radiometric systems.

The new generation of AMS system has 59 graphitized samples pressed into the holes of aluminium sample holders that are fastened to the peripheri of a carousel. They are measured fully automatically over a period of 2–3 days. The sample preparation is also automated to a high degree, except for the pretreatment of archeological and geological samples.

17.5 Comparison of dating methods

^{14}C dating by gas proportional and liquid scintillation counting with a new generation of multicounter systems give the same accuracy. The price and counting capacity of the systems is also similar. Sample preparation in gas counting is simpler and less time consuming than the synthesis of benzene. There are more factors that can interfere in the liquid scintillation method. Sample changing is, on the other hand, much simpler in liquid scintillation counting. The choice between these two methods is therefore primarily a matter of personal experience.

The main advantages of the AMS method are the small sample size, typically 1.0 mg of carbon, and high throughput, 4000 samples per year measured to an accuracy of 0.5%. If the samples were measured to 0.3%, as the radiometric methods give for one week counting time, the sample throughput would fall to about 1500 samples per year.

Before the new generation of AMS systems became operative, AMS measurements were about twice as expensive as conventional counting in radiocarbon dating. An economic comparison of the methods was made by Theodorsson (1991). A similar comparison has not been made after the new generation of AMS systems became operative. Today, the comparison should be based on improved versions of the radiometric systems discussed above. They have the added advantage that they can be established for a modest amount of money and can therefore be close to the scientists that use their services. A system with 9 detectors, including a unit for sample preparation, would probably cost less than $100,000 to build. Also, for small samples it is easy to set up an AMS sample preparation line in a conventional laboratory, sending them for measurement to some AMS laboratory.

In the opinion of the author, radiometric dating with improved systems will be used in parallel with AMS at least for the next 10–20 years. Many present day radiometric ^{14}C laboratories with old systems will, however, find it hard to compete.

17.6 Environmental tritium

17.6.1 Origin and occurrence of tritium

Tritium, ^{3}H, is a radioactive isotope of hydrogen with a half-life of 12.26 years, corresponding to a decay of 5.7% per year. It is a pure beta emitter with a maximum energy of only 18.6 keV. Libby postulated in 1946 the production of radiocarbon and tritium by cosmic rays and his group succeeded in measuring its presence in terrestrial waters in 1954. Like radiocarbon, it is produced in the upper atmosphere by the interaction of cosmic rays with atmospheric gas atoms. It has also been produced in huge quantities in tests with thermonuclear bombs in the atmosphere. Finally, it is produced in reactors of nuclear power plants, predominantly through ternary fission. A part of this tritium escapes from the fuel elements and is released from the plant into the atmosphere. The rest is released when the fuel elements are processed.

Tritium is precipitated on the surface of earth as HTO. Its concentration is usually measured in tritium units (TU) defined as a T/H ratio of 10^{-18}. One tritium unit is equivalent to 7.18 disintegrations per minute (dpm) per liter of water, or 0.120 Bq/L. Tritium in the atmosphere and in ground and surface waters has been used extensively as a natural tracer.

The concentration of tritium in precipitation and surface waters is presently 5–20 TU and a factor of 10 or more lower in ground water and in the oceans. It is usually measured, after enrichment, by gas proportional or liquid scintillation counting, but the most sensitive method is mass spectrometry, where its decay product, ^3He, is measured.

17.6.2 Electrolytic enrichment

In water electrolysis, the separation of the hydrogen isotopes is larger by orders of magnitude than in any other physical/chemical process. This process is therefore nearly always used for tritium enrichment.

The tritium separation factor β is defined as

$$\beta = \frac{T/P(\text{in water})}{T/P(\text{in } H_2\text{-gas})} \tag{17.1}$$

where T and P are the relative number of the tritium and protium (^1H) atoms, or

$$dP/P = dV/V = \beta\, d(TV)/(TV) \tag{17.2}$$

where V is the volume of the water in the cell. The solution of this equation gives us the tritium retention R factor, which is the ratio of the quantities of tritium in the final and initial volume:

$$R = \frac{V_f T_f}{V_i T_i} = \left(\frac{V_f}{V_i}\right)^{1/\beta} \tag{17.3}$$

If the tritium concentration is measured before and after electrolysis, the value of β can be found from this equation. This will, however, not represent the true separation factor, but its effective value, β_{eff}, as the tritium retention is also affected by loss through evaporation and spray droplets. These losses will, however, not decrease the retention factor seriously. β_{eff} has a value of 20–30, depending mainly on the metal used for the cathode and the type of electrolyte added to the water to secure good conductivity. Typically, 85% of the tritium in the initial volume is retained in the electrolyzed water after a volume reduction of 20:1.

Libby and his group introduced tritium electrolysis enrichment. An electrolyte must be added to the water in order to make it conductive. NaOH is normally used. It is most convenient to have it in the form of a concentrated solution in tritium-free water. Corrosion was observed when the volume reduction exceeded a factor of ten. As their goal was primarily to demonstrate tritium in terrestrial waters, intending to measure some tens of samples, they cautiously stopped the electrolysis when the volume reduction of samples had reached a factor of 10. The samples were distilled in order to remove the excessive electrolyte and the entire process repeated with a 10 times smaller volume and with the same initial NaOH concentration. Some of their tritium poor samples therefore had to be enriched in 4-5 successive stages. Later investigators followed their procedures.

Let us first look at the NaOH concentration at the beginning of the electrolysis. During the first 20 years of tritium measurements most investigators used the same initial NaOH concentration as Libby's group, 1.0% by weight. Since then, ever lower concentrations have been used. This, however, decreases the conductivity and thus increases the heat dissipation at a given current density. When the concentration is very low, corrosion problems set in.

An initial NaOH concentration of 0.3% is definitely safe, and Taylor (1982) has reported routine use of an initial concentration of 0.12% in 1.0 liter cells. These cells are presumably long, giving large cathode area, which decreases the current density and increases the cooling surface.

For about two decades scientists kept the volume concentration within a factor of 10–20 in straight electrolysis. Higher volume reduction could, however, be obtained without running the cells at too high electrolyte concentration by adding water periodically to the cells. This technique was introduced by Östlund and Werner (1962) and it has been used in many tritium laboratories. They obtained a total volume reduction of about 80 and a tritium concentration factor of 64, retaining 80% of the tritium in the electrolyzed sample.

The volume reduction in a single stage was gradually increased at various laboratories. Theodorsson (1974) showed that a volume reduction up to about 300 could be used routinely if mild corrosion was accepted. Today, practically all samples are enriched in a single stage. The samples are generally distilled before electrolysis. Here, we may be overcautious. At the author's laboratory, where a few thousand tritium samples were measured, samples of precipitation, surface water and cold ground water were not distilled but electrolyzed directly. This saved much work. All geothermal samples and sea water samples were, however, distilled. At the end of the electrolysis the enriched water is often (always in liquid scintillation counting) distilled in order to separate the NaOH from the water. A part of the tritium is bound in the electrolyte in the form of OT^-. In order to avoid this loss the sample is usually neutralized, either by bubbling CO_2 through the sample or by adding $PbCl_2$ to the water in the cell, before final distillation. The following reaction occurs:

$$PbCl_2 + NaOH \rightarrow 2NaCl + Pb(OH)_2$$

Upon heating to 150°C, $Pb(OH)_2$ decomposes to PbO and water.

Figure 17.2 shows two types of widely used cells. The most common type is the metal cell that has been made in various modifications at different laboratories. The one shown in the figure was designed at the IAEA tritium laboratory in Vienna. The teflon cylinder at the bottom of the cell ensures that all the water lies between the electrodes in the final stage of the electrolysis. The glass cells, designed by Östlund and Werner (1962), are still widely used. The cells are cooled to 2–8°C during electrolysis. The metal cells presumably ensure better removal of heat from the cells and their cathode area is much larger.

The cathode of both types of cells is usually of mild iron as it gives the highest separation factor, 35–40. The anode is usually of nickel or stainless steel. New cells give lower separation factor for the first few runs, but after that the separation factor is stable. There are various recipes for achieving a high and constant separation factor. This has been discussed in detail by Taylor (1982).

The cells, usually a set of ten, are generally connected in series. Usually one or two of the cells contain water with a known tritium concentration, "spiked" water, for monitoring tritium retention, and one with tritium free water, a background sample. A Coulomb-meter measures the total charge running through the cells and stops the electrolysis when a preset value of ampere-hours has been reached, leaving a desired end volume.

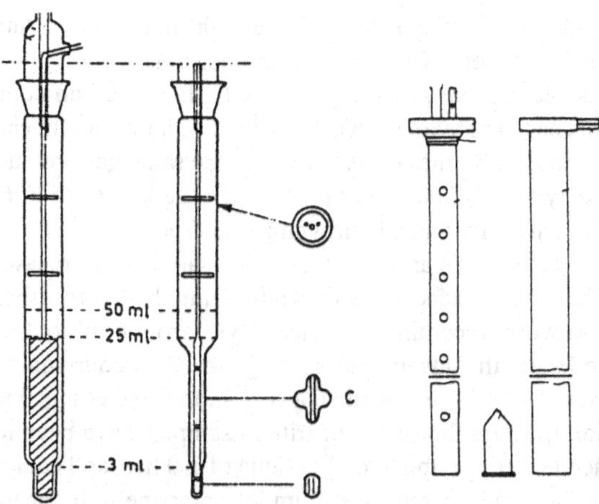

Fig. 17.2. The two most common types of electrolysis cells: the Östlund type (left) and IAEA type (right).

The cells can also be run in parallel (Theodorsson, 1974). This has the advantage that the enrichment of a new sample can be started independently of other samples and the electrolysis can be stopped automatically simply by adjusting the central electrode at a height that leaves the desired volume in the cell when the water surface drops below this level.

The initial water volume, V_i, depends on the counting method used. It is 250–500 ml when the water is measured with a liquid scintillation counter and about four times less when a gas proportional counter is used because of higher detection efficiency. Typical values for the electrolysis and measurement of the water samples are given in Table 17.1.

17.6.3 Counting tritium

Liquid scintillation tritium counting. Of 80 laboratories that took part in a tritium intercomparison study arranged by IAEA, about 80% used liquid scintillation counting (Hut, 1986). It has various advantages. The counting systems are of standard types, readily available at moderate prices. Sample preparation is simple: 8 ml of the distilled enriched sample is mixed in a vial with 12 ml of a water miscible scintillator. Sample changing is automatic. The main disadvantages of this method, compared to gas proportional counting, are its lower counting low efficiency and high background (Table 17.1).

Gas proportional counting. The gas counting systems are essentially the same as those used for radiocarbon dating, described in Section 13.15. Most of the gas counters used in the international tritium intercomparison study (Hut, 1983) had volumes of 2–3 liters and working pressure of 2–3 atm. This comparatively large volume was an advantage in the 1960s and

1970s, when tritium environmental levels were high; many of the samples could then be measured without enrichment. Out of 19 gas proportional counting laboratories that took part in the tritium intercomparison study, 14 used a hydrocarbon counting gas, predominantly ethane. This has the obvious advantage that 6 H atoms are in each molecule and the counters therefore contain 3 times more activity at the same gas pressure than when hydrogen gas is used. A typical counter can then have all the hydrogen of 6 ml of the enriched water, compared to only 2 ml when hydrogen gas is used.

A large counting sample is an advantage when samples are measured without enrichment. This is at the cost of added work in synthesizing the hydrocarbon gases. As it takes practically no extra work to continue the electrolysis from 6 ml to 2 ml and the increased tritium loss is moderate, the advantage of using ethane as counting gas is not important. Östlund and Dorsey (1977) have described a very simple and efficient tritium system at the University of Miami, where thousands of tritium samples have been measured. They use a counter of moderate size, having a total volume of 1.0 liter and count H_2 at a pressure of 2.0 atm, to which they add 0.7 atm of tritium free propane in order to get good counting characteristics. Considering that most of the samples are enriched before counting, there is little to gain in the use of counters with larger volumes as the mass of water at the end of the electrolysis can be adopted to the counter size and filling pressure. The optimum counter volume is probably be about 0.5 liters, counting H_2 at a pressure of 3 atm. This would give low background and decrease the size of the counting systems.

The main advantages of gas counting are the high counting efficiency (85–90%) and low background (0.5–1.0 cpm). A further important advantage is that the electrolysis cell can be connected directly to the filling system that produces hydrogen from the enriched water, so no distillation is needed and all the enriched water is converted to hydrogen and the final pressure gives accurately the residual water volume in the cell. Neutralization of the enriched water in order to decrease the tritium loss is not worthwhile as the loss is small and constant, typically about 4%.

17.7 Mass spectrometric tritium measurements

17.7.1 Comparison of radiometric methods

Table 17.1 lists important parameters for enrichment and counting, where either liquid scintillation or gas proportional counting is used. It is evident that the latter method, with its high counting efficiency and low background, is superior, but it has the serious disadvantage that the counting systems are not available commercially. They are, however, not much more difficult to build than the electrolysis systems, which are usually home-made.

When the samples are enriched, gas counting requires a sample size that is only about one fourth of that when the samples are measured in a liquid scintillation counter. The smaller samples save much working time and equipment expenses.

Table 17.1 compares the two methods by showing typical values for important parameters with two initial water volumes for each counting method. It is assumed that the gas proportional counting time is two times longer than in liquid scintillation counting because of the simplicity of multicounter systems.

The simple multisample liquid scintillation systems with a single phototube described in Section 14.8 will probably not give a background that is low enough in the low energy tritium window. In the two-tube systems, disparity ratio discrimination is quite effective in lowering the small pulse background originating in one of the tubes. These small pulses fall into the tritium channel, below the ^{14}C channel, and will be counted in the single tube tritium system. The relatively high quenching in the water mixed scintillator aggravates this problem. Radiopure phototubes and effective anticosmic shielding with an external guard may change this.

Table 17.1. Comparison of low-level tritium measurements using electrolytic enrichment followed by either liquid scintillation or gas proportional counting and mass spectrometry.

	Liquid scintillation		Gas proportional		Mass spectrom.
V_i, ml	250	500	50	275	50
V_f, ml	12	12	2.0	2.75	
V_i/V_f	20.8	41.7	25	100	
R	0.91	0.82	85	0.80	
Electrolysis time, d	7	10	1	7	
Counting sample, ml	8	8	2,0	2.5	
TU/cpm	70	70	100	100	
B, cpm	4	4	0.5	0.5	
Counting time, h	12	12	24	24	
Min. det. activity (3σ), TU	0.8	0.4	0.3	0.07	0.1

Notes: Columns 2–3 describe an IAEA system (Florkowski, 1981), column 4, a system at the University of Miami (Östlund and Dorsey, 1977) and column 5 an electrolysis system at University of Iceland, but assuming that a counting system equivalent to that at University of Miami is used.

Technically, a multicounter system with 4 gas proportional detectors, described in section 13.15, is of similar complexity as modern liquid scintillation counters and would probably not cost much more. Such a system would be superior to the liquid scintillation counting system. An automatic or semi-automatic system, filling the four counters serially, would be relatively simple to build and require no special skill to operate. The filling would then take similar, or less, time than distilling the enriched water samples and filling the vials in liquid scintillation counting.

17.7.2 Mass spectrometric tritium measurements

^3He is a stable, very rare, isotope of helium. Its isotopic ratio, $^3He/^4He$, in the atmosphere is 1.4×10^{-6}. The study of its concentration in hydrological systems, both in ground water and in the oceans, gives important geophysical information. Sensitive mass spectrometers were therefore developed to measure the $^3He/^4He$ ratio (Clarke et al., 1976). A study of ^3He in sea water samples in the mid 1970s showed that some of the ^3He in the oceans must have come from tritium, which decays to ^3He. Following this, a method was developed to measure tritium in water by its decay product, ^3He. 50–100 ml of degassed water were collected in glass bulbs and flame sealed, or in 10 mm inner diameter copper tubes that were closed with a pinch clamp a few cm from each end. The sample was then stored for about 6 months. After this time, the amount of ^3He formed, which is proportional to the tritium concentration, is sufficient for measurement in a sensitive mass spectrometer.

Clarke et al. found that when the water sample was 40 g and it was sealed for 6 months the detection limit was about 0.1 TU, comparable to the best that can be attained by electrolytic enrichment followed by gas counting. This sensitivity can be increased by using larger water samples and increasing the storage time. This is a simple and fast method. The degassing takes only about 5 minutes and it can be done with simple equipment in the field. The time required for the mass spectrometric measurement is 10–30 minutes. The disadvantages of the method are that it requires a waiting period of about 6 months and that the spectrometer is quite expensive.

17.8 Measurement of radon

17.8.1 Introduction

The concentration of radon in water and in air has been measured for decades in various geophysical investigations, for example in studies of flow and mixing of ground water and air masses, and in earthquake prediction research. During the last 10 years much attention has been given to radon in indoor air as its short lived alpha emitting decay products, ^{218}Po and ^{214}Po, are the largest single source of radiation dose of the general public. Furthermore, radon can be a disturbing factor, a very subtle one, in the measurement of weak radioactivity. Today, the study of radon in our environment is one of the most active sectors of low-level counting. It is therefore appropriate to devote a special section to its measurement.

Radon is the name of the element with atomic number 86, belonging to the family of noble gases. All of its isotopes are unstable and have short half-lives. The three natural radioactive series, starting with ^{232}Th, ^{235}U and ^{238}U, all include an isotope of radon:

Thorium series:	^{220}Rn	thoron	$T_{1/2} = 54.5$ s
Actinium series:	^{219}Rn	actinon	$T_{1/2} = 3.92$ s
Uranium series:	^{222}Rn	radon	$T_{1/2} = 3.82$ d

The short half-life of thoron, ^{220}Rn, limits its chances of escaping far from its thorium source. The half-life of ^{219}Rn is an order of magnitude shorter. Neither ^{220}Rn nor ^{219}Rn will be discussed here, and ^{222}Rn will be referred to simply as radon.

Because of the relatively long half-life of radon, it can be transported long distances from its source by water or air. It is therefore widely found in substantial concentrations in our environment. Half-lives and energies of the emitted particles of the radionuclides from ^{222}Rn to the long lived ^{210}Pb, are given in Table 18.2.

Table 17.2. Radioactive series from ^{222}Rn to ^{210}Pb

Nuclide	Historic name	Half-life	Particle, energy (MeV)
^{222}Rn	Rn	3.82 d	α 5.49
^{218}Po	RaA	3.05 min	α 6.00
^{214}Pb	RaB	26.8 min	β 0.65,
			γ 0.295 (37%), 0.352 (37%)
^{214}Bi	RaC	19.7	β 3.17 (23%), 1.75 (77%)
			γ 0.609 (46%)
^{214}Po	RaC'	0.16 msek	α 7.69
^{210}Pb	RaD	22 years	β 0.018

We are primarily interested in the measurement of environmental radon in air and water where its concentration is generally very low and sensitive methods must be used for its determination. Radon, or its decay products, is therefore frequently separated from a sample of air or water before the radioactivity is determined. Fortunately, radon has exceptionally good characteristics for easy separation and detection. As a noble gas it is relatively easy to separate from air and extract from water. The systems used can measure the instantaneous radon concentration, its mean value over some period or they can give a continuous record.

The literature on radon is quite extensive. Radon in our environment was discussed at two recent large meetings, at the Fifth International Symposium on the Natural Radiation Environment 1990 and a workshop in Trieste in 1991 (Furlan and Tommansino, 1993). The US National Council on Radiation Protection and Measurement has published a report on the measurement of radon in air (NCRP, 1988) and George (1990) and Harley (1992) have written short review articles on methods for measuring radon.

17.8.2 Radon in air

When radon disintegrates in air the newly formed ^{218}Po atoms will be positively charged and within seconds they become attached to ambient aerosols. In indoor air a large fraction, typically 50%, of the progeny of radon will be deposited on surfaces within about 10 minutes. Therefore, radon is generally not in equilibrium with its progeny in indoor air and the ratio of the concentrations of ^{222}Rn, ^{218}Po, ^{214}Pb and ^{214}Bi is typically given as 1.0/0.9/0.6/0.4.

The concentration of radon in air can be determined by measuring: (a) radon alone, (b) radon and its short lived progeny and (c) its progeny alone. It can be measured (usually with its progeny) directly with ZnS scintillating cells. Instantaneous values can be measured by filling an evacuated chamber with the air to be measured or the chamber can be flushed with it. The chambers can also be used for continuous radon monitoring. The decay products give a response time of about 30 minutes. This is sometimes disturbing and a faster response can be obtained by using a large chamber with an electrically negative rod in the middle of the chamber that attracts the positively charged ^{218}Po atoms when they are formed. If the distance of the rod to the ZnS layer is larger than the range of the alpha particles, the decay products give practically no contribution to the count rate. The chamber then responds practically only to radon.

Usually the concentration of radon is so low that either radon or its decay products must be collected from a large volume of air. It can be collected through adsorption on charcoal and its progeny can be collected by pumping air through a filter paper or by electrostatic deposition on a metal surface. Of these methods, the first one is used most extensively.

Adsorption of radon on charcoal. Radon can be adsorbed in practically unlimited quantities on charcoal at the temperature of dry ice and subsequently released for measurement by a flow of gas through the heated trap. This method is, however, time consuming and usually reserved for measuring very low concentrations.

Radon in air can also be adsorbed on charcoal at room temperature. When a thin layer of charcoal of mass m (grams) is exposed to air it will be saturated by radon in a rather short time. At saturation, the quantity Q of radon adsorbed is proportional to its concentration C in the air according to Henry's law:

$$Q/m = kC \tag{17.4}$$

where k is the adsorption coefficient and is measured in L/g. At room temperature k has a value of about 4 L/g. It decreases with increasing temperature. The charcoal can be considered as energy traps binding the radon atoms. At saturation there is a dynamic equilibrium of radon atoms escaping from the traps and new traps being filled. The charcoal is usually in a 2 cm thick layer in a metal canister with a diameter of 7.5 cm (Figure 18.3). The specific density of charcoal is 0.5 g/cm^3 so its mass will be 20 grams. A thicker layer will only increase the adsorbing capacity moderately because of the slow vertical radon diffusion rate in the charcoal. The canister is kept open when radon is being collected, but is closed and sealed when it is transported and measured. The canisters can be re-used by heating them to 100–120 °C for a few hours. The adsorbed radon is then expelled from the charcoal.

The concentrations of radon inside houses show large fluctuations due to variations in atmospheric conditions, mainly pressure. We are usually interested in the mean value of radon concentration during some given period, from one day to a whole year. Charcoal canisters can be used to give mean values over a period up to a few days.

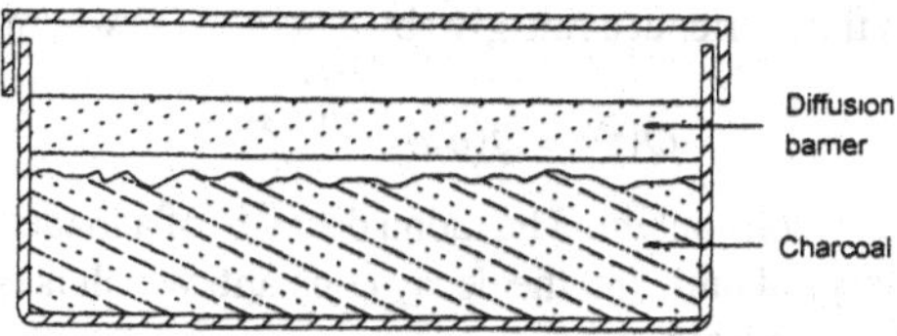

Fig. 17.3. A canister with charcoal and a diffusion barrier.

In order to obtain an integrating flow, the diffusion of radon to the charcoal must be restricted by inserting a porous barrier between ambient air and the charcoal. Radon is transported by free diffusion through this barrier at a rate that is adapted to the adsorbing capacity of the charcoal and the collection time (George 1984, Prichard and Marien, 1985, Cohen and Nason, 1986).

When charcoal of mass m (g) in a canister of this type is exposed to air with a constant radon concentration C (Bq/l) it will adsorb radon until a saturation quantity Q (Bq) is reached that is proportional to the concentration in air given by Equation (18.4). The mass of charcoal used is equivalent to an apparent volume V into which radon flows:

$$V = km \tag{17.5}$$

Water vapor will compete with radon atoms for adsorption sites and reduce the value of k (Pojer et al., 1990). The canisters are therefore weighed both before and after the collection and appropriate correction, determined experimentally, is then applied to the result of the measurement.

The rate of diffusion through the barrier is proportional to $(C_o - C_i)$, where C_o is the concentration of radon in ambient air and C_i the concentration over the charcoal given by Equation (18.4) (Cohen and Cohen, 1983, Cohen and Nason 1985). The rate of increase of the amount of the adsorbed radon, dQ/dt, is

$$dQ/dt = P(C_o - C_i) - \lambda Q \tag{17.6}$$

where P describes the apparent air flow through the barrier, for example in liters per day. λ is the decay constant of radon and λQ represents the decay loss of radon. Eliminating Q by using Equation (18.4) and rearranging the terms, we get

$$dQ/dt = -(P/V + \lambda)C_i + (P/V)C_o \tag{17.7}$$

The solution to this differential equation for a constant value of C_o is

$$Q(t) = C_o V \frac{P/V}{P/V + \lambda}(1 - e^{-(P/V+l)t}) \tag{17.8}$$

In order to find the value of P/V, describing the rate of diffusion through the barrier, the canister is initially loaded to saturation in air with a high radon concentration and then kept in air at low concentration. The process described above is now reversed, radon diffuses out of the charcoal through the barrier at a rate give by equation (4) where $C_o = 0$. The solution is then:

$$Q(t) = Q(o)e^{-(P/V+\lambda)t} \tag{17.9}$$

where $Q(0)$ is the initial radon content. The radon decay in of a loaded canister is monitored continuously by a NaI crystal unit, and the decay constant, which is equal to $1/(P/V + \lambda)$, is determined and P/V calculated.

We are naturally interested in adsorbing a large quantity of radon during the collecting period in order to get high sensitivity. However, the charcoal has limited adsorbing capacity (about 4 L/g) and, assuming a constant value of C, the rate of adsorption will decrease during the period. The quantity of adsorbed radon at the end of the collecting period will not be strictly proportional to the mean value of the radon concentration for two reasons. Each Bq of radon adsorbed at the beginning of a collecting period gives, because of decay, less contribution to the count rate than a Bq adsorbed at its end, and the difference will be larger the longer the period is. The measured radon concentration can be considered as a time weighted average. The second reason is due to the limited adsorbing capacity of the charcoal that gradually leads to a decreasing rate of adsorption. These two effects, decay and decreasing adsorption, partially compensate each other.

The rate of change of radon in the canister, dQ/dt, can be found by differentiating Equation (17.7). Let us look at the case where it is considerably faster than the decay of adsorbed radon, i.e., where P/V is large compared to λ and the latter can be neglected. We then have

$$dQ(t)/dt = C\,V\,e^{-(P/V)t} = C\,V\,e^{-t/(V/P)} \tag{17.10}$$

V/P can be considered as the diffusion time constant, τ_{dif}. Q will in this case fall exponentially with a decay time, τ_{diff}, that is equal to V/P.

In order to secure that the adsorption rate does not decrease too much during a collecting period, the sampling time must be small compared to the diffusion decay time. This, however, severely limits the quantity of radon we collect. In practice it is not uncommon that the charcoal will be filled to 60% of its saturation capacity (assuming a constant radon air concentration). In this case, the final adsorption rate will have fallen to 40% of its initial value. This is an acceptable compromise as these measurements, that give a time weighted average, are mainly used for the purposes of screening a large number of houses where high accuracy is not needed.

Let us look at an experimental case (Sensintaffar et al., 1992). The parameter P has a value of 0.030 L/min (43 L/d) and if the mass of charcoal is 20 gram, V will be 80 L and the diffusion decay time, τ_{diff}, 1.9 days. Figure 17.4 shows both the relative total amount of radon collected by the charcoal and the rate of collection as a function of time for this cell at a constant radon air concentration. For a collection time of 1.0 day the adsorption rate will fall to 59% of its initial value, and to 20% after three days, again assuming a constant external radon concentration.

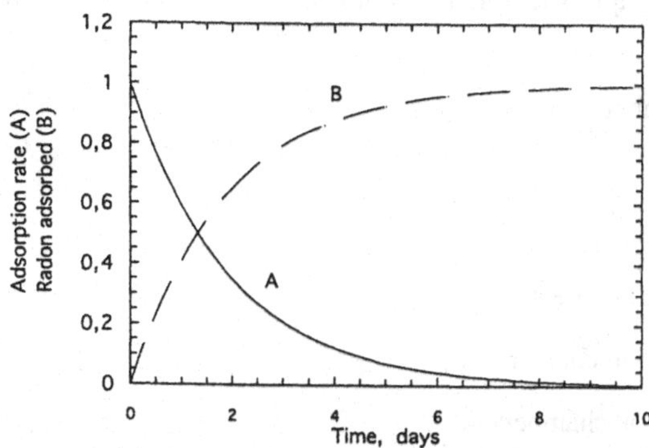

Fig. 17.4. Total amount of radon collected by a typical canister and the rate of adsorption at a constant external concentration for $\tau_{\text{diff}} = 1.9$ d.

Paper filtering of radon progeny. Most of the randon decay products in air are adsorbed by aerosol particles, which can be collected by sucking the air through a filter paper. The filter is subsequently counted with an alpha/beta gas proportional counter or a ZnS alpha counter. This method is frequently used in continuous monitoring of radon.

Electrostatic filtering of radon progeny. A 1–2 liter metal vessel with a Si(Li) diode in one end is filled with the air to be measured. This method is described in Section 17.8.4.

17.8.3 Collecting radon from water

When a mass of air is above water in a closed vessel or air is bubbled through water, the partition of radon between air and water at equilibrium is defined by the Ostwald coefficient, L_w,

$$L_w = C_w / C_a \tag{17.11}$$

where C_w and C_a are the volume concentration of radon in the water and in the air respectively. The value of L_w decreases with increasing temperature. It has a value of 4 at room temperature.

In a classical method for the measurement of radon in water, the water is first collected in a vessel with a volume of about 1 liter. Gas is then circulated through this sample and then through charcoal in a metal tube maintained at -80 °C. The cell is then heated to about 100 °C and the radon transferred to the detector unit by a flow of gas.

17.8.4 Detection systems for radon and its progeny

A large variety of detectors can be used for the measurement of radon at low levels. A general description has already been given of the detectors in Chapter 8 and the most important counting systems in Chapters 12–14. The following detectors are used for the measurement of radon:

1. Ionization chambers.
2. Gas proportional counters.
3. Scintillation cells.
4. NaI scintillation counters.
5. Liquid scintillation counters.
6. Electret ionization chambers.
7. Si(Li) alpha detectors.
8. Nuclear track foils.

Ionization chamber, the oldest type of radioactivity detector, can be quite sensitive, but it requires some skill to use. It can either be flushed or evacuated and filled with the air to be measured. Ionization chambers can also be used in a pulse mode, giving quite good alpha resolution when the chamber is filled with argon. An interesting example of the use of this method is the measurement of ^{210}Po, implanted over a long time period into glass surfaces through recoil of ^{222}Rn, ^{218}Po and ^{214}Po (Samuelsson). Due to its long half-life, the ^{210}Po concentration reflects radon concentration several decades back in time.

Gas proportional counters are sometimes used for the measurement of very low levels of radon extracted from air or water. Liquid scintillation counters are now a better choice.

Scintillation cells. The most frequently used detector for the assay of radon in air is that of the scintillation cell, frequently called Lucas cells, discussed in Section 15.3. They have been the workhorse in the measurement of radon for the past four decades. The volumes of the cells ranges from 50 ml to about 3 L. They are usually filled with air at atmospheric pressure where the ranges of the alpha particles from radon and its progeny are from 4.0 to 7.0 cm. When an atom of radon disintegrates its daughter atom, ^{218}Po, is plated out within a fraction of a minute on the wall of the chamber. The mean alpha detection efficiencies of the cells are 0.50–0.85 counts per alpha disintegration, depending on the size of the chamber. The background count rates of new cells reported in the literature varies widely, from about one count per hour to about one count per minute, depending on their volume.

NaI scintillation detectors are used extensively for the measurement of the gamma activity of radon progeny in canisters. 3"$\times$3" NaI scintillation detectors, shielded by 10 cm of lead, are generally used. The most intense gamma lines (295, 352 and 609 keV) are counted in a broad window. The radon detection efficiency for a canister with a diameter of 75 mm is then about 0.12 counts per Rn disintegration and the background is about 40 cpm.

Liquid scintillation counting. The use of liquid scintillation counters for the measurement of radon, both in air and in water, has been increasing in the last ten years. The main advantages of this method are:

1. the high solubility of radon in organic solvents,
2. a 100% alpha detection efficiency,
3. a very low background through pulse decay discrimination,
4. a good alpha resolution,
5. the wide availability of the systems.

There are variations of the method, and three of these will be described briefly, two for radon in water and one for radon in air. A new variant of radon liquid scintillation counting is also described below.

The simplest way to measure radon in water is to mix 10 ml of water with 10 ml of toluene scintillation cocktail in a vial, close it and shake it vigorously. As radon is 50 times more soluble in toluene than water, about 98% of the radon is extracted by the scintillator (Prichard and Gesell, 1977). The two immiscible liquids quickly separate gravimetrically and the sample is measured with a pulse height wide window. Bem et al. (1994) greatly increased the sensitivity by extracting radon from 500 ml of water with 20 ml of toluene and counting the separated toluene in a liquid scintillation counter with pulse decay analysis to suppress the background. Finally, Schroeder et al. (1989) have described a very attractive and sensitive method for the measurement of radon in air. They used a liquid scintillation counter to measure the radon collected by a small (2 g) charcoal trap with a diffusion barrier. After exposing the small canister, the charcoal is poured into a vial with 15 ml of scintillator. The vial is then closed and shaken. After about 10 hours, most of the radon has been extracted by the toluene and the sample is counted in a wide energy window.

Surface barrier alpha detectors. Watnik et al. (1986) have designed a continuous radon monitor where the alpha particles are counted with a Si(Li) diode. A 2 L vessel is filled with dried and filtered air. An electrically isolated Si(Li) diode with an active area of 3 cm^2 is at one end of the vessel. A voltage difference of 3000 volt is applied between the vessel's inner surface and the diode. After the decay of radon atoms the ^{218}Po daughter atoms, having a positive charge, are deposited on the negatively charged diode window. The diode resolves the ^{218}Po and ^{214}Po peaks. This method has a short response time as ^{218}Po can be counted separately. The detection efficiency for ^{218}Po is about 45% and the background 0.1 cpm.

The electret ion chamber. The electret ion chamber is a recent addition to devices for the measurement of environmental radon and nuclear radiation of various kinds. It is produced by an American firm, Ordela. It consists of a special type of ionization chamber, usually with a volume of 50–500 ml, where an electret (a specially treated plate of teflon with a semi-permanent electric surface charge) collects the ions whereby the charge of the electret is reduced (Bauser and Range, 1978; Kotrappa et al.,1988; Kotrappa et al., 1993). The initial and final surface charge is measured with a static electrometer. The electret ion chamber is a very simple and stable device that gives a true mean value of the radon concentration during the exposure period, which can be up to 3 months, even as long as a year. It has, however, a rather limited dynamic range as the ratio of full scale signal to smallest reproducible signal is about 500. This must be compensated for by proper selection of exposure time and chamber volume. A natural background radiation of 0.10 μGy/h will give the same discharge current as a radon concentration of 40 Bq/m^3. The background is therefore often measured simultaneously with a separate closed chamber. The minimum detectable level (giving three times reproducible voltage change) with an exposure time of 7 days is about 13 Bq/m^3. Considering that background gamma radiation is then 75% of the signal, a more realistic estimate of the MDL would be 20 Bq/m^3.

Delayed coincidence counting. A new method for the measurements of low levels of radon has been described by Buheitel (1993) and independently by Theodorsson (1996). In the system of Theodorsson a single phototube liquid scintillation detector counts practically only alpha pulses from ^{214}Po. This method is primarily being developed for automatic determination of radon in ground water for earthquake prediction studies, but it should be equally suitable for a variety of other studies. About 40% of the radon in 1.0 liter of water is transferred automatically by a stream of air into 25 ml of toluene in a special sample vial. Instead of reducing the background through pulse decay analysis, which is not suitable for a system operating automatically for long periods, another, rarely used, method is applied. It is based on *delayed coincidence counting* (Amano and Kasai, 1982). It uses the very short half-life of ^{214}Po (0.16 ms). Each time a ^{214}Bi atom disintegrates in the scintillator, emitting an energetic beta particle, 98% of the alpha particles from ^{214}Po will follow within 1.0 ms. This nuclide pair will thus give characteristically close pulse pairs, first a beta particle, that can be detected with an efficiency of 95%, followed by a large alpha pulse that is detected with 100% efficiency. A pulse pair of this type is rarely found in the background where the mean time between pulses is typically about 2 seconds.

The pulses must pass a gate that is normally closed. Five microseconds after the arrival of each pulse, the gate is opened for 1.0 ms. Therefore, only the second pulse of a close pair is counted. This gives a high background reduction. In a system shielded by 5 cm of lead, the background of a 25 ml sample in a broad alpha/beta counting window is about 30 cpm. The count rate of accidental pulse pairs closer than 1.0 ms will then be about 0.02 cpm. If the second pulse of a close pair is sent to a multichannel analyzer and the pulse only accepted when it lies in the ^{214}Po window, the background will be reduced by a further order of magnitude.

Radon is usually counted in a wide energy window in liquid scintillation counters, where the alpha particles are detected with 100% efficiency and the beta particles with up to 90% efficiency. Close to 5 pulses are therefore registered for each decay of an radon atom. In the present method only one of these five pulses is counted. With long counting periods this will, contrary to general expectations, only insignificantly affect the accuracy of the measurement compared to counting the five particles in a wide window. The five pulses in the series are not statistically independent, we are actually repeatedly registering the same original event, the decay of a radon atom during long counting periods. Lucas and Woodward (1962) have discussed the statistics of radon counting in an article that has not received sufficient attention in recent years.

Conventional liquid scintillation counting systems can be used for delayed coincidence radon counting by adding a small external unit that accepts amplified pulses from the system and selects for counting only close pulse pairs. Pulse shape analysis in liquid scintillation counting offers similar sensitivity as delayed pulse counting. The latter has the advantage that it does not need a special scintillation cocktail and careful setting of alpha/beta discrimination. Furthermore, the technique is specific for radon.

This method could replace NaI crystals in the measurement of radon in canisters. The liquid scintillation counting method offers automatic sample changing, 100% detection efficiency and a background less than 0.1 cpm, compared to an efficiency of 10% and a background of 40 cpm of the NaI unit. However, in this new method the radon must be transferred to the scintillation vial. It would probably not be difficult to construct an automatic manifold for this transfer, for example, with 4 samples in parallel.

Nuclear track films. Alpha particles hitting a plastic foil leave small pits in its surface. The visibility of the pits can be increased by etching the foil with NaOH solution. The number of pits per surface area is proportional to the mean radon concentration and to the exposure time, that is typically one month up to one year. This has been the most widely used method for measuring the radon concentrations in thousands of buildings all over the world.

Table 17.3. Comparison of minimum detectable radon concentration (MDL)

Method	MDL	Reference
	Air, Bq/m^3	
Ionization chamber	4	NCRP Report No.97
Lucas, 0.1L,	11	NCRP Report No.97
Lucas, 1.0L	2	NCRP Report No.97
Charcoal (20g),NaI	6	Typical system
Charcoal (2g), LSC	1	Schroeder, et al., 1989
Electret	20	Kotrappa et al., 1988
	Water, mBq/L	
Lucas, 1 L, charcoal	4	Author's laboratory
LSC, 10 ml	700	Prichard et al., 1977
LSC, 500 ml, PSA	1.5	Bem et al., 1994
LSC, 1.0L, delayed coi.	1.0	Theodorsson, 1996

17.8.5 Comparison of sensitivity

The sensitivity, the minimum detectable level (MDL), of the methods discussed above will now be compared. The comparison is based on a counting time (where appropriate) of 3 hours and a net count rate of 3σ of the background. It has also been taken into consideration that counting more than one pulse in the series of five radionuclides from radon to ^{214}Po will not appreciably increase the accuracy, as discussed above. The result of this comparison is given in Table 18.3.

Chapter 18

Statistics

18.1 Random processes

To the early investigators, radioactivity appeared to be permanent until Rutherford observed, in 1898, that thorium emanated a radioactive gas (thoron) that it lost its activity in a few minutes. He found that the activity A of thoron decayed exponentially:

$$A = A_0 \, e^{-\lambda t} \tag{18.1}$$

where A_0 is the initial value ($t = 0$) of the activity and λ is a constant with a value of about 85 s^{-1}. In the following years, a number of radioactive species were discovered which decayed in the same way, each having a characteristic decay constant. The simplest explanation of the exponential decay was that each radioactive atom had a fixed probability λ of decaying during a time period dt. According to this model, nothing could be said as to whether a given atom would disintegrate during this time interval, the process could only be described statistically. Today, it seems easy to grasp that radioactive decay is a random process, but it was difficult for scientists at the beginning of the century to accept this purely statistical description of nature.

From this very simple model we can derive the probability for the outcome of various experiments and give answers to questions such as:

1. If we have a radioactive source with a constant activity and the mean number of pulses in a given counting period is N, what is the standard deviation of N in repeated observations, and what is the probability distribution?

2. What is the distribution of the lengths of the time intervals between two consecutive pulses?

3. When a pulse is observed, what is the probability that a second one will follow within a time interval Δt?

4. A single measurement of a sample and background gives a net count rate somewhat above zero. What is the probability that this count rate represents a valid sample activity, not a statistical background fluctuation?

5. What determines the width of peaks of monoenergetic particles in various detectors?

In the present chapter, questions of this type will be addressed.

18.2 The binomial distribution

We start by asking a simple question. What is the probability that exactly n atoms out of a total number N will decay during a time interval dt? We describe the basis of the experiment in the following way:

1. We look individually at all N atoms, i.e., go through N trials.
2. Each trial either gives a decay (success, probability p) of no decay (failure, probability $q = 1 - p$).
3. The probability of decay is the same for all trials.
4. The trials are statistically independent, i.e., the outcome of one trial does not influence any other trial.

These four conditions characterize a Bernoulli process with n and N as variables. The probability $P(n)$ that exactly n atoms will decay is

$$P_n = \begin{pmatrix} N \\ \\ n \end{pmatrix} p^n \, q^{N-n} \tag{18.2}$$

$P(n)$ is called a Bernoulli distribution. The mean value μ of n can be evaluated from the equation

$$\mu \equiv \sum_{n=0}^{N} nP_n = \sum_{n=0}^{N} n \begin{pmatrix} N \\ \\ n \end{pmatrix} p^n q^{N-n} \tag{18.3}$$

Individual runs will give some number n, most of them different from the mean value μ. The scatter of these values around the mean is described by the standard deviation σ or the variance, which is equal to σ^2. The variance is defined by

$$\text{variance} = \sigma^2 = \sum_{n=0}^{N} (n - \mu)^2 P(n) \tag{18.4}$$

Calculating this sum gives

$$\sigma^2 = \sqrt{N p q} \tag{18.5}$$

18.3 Poisson and normal (Gaussian) distributions

Usually N is a very large number, $N \gg n$ and the probability that each atom will disintegrate during the time interval t is very small. That is, $p \ll 1$ and q is close to 1.0. Under these conditions the Poisson distribution gives a good approximation:

$$P_n = \frac{\mu^n e^{-\mu}}{n!} \tag{18.6}$$

The mean value of n is equal to μ. The standard deviation σ of n from its mean value μ, can be calculated from this equation and we find

$$\sigma = \sqrt{\mu} \tag{18.7}$$

The Poisson distribution describes a large number of diverse phenomena.

Let us look at numerical examples of the Poisson distribution. In low level counting the number of pulses in short time intervals is sometimes recorded in order to check the proper functioning of the system (Section 9.4). The mean number of counts, μ, is often low and the frequency distribution should fit Equation (18.6). Figure 18.1 shows the distribution when an average of 2.2 pulses are counted per time interval. For low values of μ, the probability curve is skewed. With an increasing mean number of pulses, the probability curve becomes more symmetrical around the mean value μ. Figure 18.1 also depicts the distribution when $\mu = 12$. Already at this relatively low value for μ, the distribution is nearly symmetrical. This indicates that the normal, also called Gaussian, distribution, $G(x)$, gives a good approximation for the Poisson distribution:

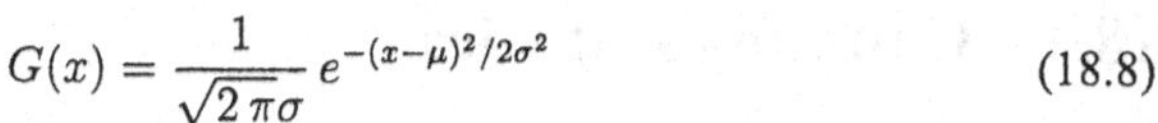

$$G(x) = \frac{1}{\sqrt{2\pi}\sigma} e^{-(x-\mu)^2/2\sigma^2} \tag{18.8}$$

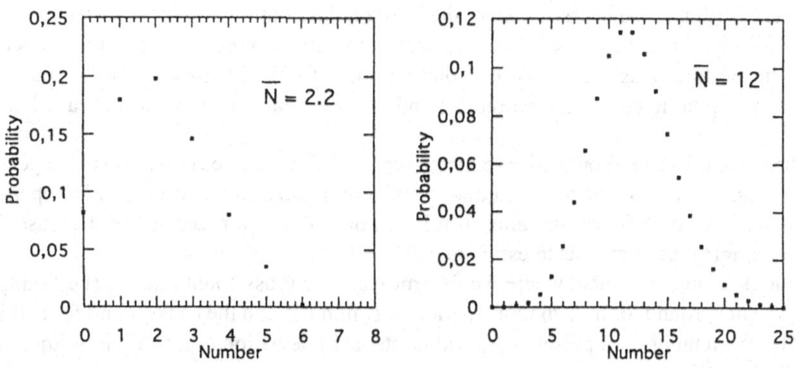

Fig. 18.1. The Poisson distribution for mean values of 2.2 (left) and for 12 (right).

The probability that the variable x has a value between x and $x + dx$ is $G(x)dx$. The area under this curve is unity. The distribution is characterized by a single parameter, μ, which is equal to the average value of x. The variance σ^2 is, as in the case of the Poisson distribution, equal to the mean value of x, $\sigma^2 = \mu$. The probability that the value of x lies between x_1 and x_2 is

$$P(x_1 \leq x \leq x_2) = \frac{1}{\sqrt{2\pi}\sigma} \int_{x_1}^{x_2} e^{-(x-\mu)^2/2\sigma^2} dx \quad . \tag{18.9}$$

This equation can be used to find the probability that x deviates more than a given value ε_0 from μ. A few sets of values are given in Table 18.1.

Table 18.1. Probability $P(\varepsilon_0)$ that the deviation ε will be less than ε_0, or $-\varepsilon_0 \leq x \leq \varepsilon_0$.

ε_0	0.67σ	1.0σ	1.64σ	1.96σ	3σ
$P(\varepsilon_0)$	0.50	0.68	0.90	0.95	0.997

This shows that the probability of the result lying within one standard deviation from the mean value is 68% and 95% lie within two σ (or more correctly, within 1.96σ). The probability that $x \leq \sigma$ is equal to $(1 - 0.68)/2$ or 0.16.

18.4 Counting statistics

We now turn to the statistics of counting radioactive samples in a fixed time interval t. The statistical distributions discussed above (the Poisson distribution for small (<10) numbers and normal distribution for large) can be used to estimate the precision of single measurements. In the following we denote the total number of pulses registered in the counting period t by capital letters (for example N) and the count rate (N/t) by lower case letters (n).

When a long-lived radionuclide is counted repeatedly under fixed conditions for a period of t minutes, we can use the mean number N of pulses found, for example, in 10 repeated runs, as a measure of the mean value μ in Equation (18.9). In practice we often use the values of single measurements to estimate σ.

We look at an experiment where we first measure the gross count rate, n_g (i.e., sample activity + background, or $n_s + n_b$) for a period of t_g minutes and the background for t_b minutes. The mean number of pulses is N_g and its standard deviation is according to Equation (18.7) $\sqrt{N_g}$. The standard deviation of the gross count rate n_g, where n_g=N_g/t_g, is equal to

$$\sigma(n_g) = \sqrt{N_g}/t_g = \sqrt{(n_g \cdot t_g)}/t_g = \sqrt{(n_g/t_g)} \tag{18.10}$$

This shows that the standard deviation of the count rate decreases as the square root of the counting time.

The background is then measured for a period t_b. The standard deviation of this count rate is

$$\sigma(n_b) = \sqrt{(n_b/t_b)} \tag{18.11}$$

The distribution in the net count rate, $n_g - n_b = n_s$, will also have a normal distribution and its variance σ_n will be the sum of the variances of the gross count rate and the background count rate

$$\sigma_s^2 = \sigma_g^2 + \sigma_b^2 \tag{18.12}$$

This can be used to find the optimum counting time if we have a given total time T to measure both the gross count rate and the background. It can be shown that the ratio of t_g and t_b, which gives minimum value for σ_s is obtained when

$$t_g/t_b = \sqrt{(n_g/n_b)} \tag{18.13}$$

In practice we usually measure the background repeatedly after a certain number of unknown samples. The mean background count rate in a number of runs is used. The background is therefore usually known to better precision than the count rate of weak samples, measured for a single period t_g.

18.5 Significant activity

Minimum significant measured activity, error of type I. We will discuss the definition of these two terms and demonstrate them through a numerical example. We assume that we have a detector with a constant mean background of 20 cpm. We need to screen a large number of samples for radioactivity by counting each sample for 10 minutes and the background once for 30 minutes and sort out all samples that are judged not active by a criterion that will now be described.

The standard deviation of a gross count rate of a blank sample, σ_g, is 1.41 cpm ($t_g = 10$ min) and for the background, σ_b, it is 0.82 cpm ($t_b = 30$ min). The net count rate, the sample count rate n_s, is $n_s = n_g - n_b$. As we are counting blank samples, the mean value of n_s is zero and the standard deviation of n_s is, assuming that we take a new background measurement each time, is given by Equation (18.12). It is equal to 1.63 cpm in the present case. n_s will give both positive and negative values and it will have a normal distribution around zero. The distribution is shown in Figure 18.2. Where shall we set the limit δ_1 (cpm) for the net count rate of a sample in order to judge that it represents a blank sample? If we claim that a blank sample is active because the counter gave a random high count rate, we say that we have made an *error of type I.* We would like to sort out weak samples, but without getting too many errors of this type. Shall we set the limit at a count rate of 21 cpm, 22 cpm, or even at 24 cpm, corresponding to a net count rate of 1, 2 and 4 cpm?

We must first decide the frequency of error, i.e., how often do we judge a blank sample to be an active one. We set this at some fixed acceptable probability, p_1, for example at

0.05, where 1 out of 20 measurements of blank samples give a false signal. The hatched area under the normal distribution curve in Figure 18.2 represents the area of high count rate that we are not willing to accept as a blank sample. This corresponds to a count rate of $k_1 \times \sigma_s$, where

$$\delta_1 = k_1 \times \sigma_s \tag{18.14}$$

According to this criterion, the minimum significant measured count rate $d-1$ corresponds to this limit:

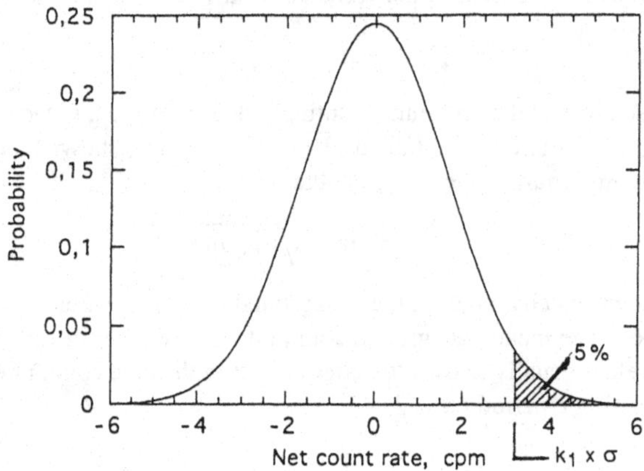

Fig. 18.2. The normal distribution for repeated counting of a blank sample and a background (actually repeated double sets of background) for the numerical case discussed in the text.

Once we have selected p_1 and found its corresponding factor k_1, we can now find what sample count rate d_1 is needed to give this minimum significant measured net count rate.

$$\delta_1 = k_1\sqrt{\sigma_g^2 + \sigma_b^2} = k_1\sqrt{\frac{\delta_1 + n_b}{t_g} + \frac{n_b}{t_b}} \tag{18.15}$$

Eliminating δ_1 we get

$$\delta_1 = \frac{1}{2}k_1^2 + \frac{1}{2}k_1\sqrt{k_1^2 + 8n_b} \tag{18.16}$$

If we select $p_1 = 0.05$ in our numerical case, k_1 has a value of 1.65 and we find that $\delta_1 =$ 2.82 cpm.

Minimum detectable true activity, error of type II. Now we turn to a somewhat different situation. We want to define a net count rate δ_2 such that we can assert with great confidence that its value shows that the sample is active. We use the same numerical example as discussed above, where $t_g = 10$ min and $t_b = 30$ cpm and $n_b = 20$ cpm. If we have a sample that gives a gross count rate of 25 cpm, what is the probability that it presents a true activity? Is the probability higher than 95%? We can also ask, is the probability less than 5% that this is not a random high count rate of a blank sample? Errors of type II occur when we say that a given excessive count rate represents a true activity, when it is actually only a high random count rate of a blank sample. More precisely, we ask, for what minimum mean count rate δ_2 is the statistical probability less than p_2 that a given count rate can be a random high count rate of a blank sample?

We look again at the numerical example of the 20 cpm. We saw that a count rate up to 22.84 cpm ($n_b + \delta_1$) can be considered as a blank sample according to the first criterion. This only tells us that at this limit the probability is 5% that it contains no activity. If we now have a true activity that gives an average net count rate of δ_2 we want to have a high probability, let us say 95%, that the count rate can be interpreted as true activity, but not as random high count rate of a blank sample. If we interpret the result of the measurement of this sample as a blank sample, we make an error of type II.

This will now be described in general terms. We first select the probability p_2 representing the maximum acceptable relative number of errors of type II. We then calculate the average net count rate δ_2 of a sample for which only a fraction $1/p_2$ of determinations will be interpreted as coming from a blank sample, i.e., a measurement giving a net count rate of d_1 or less. The net count rate of the sample will have a normal distribution with a mean value of d_2 and a standard deviation $\sigma(\delta_2)$ given by Equation (18.12). This is illustrated by Figure 18.3.

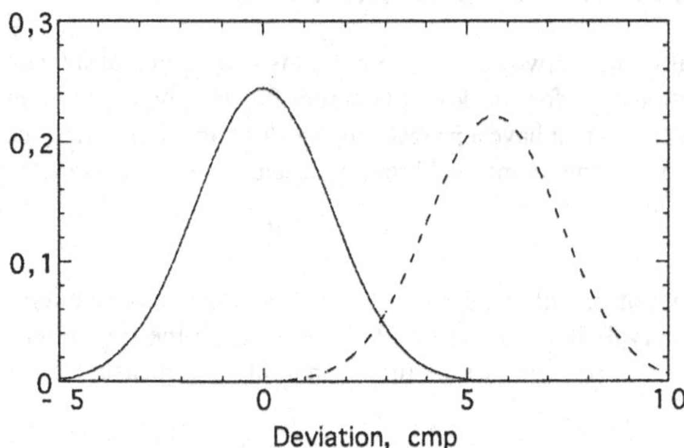

Fig. 18.3. The normal distribution for repeated determinations of the net count rate of a blank sample (curve A) and one with an average net count rate δ_2 (curve B).

The quantity $(n_g - n_b) - \delta_2$ will be normally distributed around zero. Referring to Figure 18.3 we get an equation similar to Equation (18.15)

$$\delta_2 = \delta_1 + k_2\sqrt{\sigma_g^2 + \sigma_b^2} = k_1\sqrt{\frac{\delta_1 + n_b}{t_g} + \frac{n_b}{t_b}} / t_b \tag{18.17}$$

$$\delta_2 = \delta_1 + k_2\sqrt{\frac{\delta_1}{t_g} + r_b\left(\frac{t_g + t_b}{t_g t_b}\right)} \tag{18.18}$$

which can be written

$$\delta_2 = \delta_1 + k_2\sqrt{\frac{\delta_1}{t_g} + n_b\frac{t_g + t_b}{t_g t_b}} \tag{18.19}$$

Minimum analytical activity. In analytical work we are not satisfied with knowing that a sample count rate represents true activity. We further require that the average net count rate δ_A of a sample is a factor of p_A times larger than the standard deviation of the net activity count rate:

$$\delta_A = k_A \sigma_A \tag{18.20}$$

where σ_A depends on δ_A. Solving this for δ_A as above we get

$$\delta_A = \frac{p_A}{t_g} + \frac{1}{2}k_A\sqrt{\frac{k_A^2}{t_g^2} + 4n_b\frac{t_g + t_b}{t_g t_b}} \tag{18.21}$$

If we select $p_1 = p_2 = 0.05$ and $p_A = 10$, we get, for the numerical example above, $\delta_1 = 2.84$ cpm, $\delta_2 = 5.67$ cpm and $\delta_A = 41.5$ cpm.

18.6 Distribution of time intervals

The time interval between two successive pulses is sometimes of interest, for example in dead time corrections and for checking the proper functioning of the counting system. We consider random pulses that have a probability $r \times dt$ of arriving during the interval dt. The probability $P(t)$ of finding an interval lying between t and $t + dt$ is then given by

$$P(t) = r\,e^{-rt} \tag{18.22}$$

The average length of the intervals is equal to $1/r$. It sounds like a dilemma that the mean length of time intervals is the same whether we measure the time intervals between two successive pulses or when we start the timer randomly and measure the time until the first pulse arrives.

18.7 Statistics of energy resolution

Energy resolution of particle detectors is an important characteristic. It is usually measured by the full width at half maximum of the peak (FWHM), as was described in Section 8.2. The peak width can depend on several factors, but two dominate, (1) the statistical nature of the signal, and (2) the noise of the system. The former source is frequently dominating and it will be discussed here.

Let us take as an example the gamma spectrum of ^{137}Cs measured with a NaI scintillation counter. The peak in the spectrum is from pulses when all the energy of the photons is absorbed in the crystal. Scintillation pulses will release μ electrons in average from the cathode of the photomultiplier tube. The number of electrons in individual pulses may be smaller or larger and will be distributed around the mean value according to Poisson distribution, which can be approximated by the normal distribution. The resolution R is then given by:

$$R = \text{FWHM}/E2.35\sigma/\mu = 2.35/\sqrt{\mu} \tag{18.23}$$

where E is the absorbed energy and $\sigma = \sqrt{\mu}$. For a NaI scintillation unit with a resolution of 8% for the 664 keV ^{137}Cs line, this gives that 863 photons are produced in average.

Measurements of the energy resolution of various types of detectors show that the obtained values of R can be lower by a factor of 2–4 than the Poisson distribution implies. This indicates that the formation of the charge carriers are not independent. An empirical factor F, the *Fano factor*, is used to describe the departure of observed statistical fluctuations. F is defined as

$$F = \frac{\text{observed variance}}{\text{Poisson predicted variance}} \tag{18.24}$$

The Fano factor is 0.05–0.20 for gas proportional counters and about 0.15 for semiconductor detectors.

Recommended reading

- Glenn F. Knoll, Radiation Detection and Measurement, John Wiley and Sons, second edition 1989.
- James E. Turner, Atoms, Radiation, and Radiation Protection, John Wiley and Sons, Inc., New York, 1995.
- L.A. Currie, Limits for Qualitative Detection and Quantitative Determination, Anal. Chem. 40, 586–593, 1968.

References

1. Acena, M.L., Crespo, M.T., Galan, M.P., and Gason, J.L., Determination of isotopes of uranium and thorium in low-level environmental samples, *NIM PR* **A339**, 302-308 (1994)

2. Aiginger, H., Maringer, F.J., Rank, D., and Unfried, E., A new laboratory for routine low-level measurements, (BVFA Arsenal, Wien), *MIN PR* **B17**, 435-437 (1986)

3. Al-Bataina, B. and Jäecke, J., Alpha particle emission from contaminants in counter materials, *NIM PR* **A255**, 512-517 (1987)

4. Alessandrello, A. et al., Measurement on radioactivity of ancient Roman lead to be used as shield in search for rare events, *MIN PR* **B61**, 106-117 (1991)

5. Amono, H. and Kasai, K., Abstract 1-d-10 at the 25th annual meeting of Japan Radiation Research Society (1982)

6. Anderson, E.C., Arnold, J.R., and Libby, W.F., Measurements of low level radiocarbon, *Review of Scientific Instruments* **22**, 225-230 (1951)

7. Arnold, J.R. and Libby, W.F., Age determination by radiocarbon content: Checks with samples of known age, *Science* **110**, 678-680 (1949)

8. Arnold, J.R., Scintillation counting of radiocarbon, *Science* **119**, 155-157 (1954)

9. Arthur, R.J. and Reeves, J.H., Methods for achieving ultra-low backgrounds in above ground germanium detector systems, *J. Radioanal. Nucl. Chem.* **160**, 297-304 (1991)

10. Arthur, R.J., Reeves, J.H. and Miley, H.S., Use of low-background germanium detector to preselect high purity materials intended for construction of advanced ultralow-level detectors, *IEEE Trans. Nucl. Sciences* 582-585 (1988)

11. Arthur, R.J., Reeves, J.H., and Miley, H.S., Spectrometers, *NIM PR* **A292**, 337-342 (1990)

12. Bem, H., Bakiar, Y.Y.Y., and Bou-Rabee, F., An improved method for low level ^{222}Rn determination in environmental waters by liquid scintillation counting with pulse shape analysis, *J. Radioanal. Nucl. Chem. Lett.* **186**, 119-127 (1994)

13. Boorse H.A., Mootz L., and Weever J.H., Wiley Science Editions, (1989)

14. Bötter-Jensen, L., Hansen, H.H., and Theodorsson, P., A multicounter system for scanning ultra-low-level radiochromatograms, *Nucl. Instr. Meth.* **144**, 529-532 (1977)

15. Bowman, S., Liquid scintillation counting in the London underground, *Radiocarbon* **31**, 393-398 (1989)

16. Brodzinski, R.L., Brown, D.P., Evans, J.C., Hensley, W.K., Reeves, J.H., and Wogman, N.A., An ultralow background germanium gamma-ray spectrometer, *Nucl. Instrum. Meth. Phys. Res.* **A239**, 207-213 (1985)

17. Buheitel, F., The determination of low levels of radium isotopes and radon by delayed-coincidence liquid scintillation spectrometry, *Liquid Scintillation Spectrometry, Radiocarbon*, 83-88 (1993)

18. Bunting, R.L. and Kraushaar, J.J., Short-lived radioactivity induced in Ge(Li) gamma-ray detectors by neutrons, *NIM* **118**, 565-572 (1974)

19. Chao, J.-H., Neutron-induced gamma-rays in germanium detectors, *Appl. Radiat. Isot.* **44**, 605-611 (1993)

20. Chinaglia, B. and Malvano, R., Efficiency of 3"×3" NaI(Tl) crystals, *NIM* **45**, 125-132 (1966)

21. Chung, C., Yuan, L.-J., and Chen, K.-B., Performance of a HPGe-NaI(Tl) compton suppression spectrometer in high-level radioenvironmental studies, *NIM PR* **17**, 102-110 (1986)

22. Chung, C. and Cheng-John, Lee, Environmental monitoring using a HPGe-NaI(Tl) compton suppression spectrometer, *Nucl. Inst. Meth. Phys. Res.* **A27**, 436-440 (1988)

23. Clarke, W.B., Jenkins, W.J., and Top, Z., Determination of tritium by mass spectrometric measurements of ^{3}He, *IJARI* **27**, 515-522 (1975)

24. Coccioni, G. and Coccioni Tongiorgi, V., Nuclear disintegrations induced by m-mesons, *Phys. Rev.* **84**, 29-36 (1951)

25. Coccioni, G., Tongiorgi, V.C., and Widgoff, M., Cascade of nuclear disintegration induced by cosmic radiation, *Phys. Rev.* **79**, 768-773 (1950)

26. Cohen B.L. and Cohen, E.S., Theory and practice of radon monitoring by adsorption in charcoal, *Health Physics* **50** (1983)

27. Cohen, B.L., Ganayni, M.El, and Cohen, E.S., Large scintillation cells for high sensitivity radon concentration measurements, *NIM* **212**, 403-412 (1983)

28. Cohen, B.L. and Nason, R., A diffusion barrier charcoal adsorption collector for measuring radon concentration in indoor air, *Health Physics* **50** (1986)

29. Cook, G.T. and Anderson, R., A radiocarbon dating protocol for use with Packard scintillation counters employing burst-counting circuitry, *Radiocarbon* **34**, 381-388 (1992)

30. Coursey, B.M. and Mann, W.B., Design of high efficiency liquid scintillation counting system, in: *The application of liquid scintillation counting in radionuclide metrology*, J.G.W. Taylor and W.B. Mann, eds., Serves. Bur. Internatl. Poids et Measures, pp. 23-35 (1980)

31. Croft, S. and Bond, D.S., The gamma ray detection efficiency of a 118 cm^3 Ge(Li) detector for disc-source of ^{7}Be, *Appl. Radiat. Isot.* **44**, 725-730 (1993)

32. Csongor, E. and Hertelendi, E., Low-level counting facility for ^{14}C dating, *Nucl. Inst. Meth. Phys. Res.* **B17**, 493-497 (1986)

33. Curie, L.A., Limits for qualitative detection and quantitative determination: Application to radiochemistry, *Analytical Chemistry* **40**, 586-593 (1968)

34. De Vries, H. and Barendsen, G.W., A new technique for the measurement of age byradiocarbon, *Physica* **18**, 652 (1952)

35. De Vries, H.L. and Barendsen, G.W., Radio-carbon dating by a proportional counter filled with carbondioxide, *Physica* **19**, 987-1003 (1953)

36. De Vries, H., The contribution of neutrons to the background of counters used for ^{14}C age measurements, *Nuclear Physics* **1**, 477-479 (1956)

37. De Vries, H., Further analysis of the neutron component of the background of counters used for ^{14}C age measurements, *Nuclear Physics* **3**, 65-68 (1957)

38. Dep, L., Elmore, D., Fabryka-Martin, J., Masarik, J., and Reedy, R.C., Production rate systematics on in-situ cosmogenic nuclides in terrestrial rocks: Monte Carlo approach of investigating ^{35}Cl(n,g) ^{36}Cl, *Nucl. Inst. Meth. Phys. Res.* **B92**, 321-325 (1994)

39. Draganic, I.G., Draganic, Z.D., and Adloff, J.A., Radiation and radioactivity, CRC Press, (1993)

40. Einarsson, S.A., Evaluation of a prototype low-level liquid scintillation multisample counter, *Radiocarbon* **34**, 366-373 (1992)

41. Einarsson, S.A. and Theodorsson, P., Study of background spectrum of an LSC system, *Radiocarbon* **31**, 342-351 (1989)

42. Einarsson, S.A. and Theodorsson, P., Stability of a new, multichannel, low-level liquid scintillation counter system, Kvartett, *Radiocarbon* **37**, 727-736 (1995)

43. El-Daoushy, F. and Garcia-Tenoria, R., Well Ge and semi-planar Ge (HP) detectors for low-level gamma-spectrometry, *Nucl. Instr. & Meth. in Phys. Res.* **A356**, 376-384 (1995)

44. Florkowski, T., Low level tritium assay in water samples by electrolytic enrichment and liquid scintillation counting in the IAEA laboratory, in: Proceedings of International symposium on methods of low-level counting and spectrometry, IAEA Vienna (1981)

45. Florkowski, T., Tritium electrolytic enrichment using metal cells, in: *Low-Level Tritium Measurement*, Vienna, IAEA-TECDOC 246, 133-137 (1981)

46. Freundlich, J.C., Natural radon as a source of low level laboratory contamination, Lower Hut, 538-546 (1972)

47. Gangnes et al., *Phys. Rev.* **75**, 57 (1949)

48. George, A.C., Passive, integrated measurement of indoor air using activated carbon, *Health Physics* **46** (1984)

49. Grootes, P.M., Thermal diffusion isotopic enrichment and radiocarbon dating beyond 50,000 years BP, Thesis, Groningen (1977)

50. Gulliksen, G. and Nydal, R., Further improvement of counter background and shielding, in: *Radiocarbon Dating, Proc. 9th Intern. Radiocarbon Conf. Los Angeles and La Jolla 1976* Berger, R. and Suess, H.E., eds., Univ. Cal. Press, 176-184 (1979)

51. Hansen, H.J.M., and Theodorsson, P., A multicounter system for scanning ultra-low-level radiochromatograms, *Nucl. Instr. Meth.* **144**, 529 (1977)

52. Hedberg, M. and Theodorsson, P., External radon background component of gas proportional counters, *Radiocarbon* **37**, 759-766 (1995)

53. Helmer, R.G. and Debertin, K., *Gamma and X-Ray Spectrometry with Semiconductor Detectors*, North-Holland (1988)

54. Heusser, G. and Wojcik, M., Radon suppression in low-level counting, *Appl. Rad. Isot.* **43**, 9-18 (1994)

55. Heusser, G., Gamma ray induced background in Ge-spectrometry, *Nucl. Inst. Meth. Phys. Res.* **B83** 223 (1993)

56. Heusser, G., Background in ionizing radiation detection - illustrated by Ge-spectrometry, in: Proc. 3rd Int. Summer School, Low-Level Measurements of Radioactivity in the Environment, ed. M. Garcia-Leon, R Garcia-Tenorio, World Scientific, Singapore (1994)

57. Heusser, G., Klapdor, H.V., Piepke, A., Schneider, J., Mansour, N., and Strecker, H., Construction of a low-level Ge detector, *Appl. Rad. Isot.* **40**, 393-395 (1989)

58. Houtermans, F.G. and Oeschger, H., Proportionszählrohr zur Messung schwacher Aktivitäten weicher β-Strahlung, *Helv. Phys. Acta* **28**, 464-466 (1955)

59. Hut, G., Intercomparison of low-level tritium measurements, *IAEA report*, (1986)

60. Jagam, P. and Simpson, J.J., Thorium and uranium determination in acryl by neutron activation analysis, *Journ. Radioanl. Nucl. Chem. Articles* **171**, 277-286 (1993)

61. Jagam, P. and Simpson, J.J., Measurements of Th, U and K concentrations in a variety of materials, *Nucl. Inst. Meth. Phys. Res.* **A324**, 389 (1993)

62. Johansson, L., Roos, B., and Samuelsson, C., *ARI* **43**, 119-125 (1992)

63. Kalt, P., Verteilung des radioaktiven Isotops ^{39}Ar im Nordatlantik und Pazifik, PhD thesis, University Bern (1986)

64. Kamikubota, N., Ishibashi, S., Tanabe, N., Taniguchi, T., Matsuoka, K., Okada, K., and Ejiri, H., Low-level radioactive isotopes contained in materials used for beta- and gamma-ray detectors, in: *Rare Nuclear Processes, Proc. 14th Europhysics Conf. on Nuclear Physics*, Povinec P., ed., p. 275, World Scientific, Singapore (1992)

65. Kilgus, U., Kotthaus, R., and Lange, E., Prospects of CsI(Tl)-photodiode detectors for low-level spectroscopy, *Nucl. Instr. Meth. Phys. Res.* **A297**, 425-440 (1990)

66. Kloke, F.C., Smith, E.T., and Kahn, B., The influence of radon daughter concentration in air on gamma spectrometer background, *NIM* **34**, 61-65 (1964)

67. Knoll, G.F., Radiation detection and measurements, (second edition), Publ. John Wiley and Sons, New York, p. 233 (1989)

68. Kojola, H., Polach, H., Nurmi, J., Oikari, T., and Soini, E., High resolution low-level liquid scintillation spectrometer, *Int. J. Appl. Rad. and Isotopes* **35**, 949-952 (1984)

69. Kolb, W., Background reduction of a semiplanar Ge(HP) detector, *Environm. Intern.* **14**, 367-370 (1988)

70. Kolb, W.A., Die Eigenaktivität von Blei, in: Proc. of the First International Congress of Radiation Protection, Rome 1966, 1385-1391 (1968)

71. Kotthaus, R., CsI(Tl)-photodiode detectors for spectroscopy at low radiation levels, *Report MPI-PhE/92-22*, Max-Planck-Institute für Physik, München (1992) and *Nucl. Inst. Meth. Phys. Res.* **A329**, 433-439 (1993)

72. Kromphorn, G., Messung des ^{210}Pb-Gehalts handelsüblicher Bleisorten, *PTB-Bericht* **Ra-33** 1968

73. Kulp, J.L. and Tryon, L.E., Extension of the carbon 14 age method, *Rev. Scient. Instr.* **23**, 296-297 (1952)

74. Libby, W. F., History of Radiocarbon dating, in: *Radioactive dating and Methods of low-level counting*, Proceedings of a symposium, IAEA, Vienna, 3-26 (1967)

75. Loosli, H.H. Oeschger, H., and Wahlen, M., New attempts in low-level counting and search for cosmic-ray-produced ^{39}Ar, in: Radioactive Dating and Methods of Low Level Counting, IAEA STI/PUB/152 (IAEA Vienna, 1967) 593-601 (1967)

76. Loosli, H.H., Forster, M., and Otlet, R.L., Background measurements with different shielding and anticoincidence systems, *Radiocarbon* **28**, 615-624 (1986)

77. *Low-Level Measurements of Radioactivity in the Environment - Techniques and Applications*, Proc. 3rd Intern. Summer School, Huelva, Spain 1993, Garcia-Leon and Garcia-Tenorio, eds., World Scientific, Singapore (1994)

78. Lucas, H.F., Improved low-level alpha scintillation counter for radon, *RSI* **28**, 680-683 (1957)

79. Lucas, H.F., The Argonne radon-in-air analysis system, *Radioactivity and Radiochemistry* **6**, 42-51 (1995)

80. Lucas, H.F. and Woodward, D.A., Effect of long decay chains on the counting statistics in the analysis of radium-224 and radon-222, *J. Appl. Phys.* **35**, 452 (1963)

81. Mäntynen, P., Äikää, O., Kankainen, T., and Kaihola, L., Application of pulse-shape discrimination to improve the precision of the carbon-14 gas-proportional-counting method. *Appl. Radiat. Isot.* **38**, 869-873 (1987)

82. Maushart, R., Performance of alpha/beta low-level counting systems for solid samples, *Nucl. Inst. Meth. Phys. Res.* **B17**, 501-505 (1986)

83. Mcdowell, W.J., *Alpha liquid scintillation counting: Recent applications and developments*, Vol.1, Peng, C.T., Horrocks, D.L., and Alpin, E.l., eds., Academic Press, New York, pp. 315-332 (1980)

84. Mcdowell, W.J. and Mcdowell, B.L., *Liquid Scintillation Alpha Spectrometry*, CRC Press (1994)

85. Miller, C.E., Marinelli, L.D., Rowland, R.E., and Rose, J.E., Reduction of NaI background, *Nucleonics* 40-43 (1956)

86. Minato, S., Bulk density of buildings using cosmic rays, *Appl. Rad. Isot.* **37**, 941-946 (1986)

87. Mook, W.G., International comparison of proportional gas counters for ^{14}C activity measurements, *Radiocarbon* **25**, 475 (1983)

88. Mouchel, D., A high-purity Ge detector system for the measurement of low-level radioactivity in environmental samples, in: *Low-level Measurements of Man-made Radionuclides in the Environment*, Garcia-Leon, M. and Maddurga, G., eds., p. 106, World Scientific, Singapore (1991)

89. Mouchel, D. and Wordel, R., Measurement of low-level radioactivity in environmental samples by g-ray spectrometry, *Appl. Radiat. Isot.* **43**, 49 (1992)

90. NCRP Report, National Council on Radiation Protection and Measurements, (1975)

91. Nydal, R., Proportional counting technique for radiocarbon measurements, *Rev. Sci. Inst.* **33**, 1313-1320 (1962)

92. Nydal, R., Gulliksen, S., and Lövseth, K., Proportional counters and shielding for low level gas counting, in: *Proc. Int. Conf. on Low-Radioactivity Meas. and Appl., High Tatras, CSSR* P. Povinec, and S. Usacev, eds., Bratislava, Comenius University, pp. 77-84 (1977)

93. Nydal, R., Gulliksen, S., Lövseth, K., An analysis of shielding efficiency for counters, *Radiocarbon* **22**, 470-478 (1980)

94. Oeschger, H., Beer, J., Loosli, H.H, and Schotter, U., Methods of low level counting and spectrometry, IAEA Vienna, 459-474 (1961)

95. Oeschger, H. and Wahlen, M., Low level counting techniques, *Ann. Rev. Nucl. Sci.* **25**, 423-463 (1975)

96. Oeschger, H. and Loosli, H.H., New development in sampling and low level counting of natural radioactivity, *Proc. Intern. Conf. Low Level Radioactivity Measurements and Applications*, High Tatras, Bratislava, 13-22 (1979)

97. Oeschger, H., Lehmann, B., Loosli, H.H., Moell, M., Nuftel, A., Schotterer, U., and Zumbrun, R., Recent progress in low level counting and other isotope detection methods , in: *Radioactive Dating*, Berger, R. and Suess, H.E., eds., Univ. California Press, Berkeley, Los Angeles, pp. 145-157 (1979)

98. Oeschger, H., Beer, J., Loosli, H.H., Schotterer, U., Low-level counting systems in deep underground laboratories, in: *Methods of Low-level Counting and spectrometry*, Proc. Symp. Berlin, IAEA, 459-473 (1981)

99. Östlund, H.G. and Dorsey, H.G., Rapid electrolytic enrichment and hydrogen gas proportional counting of tritium, *Proc. Int. Conf.: Low Radioactiv. Measurements and Applications*, High Tatras, Bratislava (1977)

100. Pearson, G.W., The development of high precision ^{14}C measurement and its application to archeological time-scale problems, Ph.D. thesis, The Queen's University of Belfast (1983)

101. Pei-yun, F. and Ting-kui, Z., A single photomultiplier liuid scintillation counting apparatus for ^{14}C low-level measurement, in: *Advances in Scintillation Counting*, McQuarrie, S.A., Ediss, C., and Wiebe, L.I., eds., Edmonton, University of Alberta Press, 456-467 (1983)

102. Povinec, P., Multiwire proportional counters for low-level ^{14}C and ^{3}H measurements, *NIM* **156**, 441-446 (1978)

103. Prichard, H.M. and Gesell, T.F., Rapid measurements of ^{222}Rn concentrations in water with a commercial liquid scintillation counter, *Health Physics* **33**, 577-581 (1977)

104. Prichard, H.M. and Marion, K., A passive diffusion ^{222}Rn sampler based on activated carbon adsorption, *Health Physics* **48** (1985)

105. Preusse, W., Höhenstrahlungsinduziertes Nulleffektspektrum von Gammastrahlungsdetektoren, Dissertation, TU Bergakademie Freiberg (1993)

106. Rauret, G., Mestres, J.S., and Garcia, J.F., Optimization of liquid scintillation counting conditions with two kinds of vials and detector shields for low-activity radiocarbon measurements, *Radiocarbon* **31**, 380-386 (1989)

107. Reath, C.H., Sevold, B.J., and Pederson, C.N., A multiple-anode anticoincidence ring counter, *Rev. Sci. Instr.* **22**, 461 (1951)

108. Reeves, J.H. and Arthur, R.J., Anticosmic shielded ultralow-background germanium detector system for analysis of bulk environmental samples, *J. Radioanal. Nucl. Chem.* **123**, 437-447 (1988)

109. Reeves, J.H. and Arthur, R.J., Anticosmic-shielded ultralow-background germanium detector systems for analysis of bulk samples, *J Radioan. Nucl. Chem.* **124**, 435-447 (1984)

110. Reeves, J.H. and Arthur, R.J., Anticosmic-shielded ultralow-background germanium detector systems for analysis of bulk samples, *J. Radioan. Nucl. Chem.* **124**, 435-447 (1988)

111. Review: *Measurement of Low Level Radiocarbon and Tritium*, **20**, 461 (1971)

112. Rodriguez-Pasqués, R.H., Steinberg, H.L., Harding, J.E., Mullen, P.A., Hutchinson, J.M.R., and Mann, W.B., Low-level radioactivity measurements on aluminium, steel and copper. *IJAPI* **23**, 445-464 (1972)

113. Roy, J.-C., Cote, J.-E., Durham, W., and Joshi, S.R., A Study of the indium and germanium photopeaks in the background spectra of Ge spectrometers with a passive shield, *J. Radioanal. Nucl. Chem.* **130**, 221-230 (1989)

114. Salonen, L., A rapid method for monitoring of uranium and radium in drinking water, *6th Intern. Symp. on Environmental Radiochemical Analysis*, Manchester, UK (1990)

115. Sastawny, A., Bialon, J., and Sosinski, T., Changes in the surface radioactivity of lead - the effect of the diffusion of bismuth and polonium radioistopes. *ARI* **40**, 19-25 (1989)

116. Sauli, F., Principles of operation of multiwire proportional and drift chambers, Academic Training Programme of CERN, Geneva (1977)

117. Scarpitta, S.C. and Harley, N.H., An improved ^{222}Rn canister using a two-stage charcoal system, *Health Physics* **60**, 177-188 (1991)

118. Schoch, H., Burns, M., Münnich, K.O., and Münnich, M., A multicounter system for high precision carbon-14 measurements: *Radiocarbon* **22**, 442-447 (1980)

119. Schoch, H. and Münnich, K.O., Routine performance of a new multicounter system for high-precision ^{14}C dating, in: *Methods of Low-level Counting and Spectrometry*, Proc. Symp. Berlin 1981, IAEA, 361-370 (1981)

120. Scholz, T., Untersuchung des Nulleffektes in Szintillationszählern und Zählrohren, Diplomarbeit, Heidelberg, II. Physikalisches Institut, (1961)

121. Schooch, H., Eine neue Anlage zur Präzisionsmessung des natürlichen ^{14}C-Gehaltes mit Proportionalzählrohren, Diss. A, Heidelberg

122. Schroeder, M.C., Vanags, U., and Hess, C.T., An activated charcoal-based, liquid scintillation-analyzed airborne Rn detector, *Health Physics* **57**, 43-49 (1989)

123. Sensintaffar, E.L., Chambless, D.A., Gray, D.J., and Windham, S.T., Analysis of error and minimum detection limits for ^{222}Rn measurements, *Radiation Protection Dosimetry* **45**, 33-36 (1992)

124. Shizuma, K., Iwatani, K., and Hasai, H., Gamma scattering in the low background shielding for Ge detector, *Radioisotopes* **36**, 465-468 (1987)

125. Skripkin, V., The system was described at the 15th International radiocarbon conference in Glasgow (1994)

126. Skoro, G.P., Puzovic, J., Kukoc, A.H., Vukanovic, R.B., Zupancic, M., Adzic, P.R., and Anicic, I.V., Effective depth of an underground location by single measurements of cosmic-ray intensity, *Appl. Radiat. Isot.* **46**, 481-482 (1995)

127. Smith, F.P. and Lawin, J.D., Dark matter detection, *Physics Reports* **187**, 203-280 (1990)

128. Stenberg, A. and Olsson, I.U., A low level gamma counting apparatus, *Nucl. Instr. Meth.* **61**, 125-133 (1968)

129. Stenberg, A., Reduction of the background in a NaI(Tl) scintillation spectrometer by the use of Geiger anticoincidence and massive shielding, *Nucl. Instr. Meth.* **96**, 289-299 (1971)

130. Stoop, P., Koopmana, M., de Meijer, R.J., and Put, L.W., Development of a fast radon meter for low concentrations, Report R-20, Kernfysisch Versneller Instituut, Groningen, Holland (1993)

131. Stuiver, M., Robinson, S.W., and Yang, I.C., ^{14}C dating to 60.000 years BP with proportional counters, in: *Proc. Intern. Conf. Radiocarbon Dating, California 1976*, pp. 202-215 (1979)

132. Takahashi, T., Hamada, T., and Ohno, S., Improvement in enrichment procedure for analysis of tritium in natural water, *Radioisotopes* **17**, 357 (1968)

133. Takahashi, T., Nishida, M., and Hamada, T., Improvement in enrichment procedure for analysis of tritium in natural water, (II.), *Radioisotopes* **18**, 559 (1969)

134. Taylor, C.B., Tritium enrichment of environmental waters by electrolysis, *Proc. Int. Conf. Low Radioactiv. Measurements and Applications*, High Tatras, Bratislava (1977)

135. Taylor, C.B., Current status and trends in electrolytic enrichment of low-level tritium in water, in: Proceedings of International symposium on methods of low-level counting and spectrometry, (1982)

136. Theodorsson, P., Improved tritium counting through high electrolytic enrichment, *Int. J. Appl. Radiat. Isot.* **25**, 97-104 (1974)

137. Theodorsson, P., The Geiger counter rehabilitated? *Report* **RH-P-75-B2**, University of Iceland (1975)

138. Theodorsson, P., The determination of low-level tritium in water, in: *Low-level Measurements and their Applications to Environmental Radioactivity*, Garcia-Leon, M. and Madurga, G., eds., World Scientific Publishing, Singapore, p. 571 (1988)

139. Theodorsson, P., Development of multicounter systems for low-level beta samples, in: *Low-level Measurements and their Applications to Environmental Radioactivity*, Garcia-Leon, M. and Madurga, G., eds., World Scientific Publishing, Singapore, pp. 459-470 (1988)

140. Theodorsson, P. and Heusser, G., External guard counters for low-level counting systems, *NIM PR* **B53**, 97-100 (1991)

141. Theodorsson, P. and Heusser, G., External guard counters for low-level counting *NIM PR* **B53**, 97-100 (1991)

142. Theodorsson, P., The background of gas proportional counters, *J. Phys. G: Nucl. Part. Phys.* **17**, 419-427 (1991)

143. Theodorsson, P., Gas proportional *versus* liquid scintillation counting, radiometric *versus* AMS dating, *Radiocarbon* **33**, 9-13 (1991)

144. Theodorsson, P., Quantifying background components of low-level gas proportional counters, *Radiocarbon* **34**, 420-427 (1992)

145. Theodorsson, P., Einarsson, S.A., and Olafsdottir, E.D., A combined LSC/NaI system for low-level environmental measurements, *Liquid scintillation spectrometry, Radiocarbon* (1993)

146. Theodorsson, P., Low-level counting, past - present - future, accepted for publication in *Appl. Radiat. and Isot.* (1996)

147. Turner, J.E., Hamm, R.N., Souleyrette, M.L., Martz, D.E., Rhea, T.A., and Schmidt, D.W., Calculations for b dosimetry using Monte Carlo Code (OREC) for electron transport in hot water, *HP* **55**, 741-750 (1988)

148. UNSCEAR, Sources and effects of ionizing radiation, United Nations, New York, p. 43 (1977)

149. UNSCEAR, Ionizing radiatio: sources and biological effects, p 87. United Nations, New York (1982)

150. Vojtyla, P., Beer, J., and Stavina, P., Experimental and simulated cosmic muon induced background of a Ge spectrometer equipped with a top side anticoincidence proportional chamber, *NIM PR* **B86**, 380-386 (1994)

151. Vojtyla, P., A computer simulation of the cosmic-muon background induction in a Ge γ-spectrometer using GEANT, *Nucl. Instr. Meth. Phys. Res.* **B100**, 87-96 (1995)

152. von Reden, K.F., Schneider, R.J., Cohen, G.J., and Jones, G.A., Performance characteristics of the 3 MV tandetron AMS system at the National Ocean Sciences AMS facility, *NIM* **B92**, 7-11 (1994)

153. Ward, F.A.B., Wynn-Williams, C.E., and Cave, H.M., The rate of emission of alpha particles from radium, *Proc. Roy. Soc. London* **A125**, 713-724 (1929)

154. Watt, D.E. and Ramsden, D., *High Sensitivity Counting Techniques*, Chaps. 6 and 7, Pergamon Press (1964)

155. Weller, R.I., Low level radioactive contamination, in: The natural radiation environment, eds: J.A.S. Adams and W.M. Lowder, Publ. Univ. of Chicago (1964)

156. Weller, R.I., Anderson, E.C., and Barker, J.L., Radioactive contamination of contemporary lead, *Nature* **206**, 1211-1212 (1965)

157. Wordel, R. and Mouchel, D., Low level gamma-ray measurements in a 225 m deep underground laboratory, in: *Low-level Measurements of Radioactivity in the Environment*, Garcia-Leon, M. and Garcia-Tenorio, R., eds., pp.141, World Scientific, Singapore (1993)

158. Wordel, R., Mouchel, D., Sole, V.A., Hoogewerff, J., and Hertogen, J., Investigation of the natural radioactivity of volcano rock samples using a low background gamma-ray spectrometer, *Nucl. Inst. Meth. Phys. Res.* **A339**, 322-328 (1994)

159. Yu-fu, Yu, Salbu, B., Björnstad, H.E., and Lien, H., Improvement for α-energy resolution in determination of low level plutonium by liquid scintillation counting, *J. Radioanal. Nucl. Chem. Letters* **145**, 345-353 (1990)

160. Zikovsky, L. and Kennedy, G., Radioactivity of building materials available in Canada, *Health Physics* **63**, 449-452 (1991)

Appendix A

Physical constants and conversion factors

A.1 Physical constants

Planck's constant, h	=	6.62×10^{-34} J s
Electron charge, e	=	-1.60×10^{-19}C,
Velocity of light in vacuum, c	=	2.99×10^{8}m s^{-1}
Avogadro's number, N_0	=	6.02×10^{23} mole^{-1}
Molar volume at STP (0 °C, 760 torr)	=	22.4 L
Density of air at STP (0 °C, 760 torr)	=	1.293 kg m^{-3}
Ratio proton and electron masses	=	1836.15
Electron mass, m	=	0.00054858 AMU = 0.51100 MeV = 9.1094×10^{-31} kg
Proton mass	=	1.0073 AMU = 938.27 MeV = 1.6726×10^{-27} kg
Neutron mass	=	1.0087 AMU = 939.57 MeV = 1.6749×10^{-27} kg
Alpha-particle mass	=	4.0015 AMU = 3727.4 MeV = 6.6447×10^{-27} kg
Boltzmann's constant, k	=	1.3807×10^{-23} J K^{-1}

A.2 Conversion factors

1 barn	=	10^{-24} cm^2
1 eV	=	1.6022×10^{-19} J
1 AMU	=	931.49 MeV = 1.6605×10^{-27} kg
1 Ci	=	3.7×10^{10} BQ (exactly)
1 rad	=	100 erg g^{-1} = 0.01 Gy
1 Gy	=	1 J kg^{-1} = 100 rad
1 Sv	=	100 rem
1 atmosphere	=	760 mm Hg = 760 torr = 101.3 kPa

Appendix B

Periodic table of elements

1 H 1.0079								
3 Li 6.941	4 Be 9.0122							
11 Na 22.9898	12 Mg 24.305							
19 K 39.098	20 Ca 40.08	21 Sc 44.9559	22 Ti 47.90	23 V 50.9414	24 Cr 51.996	25 Mn 54.9380	26 Fe 55.847	27 Co 58.9332
37 Rb 85.4678	38 Sr 87.62	39 Y 88.9059	40 Zr 91.22	41 Nb 92.9064	42 Mo 95.94	43 Tc (97)	44 Ru 101.07	45 Rh 102.9055
55 Cs 132.9054	56 Ba 137.34	57–71 *	72 Hf 178.49	73 Ta 180.9479	74 W 183.85	75 Re 186.2	76 Os 190.2	77 Ir 192.22
87 Fr (223)	88 Ra (226)	89–103 **	104 (257)	105 (262)	106 (263)			

*	57 La 138.9055	58 Ce 140.12	59 Pr 140.9077	60 Nd 144.24	61 Pm (145)	62 Sm 150.4
**	89 Ac (227)	90 Th 232.0281	91 Pa (231)	92 U 238.029	93 Np (237)	94 Pu (244)

								2 He 4.0026
			5 B 10.81	6 C 12.011	7 N 14.0067	8 O 15.9994	9 F 18.9984	10 Ne 20.179
			13 Al 26.9815	14 Si 28.086	15 P 30.9738	16 S 32.06	17 Cl 35.453	18 Ar 39.948
28 Ni 58.71	29 Cu 63.546	30 Zn 65.38	31 Ga 69.72	32 Ge 72.59	33 As 74.9216	34 Se 78.96	35 Br 79.904	36 Kr 83.80
46 Pd 106.4	47 Ag 107.868	48 Cd 112.40	49 In 114.82	50 Sn 118.69	51 Sb 121.75	52 Te 127.60	53 I 126.9045	54 Xe 131.30
78 Pt 195.09	79 Au 196.9665	80 Hg 200.59	81 Tl 204.37	82 Pb 207.2	83 Bi 208.9804	84 Po (209)	85 At (210)	86 Rn (222)

63 Eu 151.96	64 Gd 157.25	65 Tb 158.9254	66 Dy 162.50	67 Ho 164.9304	68 Er 167.26	69 Tm 168.9342	70 Yb 173.04	71 Lu 174.97
95 Am (243)	96 Cm (247)	97 Bk (247)	98 Cf (251)	99 Es (254)	100 Fm (257)	101 Md (258)	102 No (255)	103 Lr (256)

Index